Sexual Conflict

MONOGRAPHS IN
BEHAVIOR AND ECOLOGY

Edited by John R. Krebs
and Tim Clutton-Brock

Foraging Theory, by David W. Stephens
and John R. Krebs

Dynamic Modeling in Behavioral Ecology,
by Marc Mangel and Colin W. Clark

The Evolution of Parental Care, by T. H. Clutton-Brock

Parasitoids: Behavioral and Evolutionary Ecology,
by H.C.J. Godfray

Sexual Selection, by Malte Andersson

*Polygyny and Sexual Selection in Red-Winged
Blackbirds*, by William A. Searcy and Ken Yasukawa

Leks, by Jacob Höglund and Rauno V. Alatalo

Social Evolution in Ants, by Andrew F. G. Bourke
and Nigel R. Franks

*Female Control: Sexual Selection by Cryptic Female
Choice*, by William G. Eberhard

Sex, Color, and Mate Choice in Guppies,
by Anne E. Houde

Foundations of Social Evolution, by Steven A. Frank

Parasites in Social Insects, by Paul Schmid-Hempel

Levels of Selection in Evolution, edited by
Laurent Keller

Social Foraging Theory, by Luc-Alain Giraldeau
and Thomas Caraco

*Model Systems in Behavioral Ecology: Integrating
Conceptual, Theoretical, and Empirical Approaches*,
edited by Lee Alan Dugatkin

*Sperm Competition and Its Evolutionary
Consequences in the Insects*, by Leigh W. Simmons

*The African Wild Dog: Behavior, Ecology, and
Conservation*, by Scott Creel and Nancy Marusha Creel

Mating Systems and Strategies, by Stephen M. Shuster
and Michael J. Wade

Sexual Conflict, by Göran Arnqvist and Locke Rowe

Sexual Conflict

GÖRAN ARNQVIST
AND LOCKE ROWE

Princeton University Press
Princeton and Oxford

Copyright © 2005 by Princeton University Press

Published by Princeton University Press, 41 William Street,
Princeton, New Jersey 08540

In the United Kingdom: Princeton University Press,
3 Market Place, Woodstock, Oxfordshire OX20 1SY

LIBRARY OF CONGRESS CATALOGING-IN-PUBLICATION DATA

Arnqvist, Göran, 1961–
Sexual conflict / Göran Arnqvist and Locke Rowe.
p.   cm.   — (Monographs in behavior and ecology)
Includes bibliographical references (p.).
ISBN 0-691-12217-2 (cl : alk. paper — ISBN 0-691-12218-0 (pbk. : alk. paper)
1. Sexual behavior in animals. 2. Agonistic behavior in animals.
I. Rowe, Locke, 1959– II. Title. III. Series.
QL761.A75 2005
591.56′2—dc22   2004053520

British Library Cataloging-in-Publication Data is available

This book has been composed in Times Roman and Univers Light 45

Printed on acid-free paper. ∞

pup.princeton.edu

Printed in the United States of America

10 9 8 7 6 5 4 3 2 1

To Michelle and Elias
&
Lois and Taylor

Contents

Preface   xi

1 Sexual Conflict in Nature   1

1.1 Evolving Views of Sex and Reproduction   2

1.2 Sexually Antagonistic Selection and Sexual
     Conflict   6

   1.2.1 Intralocus Sexual Conflict   7

   1.2.2 Interlocus Sexual Conflict   10

1.3 Aims and Scope   11

2 Sexual Selection and Sexual Conflict:
      History, Theory, and Empirical Avenues   14

2.1 Darwin's Views on Sexual Selection   14

2.2 The Fisher Process   18

2.3 Indicator, or Good Genes, Mechanisms   22

2.4 The Male Trait   25

2.5 Direct Benefits   26

2.6 Preexisting Biases and the Origin of the
     Preference   27

2.7 Sexual Conflict   29

   2.7.1 Parker's Initial Models of Sexual
          Conflict   30

   2.7.2 Genetic Models   31

   2.7.3 Phenotype-Dependent and Phenotype-
          Independent Costs   34

   2.7.4 Nonequilibrium Models   35

2.8 Sexual Conflict Set in the Framework of
     Sexual Selection   35

2.9 The Roles of the Sexes in Sexual Conflict   38

2.10 Empirical Approaches to the Study
      of Sexual Conflict   40

3 Sexual Conflict Prior to Mating   44

3.1 The Economy of Mating and the Evolution
     of Resistance   45

   3.1.1 Direct Costs of Mating   45

3.1.2 Costs of Low Mate Quality 46

3.1.3 Costs of Resisting Mating 47

3.1.4 Costs to Females as a Side Effect of Male-Male Competition 48

3.1.5 Sexual Conflict and the Evolution of Sexual Cannibalism by Females 50

3.1.6 Sexual Conflict and the Evolution of Infanticide by Males 53

3.2 Adaptations for Persistence and Resistance 55

3.2.1 Harassment and Resistance 57

3.2.2 Grasping Traits 60

3.2.3 Antigrasping Traits and Other Forms of Resistance 68

3.2.4 Exploitation of Sensory Biases 71

3.2.5 Convenience Polyandry 77

3.3 Sexual Conflict and Sexual Selection 78

3.4 Mate "Screening" and Other Alternative Explanations for Resistance Traits 80

3.5 Case Studies in Sexually Antagonistic Coevolution 83

3.5.1 Diving Beetles 83

3.5.2 Water Striders 84

3.5.3 Bedbugs 87

4 Sexual Conflict after Mating 92

4.1 Female Reproductive Effort and the Conflicting Interests of the Sexes 96

4.1.1 Seminal Substances with Gonadotropic Effects 97

4.1.2 Nuptial Feeding 102

4.1.3 Male Display Traits 103

4.2 Female Mating Behavior, Sperm Competition, and the Conflicting Interests of the Sexes 106

4.2.1 Male Defensive Adaptations and Sexual Conflict 107

4.2.1.1 Costs of Delaying Remating in Females 111

4.2.1.2 Female Costs as Side Effects 116

4.2.1.3 Female Costs as a Direct Target of Male Strategies 118

4.2.2 Male Offensive Adaptations and Sexual Conflict 121

4.2.2.1 Sperm Competition and Aggressive Ejaculates 121

4.2.2.2 Direct Costs, Polyspermy, and Female Infertility 122

4.2.2.3 Indirect Costs and Deleterious Matings 128

4.2.2.4 Conflicts over Cryptic Female Choice 129

4.3 Conflicts over the Duration of Mating 132

4.3.1 Male and Female Adaptations 135

4.4 Postmating Conflicts and Male-Female Coevolution 139

4.5 Elaborated Male Ejaculates: Nuptial Gifts or Medea Gifts? 140

4.6 Are Male Postmating Adaptations Costly to Females? 146

4.7 It Takes Two to Tango: Sexually Antagonistic Coevolution in Fruit Flies 149

5 Parental Care and Sexual Conflict 156

5.1 The Basic Conflict 156

5.2 Mate Desertion 158

5.2.1 Conflict over Care and Desertion in Uniparental Species 158

5.2.2 Never Trust a Penduline Tit! 160

5.3 "Partial" Mate Desertion and Sexual Conflict over the Mating System in Biparental Species 164

5.4 Sexual Conflict over the Relative Amount of Care in Biparental Monogamous Species 170

5.5 The Dunnock: Family Life in Cambridge University Botanic Garden 174

6 Other Implications of Sexual Conflict 179

6.1 The Evolution of Genomic Imprinting 179

6.2 Sexual Conflict, Sex Ratios, and Sex Allocation 183

6.3 Dueling Worms and Stabbing Snails: Sexual Conflict within Hermaphrodites 185

6.3.1 Premating Conflict in Hermaphrodites 187

6.3.2 Postmating Conflict in Hermaphrodites 190

6.3.3 Sexual Selection and Antagonistic Coevolution in Hermaphrodites 192

6.3.4 The Love Dart in Snails—A Shot at
Paternity? 196

6.4 Sexual Conflict in Plants 200

6.5 Sexual Conflict, Speciation, and Extinction 203

6.5.1 Sexual Conflict as an Engine of
Evolutionary Divergence 207

6.5.2 Population Crosses—Inferring Process
from Pattern 210

6.6 Sexual Conflict and Sex Chromosomes 212

7 Concepts and Levels of Sexual Conflict 216

7.1 Levels of Analysis 216

7.2 Resolution of Sexual Conflict 219

7.3 Winners and Losers of Sexual Conflict? 220

7.4 Sexual Conflict over the Control of
Interactions 222

7.5 The Intensity of Sexual Conflict 223

7.6 Sexual Conflict over Mate Choice 224

8 Concluding Remarks 226

References 229

Author Index 305

Subject Index 321

Preface

Early in our Ph.D. programs, we met briefly at the XVIIIth International Congress of Entomology in Vancouver. At the time, both of us were studying the reproductive biology of water striders, so we kept in touch over the next few years. Remarkably, by the end of our degrees, we discovered we had independently come to the same conclusions about the mating system of these species—and that conclusion was sexual conflict. Since then, through the work of several laboratories, water striders have become much like *Drosophila*—a model system for the study of sexual conflict. This is a mixed blessing. On the one hand, it feels good to us to have been involved in their rise to prominence. Yet, just like fruit flies, water striders are regarded by some as poor representations of the "real" world. One aim of this book is to convince readers that this may not be true.

The idea of a book on sexual conflict was hatched in 2000, during a sabbatical leave in Umeå. At the time, growth in the field was explosive, and we felt a book was much needed. New discoveries and interpretations were accumulating rapidly as scientists increasingly considered the potential of conflict to shape male-female interactions. We felt a book would help to organize these observations, and set them in the context of established theories of sexual selection and mating system evolution. This book is written in that spirit. We are promoting the hypothesis that sexual conflict is an underappreciated force in the evolution of the sexes and their interactions. As such, many readers will feel that we have not spent enough time debating the alternatives. We acknowledge this. Instead, we point the way to interesting new systems where sexual conflict can be studied, and offer new interpretations of data that had been more or less routinely interpreted in a more classic framework. We believe that it is time to revise our view of male-female interactions, and we think that this exercise will lead to interesting discoveries. We have written this book primarily for graduate students and researchers working in evolution and evolutionary and behavioral ecology. We also anticipate that this book will have some utility for undergraduates and nonspecialists.

At the time we signed on for this book, it felt like a fairly minor endeavor—maybe like writing several papers, but without having to either do the experiments or analyze the data. It wasn't. It seemed to take a long time. And there are certainly interesting obstacles to overcome when authorship is joint. But, now that it is over, producing the book has also been more educational and more fun than we had presumed. Many people have contributed to this book. These include dozens of scientists too numerous to mention, who have provided photos and data, and who have generously shared their expertise with us. We thank them. Graduate students and other members of our laboratories have also contributed much, including ideas, references, and critical discussions, sometimes without even knowing it. Anna Gosline helped research parts of chapter 3. We are particularly grateful to our friends José Andrés, Russell Bonduriansky, Troy Day, and Ted Morrow for many hours of enlightening discussion about theory, tests, and interpretation. Drafts of various chapters were read by a number of colleagues, who all gave insightful and valuable comments that have helped us avoid most slipups (chapter): Anders Berglund (6), Mark Blows (2), Russell Bonduriansky (2,3), Tracey Chapman (4), Troy Day (2), Magnus Enquist (7), Jaco Greff (6), Dave Hoskens (4), Roger Härdling (7), Mark Kirkpatrick (2), Joris Koene (6), Hanna Kokko (2), Kate Lessels (5), Nico Michiels (6), Rolf Ohlsson (6), Leigh Simmons (4), Tamas Szekely (5), and Jon Ågren (6). Tom Tregenza and an anonymous reviewer read the entire body of the book—we are most grateful for their effort and insight. These chapters are so much better because of them. Joyce Besch did an enormous amount of editorial work on the text and references, and did it at the speed of light! Our editor Sam Elworthy and the staff at Princeton University Press were encouraging, professional, and very patient. We gratefully received financial support during the period the book was written from our main sources of research funds, the Swedish Research Council and the Natural Sciences and Engineering Research Council of Canada, as well as funds from the Swedish Foundation for International Cooperation in Research and Higher Education, the Knut and Alice Wallenberg Foundation, Premiers Research Excellence Award, and the *Magnus Bergvalls Stiftelse*. This book would certainly not have been possible without this support.

Our families never veered in their support, despite suffering our periods of distraction, and late-night writing binges, and, at times, having to live with the two of us under a single roof.

Toronto
April, 11, 2004
Göran Arnqvist
Locke Rowe

Sexual Conflict

1

Sexual Conflict in Nature

During the mating season, male robber flies roam through the vegetation in search of females. When approached by a male, a female takes to the wing, and the male pursues. If overtaken and grappled by a male, females of most species struggle violently, often successfully, to free themselves. In a few species, however, females may use another strategy to achieve the same result; if grasped by a male, they play dead! Once a female stops moving, a male apparently no longer recognizes her as a potential partner, loses interest, and releases the female, who falls to the ground and flies off.

Over the course of the egg-laying period, a male penduline tit (a song bird) makes repeated forays to his partner's nest in an apparent attempt to assess how many eggs she has laid thus far. Female partners, however, seem to make sure that this inspection job is not an easy one for males. Females bury their eggs in the bottom of the nest, perhaps to hide them, and become aggressive toward their mates. Observations in aviaries have shown that these inquisitive males are sometimes injured or even killed by their secretive mates.

As in most spiders where courting males are at risk of being cannibalized by females, a male of the funnel-web spider *Agelenopsis aperta* approaches a female slowly and very carefully. Once close, he sprays an apparent "toxin" at the female, which makes her (and occasionally him) collapse and enter an "unconscious" state. The male then hauls the cataleptic female about the web, repositions her, and finally inseminates her while she is still "knocked out."

At the time of mating, a male Malabar ricefish will dart toward a nearby female with near-lightning speed. Dashing toward the female from below, he strikes her in the genital region with a complex clublike organ. If the organ, a modified anal fin, contacts the female body, it releases a spermatophore (a sperm capsule) with a dartlike spike. This spike is pushed into the female flesh, and the spermatophore becomes firmly attached because of a whorl of recurved barbs at its tip. Females are adapted to these repeated assaults; they can be found with multiple attached spermatophores, and the skin around their genital pore is markedly thickened and hardened.

If one looks closely enough, observations such as these abound in the biological literature. But what do we make of these? It is difficult to reconcile observations of open conflict between the sexes with the common view of mating as a joint venture of two individuals that, by virtue of being the same species, share a common genome. Often, those cases where male-female interactions involve overt coercion, manipulation, deceit, or harm have been stowed away and otherwise obscured. If explanations are offered, they often rest on special circumstances.

The main message of this book is that, despite interacting males and females sharing the same genome, conflict between them is ubiquitous. In other words, some genes expressed in females will be in conflict with others expressed in males. Moreover, some genes expressed in both sexes may be favored to do divergent things when expressed in females and in males. Selection on these genes is therefore sexually antagonistic. We will argue that the robber flies, penduline tits, funnel-web spiders, and ricefish may all have something very important to tell us. That is, how males and females came to be the way they are.

1.1 Evolving Views of Sex and Reproduction

Biologists and laypeople have long regarded mating as a largely harmonious event in which males and females cooperate in producing offspring. In fact, if we assume the entire result of mating is the production of shared offspring, it is not immediately obvious why we should think otherwise. Embedded in this view may be an assumption of monogamy, where the reproductive success of each individual of the pair is equal to, and contingent upon, the other individual. In these cases, if one can elevate the reproductive success of a partner, both will reap equal benefits. If we accept all of this, it is easy to imagine that what is good for one partner is good for the other and, in fact, also good for the species. These three perceptions—harmony in sexual interactions, monogamy, and a concordance between what is good for the individual and the species—have a long history but are often incorrect.

Evolutionary views of the interactions between the sexes have been fundamentally influenced by the work of Charles Darwin. Darwin often portrays reproduction in general and sex in particular as being something essentially reserved for the monogamous and married couple. In his famous 1871 book *The Descent of Man and Selection in Relation to Sex*, he summarizes an experiment involving a small crustacean as follows: "The female, when thus divorced, soon joined the others. After a time the male was put again into the same vessel; and he then, after swimming about for a time, dashed into the crowd, and without any fighting at once took away his wife." Darwin did recognize the existence of polygamy, as evidenced in a letter to his friend Charles Lyell, where he describes female barnacles as having "two little pockets, in each of which she kept a little husband." Yet, he was clearly reluctant to entertain the idea that the "practice of polygamy" was widespread, and even saw this as the major problem for his theory of sexual selection. "Our difficulty in regard to sexual selection lies in understanding how it is that the males which conquer other males, or those which prove the most attractive to the females, leave a greater number of offspring to inherit their superiority than their beaten and less attractive rivals. Unless this result does follow, the characters which give to certain males an advantage over others, could not be perfected and augmented through sexual selection. When the sexes exist in exactly equal numbers, the worst-endowed males will (except where polygamy prevails), ultimately find females, and leave as many offspring, as well fitted for their general habits of life, as the best-endowed males." Today, thanks in part to the advent of DNA fingerprinting, we know that true genetic monogamy is in fact extremely rare (Birkhead 1997). In the absence of monogamy, as we will see, partnerships are temporary and the lifetime reproductive success of partners is no longer equal.

It is equally clear from Darwin's writings and in line with the general moral that he considered interactions between the sexes to be inherently good; an "aid to ordinary [natural] selection." By this we mean that events leading to mating, such as female choice of certain males or competition among males for access to females, work hand in hand with natural selection to improve the adaptedness of the lineage: "Just as man can improve the breeds of his game-cocks by the selection of those birds which are victorious in the cockpit, so it appears that the strongest and most vigorous males, or those provided with the best weapons, have prevailed under nature, and have led to the improvement of the natural breed or species." Darwin's own theories were likely influenced by the writings of his grandfather Erasmus Darwin. In his book *Zoonomia* (1794), he argued that the "purpose" of reproductive competition is to improve the species. "The final cause of this contest amongst the males seems to be that the strongest and most active animal should propagate the species, which should thence become improved."

This heritage is echoed in two prevalent ideas in modern evolutionary biology, which both ascribe similar utility to reproductive interactions. One concerns the hypothesis that male secondary sexual traits used in reproductive competition are "honest indicators" of male genetic quality, which enable females to select the males with the "best" genes as fathers for their offspring (Zahavi 1975). The other poses that male sexual traits function to "preserve the species," i.e., by allowing species recognition and hybridization avoidance (Mayr 1940, Lack 1968), and the linked belief among ethologists in the 1950s and 1960s that male courtship functions to allow the female to choose a male of the correct species and strengthen conspecific pair bonds.

Theoretical analyses of the consequences of male-female coevolutionary interactions confirm, in some cases, Darwin's assertion that sexual selection works in lockstep with natural selection to increase population fitness (e.g., Siller 2001, Agrawal 2001, Lorch et al. 2003). But, in other cases, the opposite is true; the outcome of coevolution between the sexes is to decrease the fitness of populations (Lande 1981, Gavrilets et al. 2001, Kokko and Brooks 2003). Yet this latter result has not been widely acknowledged. Perhaps it is simply that evolutionary change is just thought of as an inherently good thing! This underlying belief has sometimes been obvious to us when describing our own work on water striders to the public and some colleagues. In these species males grasp females without prior courtship and females struggle vigorously to get rid of males. Male mating attempts can be accounted for by selection to fertilize more eggs, and female resistance of these repeated male mating attempts can be accounted for because they are both superfluous and costly (e.g., Rowe et al. 1994, Arnqvist 1997a). Our studies suggest that both sexes have accumulated antagonistic adaptations, in a form of coevolutionary "arms race," that further their interests but at the same time are costly. The idea that mating interactions might cause the evolution of decreased fitness in a population has been particularly difficult for people to entertain, public and scholars alike, and has sometimes even been considered antiadaptationist.

The understanding among many biologists of male-female interactions has, nevertheless, changed over the last two decades. The view of reproduction as an exclusively cooperative endeavor has been challenged by the realization that the mates' interests in any interaction are often conflicting. Tracing the history of this shift is not easy. Although a few early contributions hinted at sexual conflict (e.g., Wickler 1968), the major players certainly include Robert Trivers, Richard Dawkins, and Geoff Parker. Robert Trivers (1972) was the first to provide a compelling evolutionary discussion of differences between the evolutionary interests of the sexes and the implications this might have for the evolution of parental care in particular. Richard Dawkins put the idea of sexual conflict before a wide audience—most notably in a book chapter entitled "Battle of the Sexes" in his widely read 1976 book *The Selfish Gene*, but also in a few other early contributions (e.g., Dawkins and Krebs 1978, 1979).

There is, however, no doubt that the most thorough discussion of the role of sexual conflict in the evolution of the sexes was a contribution by Geoff Parker, which was unfortunately published in a little-known book (Parker 1979). Although Parker (1979) made "no claim to originality for the suggestion that this asymmetry [in the relative interests of the sexes] can occur commonly in animals," certainly he contributed more than anyone else to the introduction and subsequent development of the field. This material was first drafted several years prior to its publication, and Parker's ideas likely influenced many other early thinkers in the field. Most ideas expressed by Parker (1979) were much ahead of their time. His landmark paper remains arguably underutilized, although it has recently begun to draw the attention it deserves (figure 1.1). Parker has continued to build the field with a steady flow of influential contributions regarding conflict in his model system, dungflies, and more generally with new theory (see the list of references). During the 1980s, the concept of sexual conflict was given some attention in the theoretical literature, but empir-

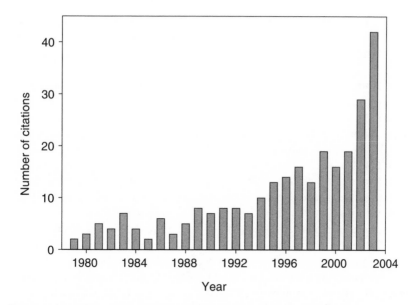

Figure 1.1. Interest in sexual conflict, and the resulting evolutionary processes, increased substantially during the mid-1990s. This point is illustrated by the number of citations/year of the classic paper by Parker (1979) that have appeared in the scientific literature since its publication. This landmark book chapter by Parker was much ahead of its time—few can hope that the citation rate of any of their contributions will continue increasing some 25 years after publication.

ical work on conflict and its consequences remained very rare. It was not until the 1990s that research on sexual conflict hit the mainstream of research in ecology and evolution, and interest has since increased dramatically. Why this sudden increase in attention?

We believe that several more or less coinciding factors have contributed to the almost explosive growth of the field. Students of animal behavior began focusing on females as active participants in coevolutionary interactions with males (e.g., Hrdy 1981, Kirkpatrick 1987a,b, Ahnesjö et al. 1993, Eberhard 1996, Gowaty 1997a, 2003, Zuk 2002). This fact may have resulted in part from philosophical shifts in thinking brought about by the feminist movement. The focus on the selective forces acting upon females in their interactions with males revealed that selection was often sexually antagonistic. Comprehensive data illustrating this point in dungflies were collected by Parker in the late 1960s (Parker 1970a,b) and began accumulating for a few other model organisms (e.g., Alatalo et al. 1981, Rowe et al. 1994, Clutton-Brock and Parker 1995). The work by Linda Partridge and coworkers on fruit flies, demonstrating male-imposed costs of mating for females, was particularly influential (e.g., Partridge et al. 1986, 1987, Fowler and Partridge 1989, Chapman et al. 1995). Despite the accumulating evidence, studies remained relatively rare, and most centered on insect behavior. One notable exception was a laboratory experimental evolution study of sexual conflict in flies (Rice 1996b).

This situation changed noticeably with the publication of a perspective by Brett Holland and Bill Rice in 1998. In this paper, they persuasively laid out the case for sexually antagonistic coevolution and did so with examples from favored taxa for studies of sexual interactions, such as birds and fishes. This paper certainly brought the evolutionary potential of sexual conflict to a much wider audience. Since then, the scope of study has increased not just taxonomically, but also in level of analysis. Molecular studies showing positive selection on, and rapid evolution of, reproductive proteins are now being interpreted as evidence of sexual conflict (see Chapman 2001, Swanson and Vacquier 2002). New theory suggests that sexual conflict has profound implications for the evolution of reproductive isolation and speciation itself (Rice 1998a, Parker and Partridge 1998). These facts reinforce the view that sexual conflict may be playing a central role in evolution within and between species. Today, the field is still rapidly expanding and filled with lively debate (Zeh and Zeh 2003). Hopefully, these accumulating data and debates will enrich our understanding of the biology of male-female interactions, and the evolution of the sexes themselves.

1.2 Sexually Antagonistic Selection and Sexual Conflict

The evolutionary interests of any two individuals are rarely, if ever, identical. This must be true, simply because they do not share the same alleles for all loci, with the possible exception of clones. Where alleles at a locus differ, there will be competition among them. Therefore, the evolutionary interests of any two individuals have the potential to conflict, even if they share much of the same genome. When two unrelated individuals of the two sexes meet to engage in sexual reproduction, conflict intensifies: while one individual needs the other to spread its genes, this dependence may be asymmetric. Further, although both have a shared interest in any mutual offspring, they usually have divergent interests in many aspects of the "economics" of reproduction. Defined as a conflict between the evolutionary interests of individuals of the two sexes (Parker 1979), sexual conflict becomes exacerbated because selection will often be tugging in different directions in the two sexes. Sexually antagonistic selection is symptomatic of sexual conflict. We expect sexually antagonistic selection to be common simply because the "roles" of the sexes in reproduction differ (Clutton-Brock 1991, Andersson 1994), and sexual conflict should be a general feature in the evolution of the two sexes.

Conflict between coreproducing individuals occurs also in isogamous species (gametes of equal size), because increased investment by one partner permits decreased investment by the other. In fact, there are good reasons to believe that "primordial" reproductive conflict in isogamous taxa has led to the evolution of anisogamy, and consequently the sexes (see Parker et al. 1972, Parker 1979, 1984, Bulmer and Parker 2002). Because males and females, by definition, produce gametes of different size (anisogamy), they typically maximize their reproductive success in more or less different ways. A classic example of this is the fact that, over much of the natural range of mating frequencies, the number of offspring produced generally increases monotonically with the number of mates in males but not in females (Bateman 1948). When this is so, there is selection favoring increased mating frequency in males, but not females. The fact that the sexes thus "play" very different roles during reproduction then sets the scene for differences in physiology, morphology, and behavior between males and females. This fact also means that the attributes or trait values favored in one sex are often not those favored in the other. Consequently, there will be sexually antagonistic selection on these traits. As such, some degree of sexual conflict will be ubiquitous in sexually reproducing taxa (Trivers 1972, Parker 1979, Lessels 1999).

The effects of sexually antagonistic selection at the genetic level can be described as intra- or interlocus conflict, depending on whether the target of selection is determined by alleles at one locus expressed in both sexes, or alleles at different interacting loci in the two sexes (Parker and Partridge 1998).

1.2.1 INTRALOCUS SEXUAL CONFLICT

Whenever selection favors different values for a phenotypic trait in males than in females, there is the potential for *intralocus sexual conflict*. For example, we could imagine that selection for elongated tails in males of some songbird occurs because females prefer to mate with males possessing long tails. We might then expect males with longer tails to be favored, even if there is some cost to efficient flight. In females, on the other hand, selection favors some intermediate tail length that maximizes flight efficiency (assuming that males are indifferent to female tail length when choosing a mate). In this case, optimal tail length depends upon which sex is assessed. If genes at the same locus (or loci) determine tail length in both sexes, then different alleles at that locus will be favored in females than in males. If expression of a given allele at that locus moves one sex toward its tail length optimum, the same allele expressed in the other sex will move that sex away from its tail length optimum. Alleles at the tail length locus are then under sexually antagonistic selection, and there is intralocus sexual conflict. Intralocus conflict will appear in all those cases where the direction of selection at a given allele depends upon in which sex it resides (i.e., whenever there is a sex x genotype interaction for fitness).

Intralocus conflict has great potential to limit adaptive evolution in both sexes. Because the sexes largely share the same genome (excluding sex chromosomes), and genes at many loci are being pulled in opposite directions by antagonistic selection in the two sexes, selection in one sex will impede adaptive evolution in the other (Rice 1984, Halliday and Arnold 1987, Lande 1987, Parker and Partridge 1998, Rice and Chippindale 2001). In the most general case, neither sex will be able to reach its adaptive peak during these evolutionary tugs-of-war. The resulting average phenotype will represent some evolutionary "compromise" between the evolutionary interests of the two sexes. However, males and females of most species do look and behave differently, they are sexually dimorphic, and this fact demonstrates that the constraint of sharing much of the same genome is not absolute. The evolution of sex-limited gene expression is one way out of the bind (Rice 1984), because it permits independent evolution in the two sexes toward their phenotypic fitness optima.

Although intralocus sexual conflict is potentially common and consequential, its evolutionary importance is debated. On the one hand, one could argue that the optimal phenotype in males and females should be different for virtually every conceivable phenotypic trait. On the other hand, one could argue that sexual dimorphism will evolve relatively rapidly and easily under selection for sex-limited expression of genes, implying that intralocus conflicts will at most have only transient effects on evolutionary dynamics. Thus, our view of the potential evolutionary significance of intralocus conflicts depends on the importance we place on evolutionary constraints in general. Some theory suggests that the evolution of sex-limited expression of sexually antagonistic genes may be slow (Lande 1980, 1987), but debate continues (see, e.g., Badyaev 2002 vs. Lindenfors 2002 for contrasting views on the evolution of sexual size dimorpism).

There is growing empirical evidence suggesting that intralocus sexual conflict may be persistent. Several studies have documented sexually antagonistic selection on traits in a variety of species (e.g., Price and Burley 1994, Merilä et al. 1997, Björklund and Senar 2001) (figure 1.2). Likewise, some laboratory studies have found genotype by sex interactions for adult fitness components (e.g., Vieira et al. 2000, Mackay 2002), and Chippindale and coworkers have provided strong evidence for intralocus conflict in laboratory stocks of the fly *Drososophila melanogaster* (Chippindale et al. 2001, Gibson et al. 2002). These authors used genetic techniques to assay the effect of sets of alleles on the fitness of males and females that otherwise shared the same genotype. When assayed in larvae, before the sexes look or behave differently, sets of alleles that increased fitness in one sex usually increased fitness in the other; good alleles were good whichever sex they were expressed in. However, when assayed in adults, when the sexes look and behave differently, sets of alleles that increased fitness in one sex generally decreased fitness in the other (figure

Females

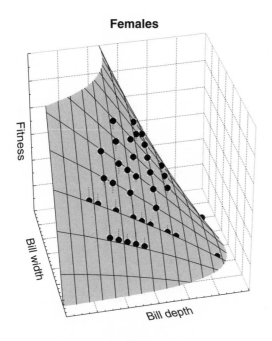

Males

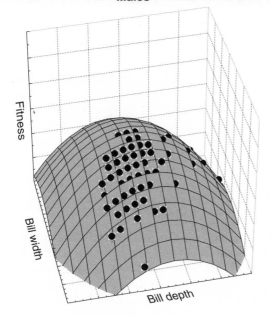

Figure 1.2. Survival selection on bill morphology differs between the sexes in a natural population of the serin (*Serinus serinus*), a carduline finch, presumably as a result of differences in food utilization between males and females. Selection on bill depth and width is directional in females but stabilizing in males. (Reprinted from Björklund and Senar 2001, with permission from Blackwell Publishing Ltd.)

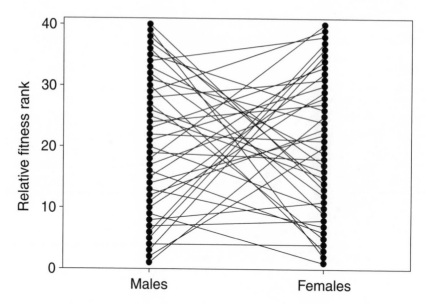

Figure 1.3. Plot of male and female adult fitness in a suite of discrete *Drosophila melanogaster* haplotypes. Note that these haplotypes tend to have opposing effect on fitness in the two sexes: haplotypes that result in high fitness in males tend to result in low fitness in females. These data demonstrate that, for at least some genes in this species, their effect on the fitness of males and females is antagonistic. (Reprinted from Chippindale et al. 2001, with permission from the National Academy of Sciences.)

1.3). These data demonstrate that there is substantial sexually antagonistic selection on these loci, and hence intralocus sexual conflict. At the moment, however, it is very difficult to evaluate how general or important intralocus sexual conflict might be, largely because of a paucity of relevant empirical studies. For this reason, we will refer to intralocus sexual conflicts to only a limited extent in this book.

1.2.2 INTERLOCUS SEXUAL CONFLICT

Whenever there is conflict over the outcome of male-female interactions, such that the optimal outcome is different for the average male and female, there is the potential for *interlocus sexual conflict*. Conflict can occur over any interaction between the sexes, including mating rate, fertilization efficiency, relative parental effort, remating behavior, and female reproductive rate. For example, imagine that there is selection for higher mating rate in males, and selection for lower mating rate in females. If mating rate is determined at the same locus, then intralocus conflict will result (Halliday and Arnold 1987). If, however, mating rate is instead determined by an interaction between a locus A expressed in males and a locus B expressed in females, the result will be very different. Alleles that are favored at locus A are those that increase the mating rate of their male bearers, and these will, therefore, spread in the population. These alleles, and their consequences for the phenotype, are adaptive for males (they increase male mating rate), but at the same time they are detrimental to females (they increase female mating rate). Consequently, their spread exerts selection at locus B for alleles that effect a decrease in female mating rate. Spread of such alleles at locus B can then affect selection among alleles at locus A. You can see that this intersexual interlocus conflict may be continuous and thereby affect a continuous stream of allelic replacement at one or many interacting loci in both females and males (see Rice and Holland 1997). Interlocus conflict is thus mediated by interactions between different loci in the two sexes and is of great interest because of its potentially central role in male-female coevolution.

When there is conflict over the outcome of any interaction between males and females, we expect intersexual interlocus conflict to result in suites of interacting and sexually antagonistic adaptations, in both sexes, that function to bias the outcome of interactions toward the evolutionary interest of their bearer (Parker 1979, Rowe et al. 1994, Rice and Holland 1997, Gavrilets et al. 2001). Parker's early work (e.g., Parker 1979, 1983a) suggested that interlocus sexual conflict should lead to complex, dynamic, and even unresolvable evolutionary chases between the sexes, and more recent theory has confirmed this. The predicted result is a coevolutionary "arms race" between the sexes, known as sexually antagonistic coevolution (figure 1.4). During these coevolutionary episodes, one set of traits in males (persistence adaptations) interacts with a different set of traits in females (resistance adaptations) in determining the outcome of a given interaction.

It is this sexually antagonistic coevolution that makes interlocus conflicts a particularly interesting field of study with important potential ramifications for diverse topics, including the evolution of gamete interactions, mate choice, social interactions, genomic imprinting, parental care, sexual dimorphism, and speciation. This book is primarily about this form of sexual conflict.

1.3 Aims and Scope

In this book we aim to illustrate the wide diversity of adaptations in both sexes that we believe are related to sexual conflict. The bulk of this book is then a series of chapters dealing with the major components of male-female interactions. These chapters highlight the ways in which the interests of the sexes differ in an interaction, and the adaptations in both sexes that have resulted from this conflict. For each component of male-female interaction, we discuss examples that we believe are particularly compelling, illuminating, or otherwise interesting. Our hope is that this survey of the natural history of sexual

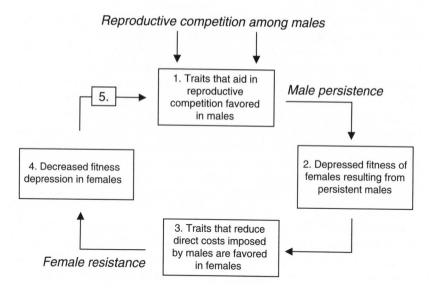

Figure 1.4. Interlocus sexual conflict can result in perpetual cyclical or escalating coevolution of reproductive traits in the two sexes, in a process known as sexually antagonistic coevolution. This process is generally seen as being driven by reproductive competition among males, by male-male interference and/or exploitation competition, simply because males have a higher potential reproductive rate in most taxa. Selection thus continually probes males for new mutations that are beneficial to their bearers in this competition (1). A subset of novel male adaptations will depress fitness in females that interact with the bearers of these adaptations (2). Such traits are referred to as male persistence adaptations. The spread of male persistence adaptations will result in selection among females to reduce the direct costs imposed by persistent males (3). Female traits, or alleles, that in any way reduce such costs are referred to as female resistance (4). The evolution of female resistance adaptations may then feed back and change the strength or form of competition among males (5), causing selection for altered persistence.

conflict will convince readers that sexually antagonistic coevolution may be, or has been, involved in the evolution of a great variety of these traits that distinguish the sexes.

A second aim is to set sexually antagonistic coevolution into the well-established theoretical framework of coevolution between the sexes. There has been a great burst of empirical research into sexual conflict in the last decade. Yet, as in any young and rapidly expanding field, there is a risk that collection and interpretation of data gets beyond the development of a solid theoretical foundation. At present, there is considerable variation in how these new data are being interpreted with respect to competing theories of coevolution between the sexes. Documenting coevolution between the sexes has been relatively straightforward, but distinguishing among alternative mechanisms that may be responsible for this coevolution has been much more difficult. For example, there is still little consensus among students of sexual selection about the extent to which females choose mates on the basis of the vigor their off-spring will have or the sexiness of their sons, despite more than two decades of intense research (Kirkpatrick and Ryan 1991, Andersson 1994, Kokko et al. 2003). Addition of a new mechanism, sexually antagonistic coevolution, certainly does not make this task any easier. Yet we hope that by providing evidence for sexually antagonistic coevolution in nature and setting this newer idea within the established framework of sexual selection, some progress can be made.

Finally, in this emerging field, conceptual ambiguities and inconsistent ter-minology are holding back progress. Ambiguities include finer distinctions (e.g., what is indirect selection?) but also extend to the core of the field (e.g., what is sexual conflict?). Much of the ambiguity results from the distinct lexi-cons and conceptual backgrounds of contributing scientists. We hope to bring this field a bit forward by being more precise in discussing concepts, or at least by identifying concepts that remain fuzzy.

We first review the development of the theories of sexual selection and sexual conflict, and attempt to integrate them (chapter 2). This chapter neces-sarily assumes some background in evolutionary biology, and consequently will be of more interest to students of evolutionary biology than to the general reader. The next four chapters constitute the natural history of conflict in male-female interactions among animals. These interactions are broken into those occurring between individuals prior to mating (chapter 3) and those occurring after mating (chapters 4 and 5). In chapter 6, we discuss a series of interesting issues that fall somewhat outside the three previous chapters. For example, we discuss conflict in hermaphrodites and in plants, and the consequences of con-flict for divergence of lineages. Finally, in chapter 7 we discuss some ambigu-ities that have arisen as a consequence of the rapid growth of this field, and the merging of concepts and terminology from game theory, evolutionary ge-netics, and behavioral ecology.

It is not our intention to present a complete treatise on sexual conflict, an endeavor that is probably better left for a time when the field has matured. We have also emphasized some aspects over others (e.g., interlocus over intralocus conflict), in part because these are richer in examples, but also, admittedly, because of our own research interests. If our presentation comes across as being biased, speculative, and/or promotional at times, it is because our aim is to make the case for the ubiquity of sexual conflict rather than provide a balanced review of all aspects of male-female coevolution. We hope to encour-age many readers to think in novel ways about those interactions involved in reproduction.

2

Sexual Selection and Sexual Conflict: History, Theory, and Empirical Avenues

Since Darwin (1871), at least, the notion of conflict between the sexes has been present in discussions of sexual selection, sexual dimorphism, and the sex roles. However, it was a century after Darwin before sexual conflict, and its potential to shape the sexes and their interactions, was fully recognized. As mentioned earlier, we see two landmarks in this early period. The first was R. L. Trivers' (1972) recognition that asymmetries between the sexes in parental investment, beginning with anisogamy, led to distinct roles for the sexes, and distinct evolutionary interests. Trivers' fundamental insight offered a single explanation for widespread patterns in mating systems, sexual selection, and sexual conflict. Second was G. A. Parker's (1979) conversion of his own observations of conflict in nature into a formal theory of sexually antagonistic coevolution. These two book chapters launched the field and remain essential reading today.

In this chapter, we aim to review the development of current theory in sexual conflict, and this necessitates a review of the theory of sexual selection. Our review of sexual selection is brief, but we note that several recent reviews exist (Bradbury and Andersson 1987, Kirkpatrick 1987a, Pomiankowski 1988, Kirkpatrick and Ryan 1991, Kokko et al. 2003), including Malte Andersson's (1994) monograph on the topic. We will argue that much of the research on sexual conflict, at least interlocus conflict, fits nicely within the broader theoretical structures of the modern theory of sexual selection. In fact, the fit is so good that it is entirely unclear whether a distinct boundary even exists between what we refer to as sexual conflict and the body of theory on sexual selection. Yet it is clear that novel insights and hypotheses and new data have been generated by the recognition of intersexual conflicts involved in male-female interactions.

2.1 Darwin's Views on Sexual Selection

In the *Origin of Species by Means of Natural Selection*, Darwin briefly outlined the problem of sexual dimorphism and offered his theory of sexual selection as an explanation (Darwin 1859). His definition of sexual selection pointed to the central issue his theory intended to address. Sexual selection "depends, not on a struggle for existence in relation to other organic beings or to external conditions, but on a struggle between the individuals of one sex, generally the males, for the possession of the other sex." The problem sexual selection solved was the widespread differences between the sexes in morphology and behavior, which were not readily explained by his theory of natural selection. These differences included "the greater size, strength, and pugnacity of the male, his weapons of offence or means of defence against rivals, his gaudy colouring and various ornaments, his power of song, and other such characters" (Darwin 1871). Natural selection, "the survival of the fittest," concerned the struggle for existence. Yet there were many traits, particularly the extravagant traits of male birds and armaments of male insects, that seemed ill suited to "the struggle." His major treatment of sexual selection was saved until 1871, when it occupied the second half of his book *The Descent of Man and Selection in Relation to Sex*. There the definition of sexual selection appeared as "the advantage which certain individuals have over others of the same sex and species solely in respect of reproduction" (Darwin 1871). Thus the study of sexual selection, introduced by Darwin and adhered to today, is the study of those traits that yield an advantage in reproductive competition and, therefore, the fitness of their bearers.

On the face of it, there is no obvious reason to see sexual selection as truly distinct from natural selection (Wade and Arnold 1980, Endler 1986, Andersson 1994). Indeed, many or most evolutionary biologists today view sexual selection as a component of natural selection (figure 2.1). Studies of sexual selection concern the causes of variance in mating success, or more precisely fertilization success, whereas studies of natural selection concern the causes of variance in the remaining components of fitness (survivorship and fecundity). Yet we still typically draw a line between studies of natural and sexual selection. There are probably a number of reasons for this, foremost being history. But sexual selection also appears to be a particularly diversifying and strong force of evolution. Much of the diversity that is evident in nature consists of traits related to reproduction that differ between the sexes in their degree of expression. This is illustrated with a quick perusal of your field guide to the local birds. Species are usually recognized by plumage patterns or song, which are usually exaggerated in only one sex, usually the male, and these traits are usually shaped by sexual selection. Sexual selection also appears to be a particularly strong force of evolutionary change. At the time, Darwin saw sexual selection as a weaker or "less rigorous" force of evolutionary change than natural selection, because in contrast to natural selection "the result is not death to the unsuccessful competitor, but few or no offspring" (Darwin 1859). However, this is not a viewpoint to which many subscribe today. Variance in mating success is often greater than that of other fitness components, including

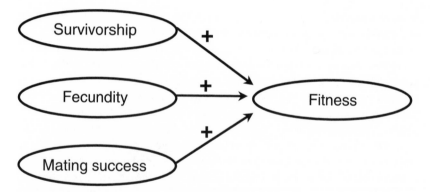

Figure 2.1. Fitness can be divided into three fitness components: survivorship, fecundity, and mating/fertilization success. Sexual selection describes variance in mating/fertilization success that is attributable to variation among individuals in some underlying metric trait (behavioral, physiological, or morphological). For example, males with brighter plumage markings may have higher mating success, and therefore fitness, than males with drabber markings. Natural selection describes variance in all other fitness components. Thus sexual selection can be viewed as the subset of natural selection that concerns variance in mating/fertilization success. Other visual schemes for distinguishing sexual from natural selection are available (e.g., Arnold and Duvall 1994), but the broad picture does not change.

survivorship, and consequently sexual selection is relatively strong (Hoekstra 2001, Kingsolver et al. 2001).

Not all differences between males and females are attributable to sexual selection. Most obvious among these are certain primary sexual traits, such as the mammary glands of mammals, which are necessary for offspring survival. Such traits, according to Darwin, do not require sexual selection as an explanation because natural selection suffices. In contrast, secondary sexual traits, such as the gaudy plumage of birds of paradise, have no apparent function other than competition for mates or fertilizations. Darwin (1871) did not recognize the importance of postmating sexual selection (i.e., sperm competition and cryptic female choice) (Birkhead 1997), and therefore struggled with the distinction between primary and secondary sexual traits: "Unless indeed we confine the term primary to the reproductive glands, it is scarcely possible to decide which ought to be called primary and which secondary." We know today that many traits traditionally considered to be primary sexual traits subjected to natural selection have actually been sculpted by postmating sexual selection, for example, male genitalia (e.g., Eberhard 1985, Arnqvist 1998) and reproductive glands (e.g., Hosken 1997, Dunn et al. 2001), and there are several good reasons to abandon the dichotomy between primary and secondary sexual traits. The problems with distinguishing between primary and secondary sexual traits are discussed in, for example, Darwin (1871), Eberhard (1985, 1996) and Andersson (1994). As we will see throughout this book, drawing sharp distinctions between classes of traits or evolutionary processes in sexual selection is often futile.

Darwin divided sexual selection into two classes of competition for mates, where "in the one it is between individuals of the same sex, generally the males, in order to drive away or kill their rivals, the females remaining passive; whilst in the other, the struggle is likewise between the individuals of the same sex, in order to excite or charm those of the opposite sex, generally the females, which no longer remain passive, but select the more agreeable partners." (Darwin 1871). The first we now refer to as intrasexual competition for mates (male-male competition) and the second intersexual competition (female choice). This distinction was fundamental to Darwin. At one end of the pole one might imagine a species in which males, armed with horns, fight among one another to control territories occupied by receptive females. Here the horns are under intrasexual selection, and no coevolution of horns and some other trait in females is required to understand their evolutionary exaggeration. At the other end, one might imagine a species where males gather and display their ornaments to passing females, who first inspect the lot, and then choose the most ornamented for a mate. Here, male ornaments are under intersexual selection by female choice and there is an expectation of coevolution between ornament and choice (see below). However, it is easy to find cases where the distinction is not so clear and the boundary is hotly debated (e.g., Berglund et al. 1996, Wiley and Poston 1996, Birkhead 1998, Pitnick and Brown 2000).

Darwin noted that choice was typically exercised by females, and intrasexual competition occurred among males, but was aware that these "roles" were not universal. More recent research suggests that sex role reversal or partial reversal may be much more common than once thought (reviewed by Gwynne 1991, Bonduriansky 2001). In this book, we will focus primarily on direct interactions between males and females. This is not to say that male competition for mates does not have positive and negative side effects on female fitness. It sometimes does, and we will discuss these cases. Instead we focus on intersexual selection because sexual conflict is most evident in cases where interactions occur between the sexes, and these interactions have garnered the most theoretical and empirical attention.

Our focus requires a working definition of choice, or preference. Defining preference in the context of sexual selection has created a lot of debate (Halliday 1983, Maynard Smith 1987, Pomiankowski 1998, Andersson 1994, Kokko et al. 2003). We will follow Halliday (1983), Maynard Smith (1987), and Pomiankowski (1988) and use a broad definition of female preference: any female trait (behavior, structure, etc.) that biases conspecific male mating/fertilization success toward certain male phenotypes. This definition of preference has the virtue of implying no higher mental capabilities in those exercising preference, and no particular evolutionary force in its origin. We will also use the terms preference, choice, and mating biases interchangeably.

For Darwin, the currency of sexual selection was the number of matings achieved by a competitor. This currency assumes that there is some tight and consistent relationship between mating success and fertilization success. This need not be the case. It was one of Parker's (1970c) fundamental insights to extend sexual selection to interactions among competitors that occur after mating. For example, males mating with already mated females may displace the sperm of the first male and replace it with their own. In this case, both first and second males have mated, but the second male has a disproportionate success in fertilization and hence fitness. These postmating processes are now known to be a powerful component of sexual selection (reviewed by Eberhard 1996, Birkhead and Møller 1998). We will consider sexual selection to comprise all of those pre- and postmating events that result in variance in the rate at which males achieve fertilizations.

Finally, Darwin viewed sexual selection primarily as a positive force that worked alongside natural selection to promote those variants that were most adapted to their environments. "Amongst many animals, sexual selection will give its aid to ordinary [natural] selection, by assuring to the most vigorous and best adapted males the greatest number of offspring" (Darwin 1859). The conviction that sexual selection is "adaptive" in this sense, is still a common view today. However, a debate exists between this adaptive school and those that believe sexual selection can be essentially neutral or even "maladaptive" to the fitness of populations (e.g., Agrawal 2001, Siller 2001, Kokko and Brooks 2003, Lorch et al. 2003). Much of the intuition underlying the claim that sexual conflict is distinct from sexual selection hinges on whether male-female mating interactions are "adaptive" or "maladaptive" for females. For this reason, it is a point to which we will return in this chapter and elsewhere. Suffice it to say that the debate also occurred in Darwin's own writings. On the one hand, Darwin created the theory of sexual selection to explain particular extravagant traits that appeared maladaptive to males in terms of natural selection, yet at the same time he apparently believed that sexual selection improved overall adaptation of the lineage.

2.2 The Fisher Process

Male competition for mates was accepted by most evolutionists after Darwin, perhaps because it was relatively easy to see in action and therefore to understand (Andersson 1994). Female choice, however, was not widely accepted in the years following publication of the *Descent*. Darwin's description of the process of choice was not particularly helpful, consisting primarily of the observation that females discriminate among males and the speculation that this drives the evolution of male characters. On the one hand, it is easy to see why those male traits favored by females are exaggerated over evolutionary time.

But it is not immediately obvious why females should prefer those traits in the first place. The origin and exaggeration of female mating biases are at the heart of the modern problem of sexual selection. This problem largely reduces to an understanding of the details of selection acting upon the preference, now and in the past. Indeed, the myriad of extant models of sexual selection can be categorized according to the selective forces that are assumed to shape the preference itself (Kirkpatrick 1987a, Kirkpatrick and Ryan 1991). None of these models were present for several decades after the *Descent*, and it is likely for this reason that sexual selection by female choice languished until the first verbal model by Fisher (1915, 1930) was formalized by O'Donald (1962), Lande (1981), and Kirkpatrick (1982).

In a short passage of his book *The Genetical Theory of Natural Selection*, R. A. Fisher (1930) outlined an explanation for the evolution of preference and subsequent coevolutionary exaggeration of both the preferred trait and the preference itself. Curiously, his logic may have its origin in a sarcastic comment by T. H. Morgan (Andersson 1994). Fisher imagined first that females preferred some unremarkable male trait that was correlated, even weakly, to fitness. At this stage natural selection and sexual selection, through the preference, would work together. A mutation causing females to prefer these males would be favored because male (and perhaps female) offspring from this union would be more fit themselves, owing to carrying the genes of their more fit fathers. These offspring would also carry their mother's preference genes, and through this statistical association, the preference itself would spread in the population. The "initial advantage" was not the result of the preference itself, but the result of the higher nonmating fitness of preferred males. This piece of his argument has the ring of current "good genes" models (section 2.3; see also Fisher 1915), and offers an explanation for the early spread of the preference. It is also true that it is often overlooked in explaining Fisher's contribution to the field, where instead the second half (below) of the argument is emphasized.

As the preference spreads, males gain an additional advantage because an increasing proportion of the population of females prefers to mate with them over other males. In this way, the coevolutionary process becomes self-reinforcing. When the preference increases in frequency or intensity, exaggeration of the male trait yields an increasing advantage to males. Male offspring from these matings are increasingly favored because they express their father's preferred trait. These same males carry the genes for the preference, and therefore the preference itself is increasingly favored. As the male trait is exaggerated, it eventually becomes a hindrance to survivorship, and its further exaggeration can be checked by natural selection. If evolution of the male trait is halted, the female preference is likewise halted, because it evolves only through its statistical association with the male trait. This statistical association between genes for the preference and for the male trait is the central element of this process. It forms naturally as a consequence of assortative mating between individual females with the preference and males with the trait, and appears as a genetic correlation between male trait and female preference.

We will refer to the self-reinforcing effect resulting from a genetic correlation between preference and trait as the Fisher process. The term "sexy sons" is also often used to describe the process. Although its origin was decades later and was meant for a more specific scenario than Fisher's (Weatherhead and Robertson 1979), it appears to capture his thoughts—"hens choose the fashion of their sons' ornaments." Fisher's argument addresses the potential for rapid joint or coevolutionary exaggeration of both the preferred trait and the preference itself, and thus offers an explanation for Darwin's observation of often wildly exaggerated male traits that would otherwise be lost due to natural selection.

Fisher's intuition was largely confirmed in a series of models a few decades later (e.g., O'Donald 1962, Lande 1981, Kirkpatrick 1982; reviews by Arnold 1983, Kirkpatrick 1987a, Andersson 1994). These early models demonstrated that sexual selection through female choice could easily account for the presence of sexual dimorphism, and the costly nature of preferred traits. Interestingly, Lande (1981) and Kirkpatrick (1982) showed that there is no single point of equilibria in these genetic models. Instead, there is a line of equilibria along which the forces of sexual and natural selection on the male trait are balanced. Toward one end, both sexual selection and opposing natural selection are strong, with both diminishing as you move toward the other end. Where a population comes to reside along this line depends largely on the values of trait and preference when the process was initiated. This fact alone offers an explanation for the extraordinary variation we see among species in sexually selected traits (Lande 1981).

Several elements of the assumptions and results of these collective efforts should be emphasized. It is assumed that there is no *direct* selection on the preference itself: females that have the preference share the same values for survival and fecundity as those females with no or a less developed preference. If there were some direct selection on the preference, the equilibrium value of preference would collapse to its natural selection optimum. An association, or genetic correlation, between genes for the preference and those for the male trait is a prediction of the model, and is critical to the dynamics of their co-evolution. In the absence of direct selection, the preference is evolving entirely because of its genetic correlation with the male trait, which is under direct selection by female choice. That is, preference is evolving under *indirect* selection. Moreover, the rate of evolution of the preference depends upon the degree of genetic covariance with the male trait. If the covariance is great enough, then a runaway of trait and preference can occur. In this case, one to which Fisher alluded, evolution can be very rapid, and under this assumption neither trait nor preference reaches an equilibrium.

The feeling one gets from these models is that sexual selection is not promoting adaptation in the Darwinian sense. Instead, sexual selection is "dragging" male traits off their natural selection optima. In fact, if there is not perfect sex limitation of favored male traits, then females themselves will be knocked off their natural selection optima. In other words, as female preference pulls the male trait off the male natural selection optimum, it will simultaneously be pulling the female trait values off their optimum, to the degree that there is a genetic correlation between males and females for the trait (Lande 1981). Such a genetic correlation could arise for a number of reasons, one being that the loci underlying the preferred trait are shared by the sexes. We may expect sexual selection to often lead to intralocus conflict simply because it tends to drive male and female optima apart. This "non-adaptationist" view of sexual selection is at odds with both Darwin's belief that "sexual selection will give its aid to ordinary [natural] selection," and the intuition and evidence of biologists that sexual selection often favors the highest-"quality" or most "vigorous" males and that this will necessarily benefit the female partner (e.g., Maynard Smith 1956, Williams 1966, Trivers 1972, Zahavi 1975, Borgia 1979, Thornhill and Alcock 1983; for reviews of the evidence, see Johnstone 1995, Møller and Alatalo 1999, Jennions et al. 2001).

An important shortcoming of the Fisher process, as a singular explanation for sexual dimorphism, is its assumption that the preference is under no direct selection at all. Yet costs of mate choice appear to be common (Pomiankowski 1987b, Reynolds and Gross 1990, Andersson 1994). Such costs may include increased risk of predation, reduced foraging, delayed production of offspring, and a variety of other costs to females. These costs amount to direct selection on the preference, because the expression of preference affects that female's fecundity or viability. Several modeling efforts have demonstrated that costly preferences cannot be maintained by the Fisher process alone (e.g., Kirkpatrick 1985, Pomiankowski 1987a, Bulmer 1989, Cameron et al. 2003). When direct selection on the preference is included, the "line of equilibria" usually collapses to a single point where female fecundity and survivorship are jointly maximized. In other words, when the preference is under natural selection, the Fisher process has no effect on the equilibrium value of preference and trait. There are some interesting exceptions to this rule (Pomiankowski et al. 1991, Day 2000, Cameron et al. 2003), but, in each case, other relatively improbable factors are required (e.g., weak costs of preference combined with biased mutation keeping the male trait off its optima). The fact that the Fisher process is so sensitive to direct selection on the preference severely limits the likelihood that it alone explains much of the prevalence of elaborate male traits.

Another once controversial set of models, the indicator or good genes models, address two potential shortcomings of the Fisher process as modeled by Lande (1981) and Kirkpatick (1982); its extreme sensitivity to direct selection on the preference, and its apparent inability to account for the observation that females often prefer to mate with high-quality males.

2.3 Indicator, or Good Genes, Mechanisms

Fisher (1915), and later Williams (1966), reasoned that females would be favored to prefer the most fit males to father their broods, and only the most fit males would be able to fully express secondary sexual characters. Thus, it would pay females to prefer exaggerated male traits as a means of identifying fit males. This hypothesis became known as the "good genes" or "indicator" model of preference evolution. It is also sometimes referred to as the "handicap" model because of Zahavi's (1975, 1977) emphasis on the handicapping function of preferred traits. He argued that preferred traits, by virtue of their costs, act as handicaps to their bearers, where only males of the highest genetic quality could bear the costs of expressing these traits to their fullest. Therefore, Zahavi reasoned that trait expression acts as an honest indicator of male genetic quality, and female preference for the indicator trait is favoring the same males that are favored by natural selection. This model accounts for the widespread observation that females tend to mate with high-quality males, and is in accord with Darwin's view that sexual selection would "give aid" to natural selection. Yet the hypothesis met stiff opposition from theorists of the day (see Maynard Smith 1985, Kirkpatrick 1986, Pomiankowski 1988, Andersson 1994). The argument appeared at one point to degenerate into two camps populated by the adaptationists (good genes school) and nonadaptationists (Fisher school) (e.g., Borgia 1987, Kirkpatrick 1987b). Notably, a similar argument is occurring today in the field of sexual conflict (section 2.10).

Subsequently, several models were produced that demonstrated that the indicator process could account for the coevolution of female preference and male traits (e.g., Bell 1978, Andersson 1982, 1986, Pomiankowski 1987a, Iwasa and Pomiankowski 1991). Moreover, the process worked in the absence of linkage disequilibrium between trait and preference (the Fisher process). Unlike the lines of equilibria common to Fisher models, these indicator models had single point equilibria under most assumptions. Nevertheless, handicap models are closely related to Fisher models in that coevolution is driven by *indirect* selection on the preference. Male trait expression is assumed to be an increasing function of male genetic quality, or nonmating fitness (e.g., viability), and quality is always under directional selection to increase. Here, assortative mating between females with the preference and males with the trait leads to a genetic correlation between alleles for the preference and alleles that confer nonmating fitness. Male (and perhaps female) offspring of matings between females with the preference and males with the trait inherit high fitness from their father. These same offspring carry the preference alleles, and for this reason the preference spreads.

Indicator models have one substantial advantage over Fisher models: they can bear costs of choice (e.g., Andersson 1986, Kirkpatrick 1987a, Pomiankowski 1987b, Houle and Kondrashov 2002). Nonmating fitness is always expected to be under directional selection to increase, and because preference alleles become associated with those determining nonmating fitness, indirect selection is always pushing the preference toward greater values. This force can offset some direct selection (costs) on the preference itself. At equilibrium the female preference settles at the point where the costs to females of the preference are balanced by the indirect benefits of the preference. The preferred male trait, as in the Fisher process, settles where the upward force of sexual selection is balanced by the downward force of natural selection. Therefore, indicator models, like Fisher models, offer an explanation for the costly exaggeration of male traits by female preference. But, unlike Fisher models, indicator models can account for the presence of costly preferences. There are two related assumptions, however, that have the potential to greatly limit the explanatory power of indicator models. First, the process requires that genetic variance in nonmating fitness of males, reflected in some indicator trait, is maintained at equilibrium, and second, this variance has to be very substantial to account for all but a minute cost of preference.

Many have argued that the joint forces of natural and sexual selection, assumed in indicator models, would deplete genetic variance to the extent that there would be no indirect benefits to female choice. This problem has been called the "lek paradox," and its resolution was required before the widespread evolution of preferences could be attributed to the indicator process (Borgia 1979, Taylor and Williams 1982). Fortunately, the paradox is resolved, in one sense, by the empirical observation that additive genetic variance in fitness (Houle 1992, 1998, Burt 1995, 2000) and in sexually selected traits (Pomiankowski and Møller 1995) is abundant in nature. There are a great variety of explanations for the maintenance of variation in fitness, ranging from those based on the input of deleterious mutation to variable environments, or an interaction between the two (Hamilton and Zuk 1982, Charlesworth 1987, Burt 1995, Houle 1998). Hypotheses for the maintenance of relatively high genetic variance in indicator traits are fewer (Pomiankowski and Møller 1995, Rowe and Houle 1996). One hypothesis is based on the two premises that the costly nature of sexually selected traits leads to the development of covariance between the trait and condition (quality), and that condition is a large target for mutations. The development of condition dependence of these traits means that the abundant genetic variance in condition is captured in the trait, to the extent that they are correlated (figure 2.2). The dependence of many sexually selected traits on phenotypic condition is well established (Price et al. 1993, Andersson 1994, Johnstone 1995, Cotton et al. 2004). Much less is known about the genetics of condition, although complex traits like condition appear to be large targets for mutation (Houle 1998). Notably, this hypothesis is concordant with one key feature of indicator models: male traits are honest indicators of nonmating fitness (Zahavi 1975, 1977).

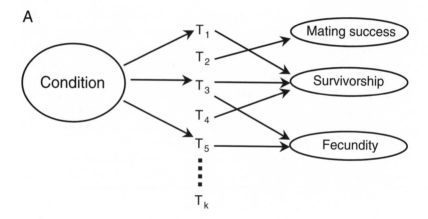

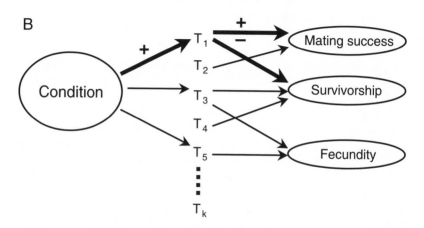

Figure 2.2. The evolution of condition-dependent expression of a sexually selected trait is illustrated in the path diagrams shown in panels A and B. These show the relationship between condition, a variety of traits $(T_1 \Pi \ T_k)$, and the components of fitness. We have omitted the sign of the pathways for ease of interpretation. Before sexual selection is imposed (panel A), some of these traits have some dependence on condition, which could be large or negligible. All traits have effects on one or more fitness components. The trait of interest, T_1, is under some form of natural but not sexual selection, indicated by a single pathway from it to survivorship. After sexual selection is imposed on T_1 (panel B), we would expect it to be dragged off its natural selection optimum. This is indicated by the positive sign of the pathway between T_1 and mating success and the negative sign of the pathway between T_1 and survival. Hence exaggeration of the trait is *costly*, and such costly traits are expected to evolve condition dependence. As the trait is exaggerated we expect the covariances between it and condition, mating success, and survivorship to grow. The strength of these covariances is indicated by the thickness of the paths.

Current estimates of additive genetic variation in fitness in natural populations (Burt 1995, 2000), accessible through indirect benefits, suggest that it might be too low to offset much direct selection on female preference (Kirkpatrick 1996, Kirkpatrick and Barton 1997). First, coefficients of additive genetic variance in total fitness appear to be on the order of one or a few percent. Moreover, by virtue of being indirect, these benefits are discounted by the genetic covariance between the alleles for nonmating fitness and the alleles for the preference. The theory of Kirkpatrick and Barton (1997) suggests that good gene effects are likely to account for exaggeration of male traits by only a few standard deviations of the mean. However, this conclusion may be too conservative. They assumed that costs of the preference accelerate as the mean population male trait value increases. An alternative assumption is that females choose to mate with the best of n sampled males (Janetos 1980). In this case, the cost of preference may increase with n, but not the mean male trait value. A recent model using an analogous assumption about the form of preference, and realistic assumptions about the magnitude of mutation and the costs of preference, suggests that good genes models can account for a quite dramatic exaggeration of male traits (Houle and Kondrashov 2002). Nevertheless, even with this particular form of mate choice, the cost to fitness of rejecting any one male must be relatively low (~1%), and this may limit its applicability.

We are still some distance from drawing quantitative conclusions about the extent to which "good genes" mechanisms can account for the evolution of female preference in nature (Kokko et al. 2003). A recent review of empirical studies, although supportive of good genes, reveals relatively small indirect benefits in nonmating fitness (Møller and Alatalo 1999). However, this too may be an underestimate, because allocation of resources by males into mating traits may reduce estimates of fitness that exclude mating success (Kokko 2001). One new and promising approach is to use experimental evolution to evaluate the effect of the opportunity for sexual selection on the evolution of fitness. Strong effects of good genes would be seen as increased fitness or rate of adaptation in those lines with sexual selection present. Results of the relatively few of these experimental evolution studies that have been conducted are mixed (Rice 1996, Promislow et al. 1998, Holland and Rice 1999, Holland 2002, Martin and Hosken 2003).

2.4 The Male Trait

At this point, it is worth considering in more detail the evolution of the male trait and its covariance with other traits under models of sexual selection based on indirect effects. The indicator models assume that the degree of trait expression is an increasing function of male genetic quality. Although this is not a required assumption of Fisher models, it is nevertheless a likely outcome.

This statement is based on the premise that female preference, no matter what its origin may be, will drive male traits to levels of exaggeration that are costly to the male and consequently result in their dependence upon condition. As we have seen, both Fisher and indicator processes predict equilibria where male traits are costly under natural selection: in both, the male trait settles at the point where sexual selection is balanced by natural selection. Several models have demonstrated that condition dependence of costly traits is expected, when variance in condition has either an environmental origin (e.g., Parker 1983b, Nur and Hasson 1984, Grafen 1990a,b, Price et al. 1993, Getty 1998) or a genetic origin (Iwasa and Pomiankowski et al. 1991, Lorch et al. 2003). In short, when the marginal cost to fitness of a given increment of trait exaggeration is higher for low- than high-condition males, traits will evolve to reflect condition (see also Getty 1998, Kokko 1997, Proulx et al. 2002). As expected, preferred traits do appear typically to be condition dependent (Price et al. 1993, Johnstone 1995, Jennions et al. 2001), although it is not yet clear whether they are generally more or less so than other types of traits (e.g., Møller and Petrie 2002, Bonduriansky and Rowe 2004).

One inference from these data and analyses is that preference, whether driven by Fisherian or indicator processes, will lead to condition-dependent male traits, at least near equilibrium, and therefore "good genes" will be contributing, to some extent, to the evolution of these preferences (Rowe and Houle 1996). This conclusion is similar to one made earlier: a genetic covariance between trait and preference, the key force in the Fisher process, is also expected whenever there is assortative mating (Lande 1981). A general conclusion, then, is that information about the genetic covariance between male trait and condition or female preference is not helpful in distinguishing the Fisher process from the indicator process. We expect both to be present to some extent in both cases. Instead, a quantitative understanding of the forces of selection acting on the preference is required (Kirkpatrick and Ryan 1991).

2.5 Direct Benefits

Up to this point, we have been assuming either that there is no direct selection on the preference or that direct selection works against the preference. However, there is a large number of cases where the preference is actually favored by direct selection. For example, females may simply prefer male traits that allow females to reduce mate search costs (e.g., Reynolds and Gross 1990, Westcott 1994) or the preference may be favored in some context other than mating (see section 2.6). A second group is known as "direct benefits," and here the evolution and maintenance of the preference are much simpler to understand than for those predicated on indirect selection (e.g., Heywood 1989, Grafen 1990b, Reynolds and Gross 1990, Price et al. 1993). These describe cases where males contribute materially to the fecundity or viability of the female. Examples include males that pass food resources to females (e.g., nuptial gifts), parental care for her offspring, sufficient sperm to fertilize her eggs, or a reduced probability of parasite transfer at mating (Andersson 1994). In some instances the resource value can be assessed directly by the female, where in other cases she must use an indicator trait. In the latter, the models behave much like the indicator models discussed above (section 2.4), except that all the initial forces of selection acting on the preference are direct. The equilibrium value for female preference represents a compromise between any costs of the preference and those material benefits. Preferred male traits evolve to a balance between the forces of sexual selection and natural selection. Because upward selection on the preferred trait is direct rather than indirect selection, any costs of preference are much less constraining. As in previous models, we expect a genetic correlation both between trait and preference and between trait and condition. Therefore, indirect effects from both the indicator and Fisher processes are likely to be present if the requisite genetic variance is present. One important difference between these direct benefit models and the *indirect selection* models is that these benefits to females could be determined entirely by environmentally generated differences among males. For example, all males may share the same genotypes, but some had the good fortune to acquire a resource preferred by the female.

2.6 Preexisting Biases and the Origin of the Preference

Direct selection on the female sensory system, morphology, or behavior, in some context other than mate choice, may have a side (pleiotropic) effect of biasing the mating or fertilization success of males (e.g., West-Eberhard 1979, 1984, Kirkpatrick 1987a, Ryan 1990, Endler and Basolo 1998, Boughman 2002). For example, a female's sensory system may be shaped to be sensitive to red color if a major food source in her diet is red. In such a case, females will be more sensitive to males that have red color markings, and this may result in higher mating frequency for these males (Rodd et al. 2002). Given that sensory systems are shaped in complicated ways by natural selection, there is good reason to believe that there is the potential for preexisting biases toward a great variety of colors, sounds, scents, and even behaviors and morphologies (Arak and Enqvist 1993, 1995, Enqvist and Arak 1993). Moreover, the empirical evidence for these preexisting biases playing a central role in the evolution of mate choice is growing rapidly (reviews by Ryan 1990, Basolo and Endler 1995, Endler and Basolo 1998, Boughman 2002).

Sensory bias differs in significant ways from the previously discussed models of female preference evolution. First, preference is a pleiotropic effect of direct selection on females in some context other than mating interactions. The model implies that preferences can arise prior to male traits, and this prediction has been supported with phylogenetic inference (e.g., Basolo 1990, Ryan et al. 1990, Proctor 1991, all reviewed by Shaw 1995). Second, sensory bias, in a strict sense, is not a model of male-female coevolution, since it does not address the evolution of female mating biases in response to an evolutionary change in male traits. The fact that some preferences can be explained in a context outside of sexual interactions, however, does not imply that other processes do not come into play once male traits have evolved to exploit such preference (Basolo and Endler 1995, Sherman and Wolfenbarger 1995, Gavrilets et al. 2001). Yet preexisting biases may offer an explanation for the presence of "preference" in the first place (Kirkpatrick 1987b, Kirkpatrick and Ryan 1991, Payne and Pagel 2001).

It is an interesting exercise to consider what will happen to preferences and traits when a male trait invades in the presence of some preexisting bias. Consider a case where females are attracted to red colors because of some preexisting sensory bias, there is stabilizing natural selection on the bias, and females mate once during a season. Then a "red spot" trait invades the population. First, if there is genetic variance in the bias, then a genetic correlation between trait and bias (preference) will form through assortative mating. Both trait and preference now have the potential to be exaggerated by the Fisher process once this genetic covariance is formed, but because there is direct natural selection on the preference, it will not move off its natural selection optimum. Nevertheless, that optimum preference favors red spots, so we expect red spot to spread or be exaggerated in males. It is also reasonable to assume that the size and intensity of the red spot will increase until it is checked by natural selection on the male (perhaps red also attracts predators). Given that the male trait is thus costly at equilibrium, we may then expect it to evolve into an indicator of male quality, and this indirect selection can move the preference off its previous natural selection optima. In essence, the females are paying some direct cost (the preference is moved off its previous natural selection optima) for the preference, and this is offset by the indirect benefits of mating with high-quality males. Thus, despite the central role of preexisting bias in starting this process, the preference at equilibrium appears as one would under the indicator model. Similarly, the male trait and its covariance with preference and nonmating fitness have similar characteristics to those predicted under Fisher, indicator, or direct benefits processes.

The scenario above is only one alternative for the coevolution of preference and trait under the preexisting bias model. Another has a quite different trajectory and consequences for female fitness. Imagine a similar system, except in this case females may mate multiply and there is stabilizing selection on female mating frequency that is greater than one. Now when red spot invades, it will similarly become exaggerated because of the preexisting bias in females. However, in this case the exaggerated red spot in males may stimulate females to mate suboptimally (e.g., beyond their optimum mating frequency). Thus, exaggeration of the male trait is costly to females, and males can be viewed as *exploiting* the preexisting bias in females with a *manipulative* trait (the red spot). Females are then under direct selection to *resist* this manipulative male trait. If females effectively evolve to resist, then there is sexual selection on males to further increase the expression of the manipulative trait, causing further selection on females to resist.

This second coevolutionary scenario is analogous to the one Rice and Holland (1997, Holland and Rice 1998) used to introduce the theory of sexually antagonistic coevolution (the chase away model; see also Gavrilets et al. 2001). Rice and Holland further reasoned that much or all of male-female coevolution may be shaped to some extent by such conflict. Their papers had a remarkably large and rapid effect on the field of sexual selection. It came as a surprise to many, however, that formal analysis of sexually antagonistic coevolution (SAC) predated Rice and Holland's verbal model by two decades. Already in 1979, Parker reasoned that sexual conflict was widespread and outlined a comprehensive formal analysis of SAC (Parker 1979).

2.7 Sexual Conflict

Parker (1979) built on Trivers's (1972) thesis on the origin of the sex roles to argue that the sexes would often be in conflict over reproductive decisions. Trivers noted that the sexes typically differ in their investment in offspring, and that this difference leads to different sex roles. The asymmetry begins with the ancient state of anisogamy (unequal gamete size between mating partners), but may extend to parental investment in unborn and born offspring (see also Bateman 1948). He reasoned that the sex that invested more in offspring would be more selective, on account of having a larger investment at risk, than the sex that invested less. Females, on the basis of larger gametes (eggs vs. sperm), typically invest more and, therefore, are the more discriminating sex. These sex roles are a reflection of divergent interests between the sexes. Trivers' reasoning provided an ingenious explanation for Darwin's observation that males typically competed for females, and females typically chose among males. He bolstered his argument by noting that, when relative parental investment is greater in males than females, the sex roles are reversed (reviewed by Gwynne 1991). More recent discussion of the sex roles can be found in Clutton-Brock (1991), Andersson (1994), and Queller (1997).

Parker (1979) recognized that the fact that the mates' genetic interests in mating were different meant that there would often be conflict between the sexes over mating and other reproductive decisions. Thus, "selection can act in opposing direction in the two sexes," and sexual conflict will occur whenever differences occur in the "evolutionary interests of individuals of the two sexes." All reproductive interactions between the sexes are likely to include conflict, except in the rare case of lifelong true genetic monogamy where the interests of the sexes after pair formation necessarily coincide (chapter 7). Such interactions include, for example, mating rate, female remating behavior, infanticide, sperm usage, and parental investment. Parker (1979) modeled a series of these evolutionary conflicts between the sexes (see also Parker 1983a,b, Hammerstein and Parker 1987). Although his language and modeling techniques differed (e.g., he largely relied on game theory rather than genetic models), his theory can be articulated in the terms of those models discussed so far.

2.7.1 PARKER'S INITIAL MODELS OF SEXUAL CONFLICT

Parker (1979) initially considered the invasion of a sex-limited autosomal allele that gave an advantage in mating for males, but decreased female fecundity and, therefore, had a cost to both sexes. He demonstrated that the trait could invade despite its direct costs to females, but that this depended upon just how costly the trait was, and whether it was dominant or recessive in action. Andrés and Morrow (2003) extended this model, and suggested that such costly male traits invade even more readily when sex linked. Parker's (1979) model clearly shares some features with the preexisting bias models above, in that there is sexual selection on the male trait as soon as it arises, and that it can have direct costs to females. The model also includes a "sexy son" benefit to females, and implies that this benefit could alter the optimal strategy of females during the invasion of a male trait. Yet, as noted above, in the presence of natural selection on the mating bias, as assumed by Parker, genetic models demonstrate that the sexy son effect will play no role in affecting the evolutionary equilibrium.

Parker (1979) also considered two other scenarios of male-female coevolution that have direct relevance to sexual selection. First he considered a game concerning mating encounters between the sexes where males have two states (persist or give up when females resist) as do females (accept mate or resist). It is assumed that the female has already mated and therefore does not need sperm to fertilize her eggs. All selection in this model is direct (i.e., there are no sexy sons or good genes). There are direct costs to females of both mating and resisting, and there are costs to males of persisting (when females resist), but benefits of mating. These costs depend upon the other "player" (i.e., a "war of attrition"). For example, persistence costs are paid only if the male encounters a resistant female. Parker concludes that the outcome of the game depends on the strength of selection on the two sexes, the starting conditions, and the asymmetries in cost functions. Yet he also notes that one likely result is an "evolutionary chase" in which the frequencies of both male persistence and female resistance increase.

In another model, Parker (1979) examines a similar game, but one in which the costs of the persistence and resistance traits are independent of who is matched (i.e., an "opponent-independent cost game") (see also Parker 1983a). An example of such a case would be where the persistence and resistance traits are morphological traits that are built (and paid for) at maturity. Parker concludes that unresolvable evolutionary chases are particularly likely in this case (see also Härdling 1999, Härdling et al. 2001). This model appears similar to the "chase away," in more than just its name. As in the chase away, no indirect selection on females is assumed; male persistence is favored by sexual selection, but has a direct cost to female fitness. It is easy to imagine that Parker's male persistence trait is a manipulative red spot! This chase also captures Dawkins' idea of the sexual arms race, where the armaments are persistence and resistance (Dawkins 1976, Dawkins and Krebs 1979).

These and other models in Parker's formal treatments of sexual conflict (e.g., 1983a,b) were groundbreaking. They captured the main ideas reflected in subsequent contributions and they remain current today. Nevertheless, these ESS models are not strictly comparable to the genetic models upon which the theory of sexual selection is grounded, and their applicability in this context has shortcomings (Maynard Smith 1978, Enquist et al. 2002, Wade and Shuster 2002). Gavrilets et al. (2001) have produced a model that is directly comparable to the typical genetic models of sexual selection and captures the central scenario of a conflict over mating rate. Their results verify the intuition of Rice and Holland and the earlier ESS models by Parker and others (reviewed by Clutton-Brock and Parker 1995).

2.7.2 GENETIC MODELS

Gavrilets et al. (2001) analyze a conflict over mating where males have an optimal mating rate that is very much greater than that of females, and therefore there is direct selection for higher male persistence and female resistance traits. Although specified for a conflict over mating rate, its general conclusions are applicable to conflict over other variables. In the model, male fitness is assumed to be a monotonically increasing function of mating rate so that male fitness is always greatest with the highest rates of mating. Conversely, female fitness is modeled so that it is maximized at an intermediate rate of mating, declining both above and below the maximum. All forces of selection in the main model are direct (although indirect "sexy son" effects were explored in an appendix). The mating rate is a function of the difference between the level of male stimulation (the persistence trait) and female resistance; in this model both male stimulation and female resistance are each determined by a single quantitative trait. Females are characterized by preference functions that are fixed in shape but variable in location across the distribution of male traits. This means that females may resist the increased stimulation associated with increases in the male trait by adjusting the amount of male stimulation they require for mating. The male trait, female preference, and female mating rate each had natural selection acting on them. In their analysis of the dynamics of the model, Gavrilets et al. (2001) alter the strength of selection on each of these elements.

The main conclusion of the Gavrilets et al. (2001) model was that exaggeration of male traits that facilitate overcoming female resistance to mating can result from direct selection on females to elevate resistance to mating. When Gavrilets et al. (2001) removed natural selection on both the male and female traits, no fixed equilibrium value for the male and female traits were found. Instead, a perpetually increasing runaway between the male and female traits resulted. More in tune with nature, when costs to males or both sexes were incorporated, joint evolution of costly female mate choice and exaggerated male traits occurred under a wide range of circumstances. The system then evolved toward either a stable equilibrium or, more rarely, a stable limit cycle where persistence and resistance became locked in a stable coevolutionary oscillation.

The model of Gavrilets et al. (2001) contrasts with the more typical models of sexual selection (good genes, Fisher, direct benefits) in two important ways. Indirect selection, through either good genes or sexy sons, was excluded from the main model. Second, unlike in any of these models, mating with "preferred males" had exactly the same consequences for females as mating with nonpreferred males. Fitness of a female derived from her mating rate, rather than from her mating partner. This model successfully captured the verbal theory of sexually antagonistic coevolution outlined by Holland and Rice (1998), showing clearly how resistance in one sex can drive the evolution of "coercive" or "manipulative" traits in the other. However, the constraints that were imposed on the evolution of the female preference function may overestimate the probability that "resistance" drives the exaggeration of antagonistic traits.

In this model, the preference function (figure 2.3) is fixed in shape, so that it can evolve only by shifting its location along an axis describing the value of the male stimulus (or its intercept). As such, females could minimize the costs of mating only by increasing the degree of male trait expression required for a given mating rate. This assumption makes sense only if natural selection places constraints on the evolution of female response to the type of stimuli produced by males, as would be the case if, for example, responsiveness to red colors in females would be required to successfully forage for red food items (e.g., Rodd et al. 2002). This constraint, however, makes it more likely that male trait values and the female "thresholds" for mating will increase during the coevolutionary process. An alternative model would allow the shape, or the slope, of the preference function (figure 2.3) to evolve as well as

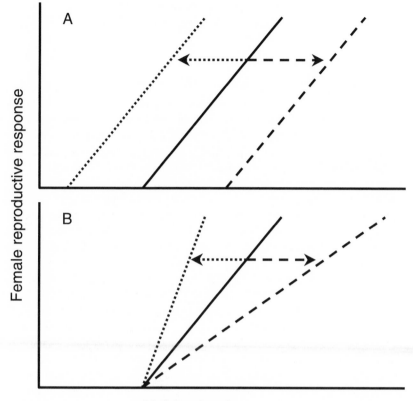

Figure 2.3. When males stimulate females to mate at higher rates than optimal, females may reduce the mating rate in one of two ways. Females may either (A) increase the threshold amount of male stimulation (male trait) required to initiate mating or (B) adjust their sensitivity (slope of the preference function) to the male trait.

the threshold (Rosenthal and Servedio 1999). Evolutionary and plastic change in the slope of preference functions is not uncommon (Berglund 1993, Hedrick and Dill, 1993, Morris et al. 1996, Wagner 1998).

A model of Rowe et al. (2004), based on the Gavrilets et al. (2001) framework, allowed both the threshold and shape of the preference function to evolve. In those simulations where the slope of the preference function was relatively free to evolve, females often reduced the slope of the preference function, which reduced selection for exaggeration of costly male traits. In essence, females may evolve insensitivity to the type of stimuli produced by males and, in some circumstances, even discrimination against more exaggerated traits. When this occurs exaggeration of the male trait stops or is reversed.

These results suggest that arms races between the sexes may be less common than would be predicted by Gavrilets et al. (2001).

2.7.3 PHENOTYPE-DEPENDENT AND PHENOTYPE-INDEPENDENT COSTS

When Parker (1979) first set out to illustrate sexual conflict, he sketched a scenario in which a sexually antagonistic mutation in males conferred a mating advantage on males, but a direct cost to females mating with males bearing this mutation. We will refer to this class of costs of antagonistic adaptations as *phenotype dependent*, because the cost to females depends upon the phenotype of the male to which they mate. Examples of phenotype-dependent costs may include decreased fertility experienced by females mating with more attractive males (Warner et al. 1995, Baker et al. 2001), decreased survival of females mated with large males (Pitnick and García-González 2002, Friberg and Arnqvist 2003), decreased amounts of paternal care received by females pairing with more attractive males (de Lope and Møller 1993, Slagsvold and Lifjeld 1994), and decreased survival of offspring fathered by such males (Brooks 2000).

However, per mating payoffs need not depend on male phenotypes for sexual conflict to generate sexually antagonistic coevolution. For example, imagine that mating at a high rate is costly for females but that per mating costs are essentially independent of the male phenotype to which females mate, as in the red spot example (see section 2.6). Similarly, females may be induced to mate at suboptimal times or places. We will refer to these costs of antagonistic adaptations as *phenotype independent*. Such costs may be very common because multiple mating per se, rather than with whom, is often costly to females (reviewed by Arnqvist and Nilsson 2000), and any persistence trait in males that increases mating rate could qualify. For example, in water striders (Gerridae), where mating is costly to females, various male adaptations allow them to overcome female resistance to mating, yet there is no evidence that the costs of mating to females depend upon these adaptations (Rowe et al. 1994, Arnqvist 1997a).

This dichotomy between phenotype-dependent and phenotype-independent payoffs is reflected in an analogous dichotomy between "wars of attrition" and "opponent-independent cost games" in ESS models of sexual conflict (e.g., Parker 1979, Hammerstein and Parker 1987, Härdling et al. 2001; section 2.7.1). The two genetic models discussed above (section 2.7.2) have assumed that costs to females were phenotype independent, yet it seems reasonable to assume that their dynamics would be similar if costs were phenotype dependent. This is an issue that warrants further analysis. Nevertheless, whichever is the form of the cost, females are expected to evolve resistance or, if possible, insensitivity to the trait.

2.7.4 NONEQUILIBRIUM MODELS

All the sexual conflict models we have discussed so far consider the evolution of only one male trait and one female preference function, and use an "equilibrium" approach. This approach fits with the bulk of models in sexual selection, but does not fit as well with many of the verbal descriptions of sexually antagonistic coevolution. For example, Rice and Holland (1998) reason that, as females evolve resistance to one manipulative trait, males respond by evolving a new trait that manipulates some other undefended pathway in females, and so on. This scenario describes a particularly diversifying, nonequilibrium, and potentially endless evolutionary chase. The chase is limited only by the number of exploitable pathways in females and the evolutionary potential of males to exploit them. Interestingly, precisely this reasoning was outlined by West-Eberhard more than ten years earlier (West-Eberhard 1983).

Iterated chases, involving new pathways for each iteration, have not been studied with genetic models, but recent work employing neural networks has addressed these chases (Arak and Enquist 1993, 1995, Wachtmeister and Enquist 2000). These models capture the reasoning outlined by Rice and Holland (1998), and confirm that chases incorporating many signal-receiver pathways along their span are quite probable. It remains an open and general question whether the equilibrium approaches we have been relying upon to understand coevolutionary interactions are reasonable, and this may be particularly questionable for sexually antagonistic coevolution (Enquist et al. 2002)

2.8 Sexual Conflict Set in the Framework of Sexual Selection

It should be clear from the models of sexual conflict discussed above (section 2.7) that sexual selection is an inevitable facet of sexually antagonistic coevolution. Male traits that decrease female fitness spread and are exaggerated because females "prefer" them (i.e., bias mating or fertilization success toward males with exaggerated or more manipulative traits). The fact that these traits are costly to females leads to selection for resistance or the evolution of insensitivity to them. In the former case, resistance can favor further exaggeration (persistence), and this cycle of persistence and resistance describes sexually antagonistic coevolution (figure 1.4). In the latter case, sexual selection on the male trait is reduced or halted. How, if at all, does this process differ from other forms of coevolution by sexual selection (e.g., good genes, Fisher, and direct benefits)? As in our discussion of these other models, we believe it is most useful to look at the evolutionary forces acting on the female preference/resistance that are driving the evolution of male traits.

Following Kirkpatrick (1987a, Kirkpatrick and Ryan 1991) the forces acting on the preference can first be divided into *indirect* and *direct selection* as we have done above. The good genes and Fisher, or sexy sons, models describe evolution of the preference through indirect selection (sections 2.2 and 2.3), whereas direct benefits and the original preexisting bias models describe evolution of the preference through direct selection (sections 2.5 and 2.6). Our view is that sexually antagonistic evolution is characterized by direct selection on the preference, where it is distinguished from these other models by the sign of selection (Gavrilets et al 2001, Chapman et al. 2003a, Cameron et al. 2003). In sexually antagonistic evolution, the trait that gives an advantage to males in mating is *disadvantageous* to females. Male trait exaggeration is then impelled by evolutionary responses in females to these costs. In contrast, in direct benefits models the trait that gives an advantage to males in mating is *advantageous* to females, and it is this advantage that impels coevolution of male and female traits. In the early models of sexual selection resulting from preexisting bias, although it was not precluded, there was no discussion of an evolutionary response of female preference to the spread of preferred traits. Nevertheless, as argued above, the preference could evolve from either indirect or direct selection (section 2.6). If the spread of the preferred male trait reduces female fitness and this leads to an evolutionary modification of the female preference, we can refer to the coevolutionary process as sexually antagonistic.

Although our focus will be on direct effects, the presence of these does not preclude indirect effects, including sexy sons and good genes, occurring simultaneously during phases of sexually antagonistic coevolution. This issue has received a lot of attention in debates on the topic (e.g., Cameron et al. 2003, Chapman et al. 2003b, Cordero and Eberhard 2003, Córdoba-Aguilar and Contreras-Garduño 2003, Eberhard and Cordero 2003, Kokko et al. 2003) and we have touched on these previously in this chapter. First, assortative mating leads naturally to a genetic correlation between male traits and female preference (i.e., sexy sons; see section 2.2). Therefore, its presence says nothing about the forces of selection acting on the preference—the selective agent at the root of assortative mating. Second, if antagonistic male traits evolve to the extent that they are costly under natural selection, we expect them to become condition dependent, and therefore come to indicate some aspect of genetic quality. If so, good genes may play some role in the coevolutionary process (see sections 2.3 and 2.4). Thus, one conclusion is that there is no expectation that only a single process (i.e., Fisher vs. good genes vs. sexual antagonism) should be responsible for coevolutionary dynamics of trait and preference, a point made several times before (e.g., Kirkpatrick and Ryan 1991, Rowe and Houle 1996, Chapman et al. 2003a, Kokko et al. 2003). A related conclusion is that the relative importance of different processes will likely vary during different phases of male-female coevolution (see section 2.6). For example, we would often expect sensory exploitation to be important in early stages, whereas antagonistic coevolution could be significant during subsequent escalation, and indirect effects generated by good genes mechanisms coming into play as equilibrium is approached.

A quantitative assessment of the role of each in the evolution of female preference has been a difficult empirical issue, and the inclusion of sexually antagonistic coevolution in the array processes has not made it any easier (Chapman et al. 2003a, Cordero and Eberhard 2003, Kokko 2001, Kokko et al. 2003). Nevertheless, some conclusions, drawn from theory, are possible. For example, we do not expect the production of sexy sons to affect the equilibrium value of female preference (resistance) when there is direct selection acting on the preference. The easiest way to see this is to imagine an equilibrium of trait and preference that was reached in the absence of any direct selection acting on the preference (as Fisher did in his verbal models; see section 2.2). At equilibrium, male trait values are held in place by the upward force of female preference and the downward force of natural selection acting on the exaggerated trait. Thus, the average male has the highest fitness. Females mating the average male will produce the most fit sons, and this holds the preference in place. Females mating males with relatively high trait values will produce sons with relatively high mating success, but this advantage will be more than outweighed by their low nonmating fitness. Likewise, females mating males with relatively low trait values will produce sons with high nonmating fitness, but this advantage will be more than outweighed by their low mating success. Now imagine that the preference has some direct selection on it and ask what happens to trait and preference. To be most relevant, we can introduce a cost to female resistance, assuming that resistance biases mating success toward the most persistent male types, and then ask whether the production of persistent sons can compensate for the costs of resistance. Given that there is now downward selection on the preference, the preference will decline, and consequently upward sexual selection on the male trait will be reduced and male trait values will be driven down by natural selection on that trait. Nothing will stop this downward slide of preference and trait until the preference reaches its natural selection optima. The conclusion is that the existence of costly female resistance to mating cannot be accounted for by the benefits of resistant females producing persistent sons. As could be expected from this verbal reasoning, and from earlier formal analyses (see sections 2.2 and 2.3), inclusion of these sexy son effects into formal models of sexually antagonistic coevolution has no effect on the equilibrium values of trait or preference (Gavrilets et al. 2001). In short, the equilibrium in these models can be described entirely by the forces of direct selection acting on the preference (resistance).

The same argument does not apply to indirect effects generated by good genes mechanisms. The critical difference is that in good genes models, because of the positive covariance between trait and nonmating fitness, males with exaggerated traits are those with relatively high nonmating fitness. Nonmating fitness is always under directional selection to increase, whether or not the preference is present in the population. This is distinct from sexy sons, where selection for exaggeration of the male trait is entirely dependent upon the preference itself. In the former case, females mating males with more exaggerated traits will produce sons (and perhaps daughters) with higher nonmating fitness. This indirect selection favoring the preference can counterbalance some degree of direct selection against the preference (see section 2.3 for further discussion).

It is important to note that the arguments of the last four paragraphs are based on equilibrium assumptions. As noted in section 2.7.4 many views of sexually antagonistic coevolution are decidedly nonequilibrium, despite the fact that the vast majority of theory on sexual selection is equilibrium based. If there is no assumption of equilibrium, all bets are off. We know, for example, that sexy son effects can be important in the transient coevolutionary dynamics of trait and preference, whether there is direct selection on the preference or not. In contrast, there is no particular reason to expect the evolution of condition dependence of male traits during the early phase of trait and preference coevolution. Consequently, there is no particular reason to expect that good genes mechanisms will play any role at all in these coevolutionary dynamics.

2.9 The Roles of the Sexes in Sexual Conflict

Although conflicts between the sexes are usually discussed as directional with male traits exploiting or manipulating female biases, these roles are not in any way fixed. The roles of the sexes in sexual conflict are not determined by gender per se, but are dictated by the extant pattern of intrasexual competition. In general, females make a higher premating investment in offspring production, ultimately because of the asymmetry in gamete size of the two sexes (Bateman 1948, Trivers 1972). As a consequence, males typically have a higher potential rate of reproduction than females (Clutton-Brock and Parker 1992, Kvarnemo and Ahnesjö 1996, Parker and Simmons 1996), leading to a male bias in operational sex ratio, and more intense intrasexual reproductive competition among males. Premating asymmetries in offspring investment can then promote similar postmating asymmetries for two related reasons. First, the benefits of parental care are smaller for males simply because they are, on average, less related to mutual offspring than are females (because of polyandry and extrapair matings) (Queller 1997, Wade and Shuster 2002). Second, males that actually do reproduce successfully (a subset of all males) will be less selected to invest intensely in offspring compared to their mates (Queller 1997, Ahnesjö et al. 2001). In this way, sexual selection acts to reinforce asymmetries in investment in offspring, by a positive feedback loop involving the premating asymmetries that form the basis for sexual selection in the first place

Figure 2.4. Sex roles are not inherent to the genders, but depend on the relative invest-ment of the sexes. In the dance fly *Rhamphomyia longicauda*, males offer "expensive" nuptial gifts to females at mating and females have evolved secondary sexual traits which apparently deceive males. As shown here, inflatable abdominal airsacs and pinnate legs in females together exaggerate abdominal swelling, and give the false impression of an abdomen full of mature eggs. (Reprinted from Funk and Tallamy 2000, with permission from Elsevier; photo by D. H. Funk.)

(Queller 1997, Kokko and Jennnions 2003, Andersson 2004). While males most often have higher potential reproductive rates, this is by no means always true, and cases of sex-role reversal are common (Clutton-Brock 1991). In these taxa, females compete among themselves for access to males (e.g., Gwynne 1991, Owens and Thompson 1994, Funk and Tallamy 2000, Bonduriansky 2001), and we would expect females to express traits that exploit or manipulate males in order to access limiting resources provided by males.

One study nicely illustrates this point (Funk and Tallamy 2000; see also section 3.2.4). In the dance fly *Rhamphomyia longicauda*, males carrying ar-thropod prey items enter a swarm of females. A male appears to choose a female from the swarm, donate his prey item, and fly off with her to copulate while she feeds on the "nuptial gift." In this species, females compete for such gifts and are equipped with inflatable abdominal sacs and conspicuous leg scales that seemingly function to exaggerate their apparent size (see figure 2.4). Male *R. longicauda* prefer to pair with large females, because they are more visible and/or because large females tend to be more fecund. Funk and Tallamy (2000) suggested that females have evolved deceptive secondary sexual traits (abdominal sacs and leg scales) that exploit an adaptive male preference for large females and thereby manipulate males into allocating their paternal effort in a suboptimal manner. Whether males have responded by evolving resistance to these manipulative displays is not known (see Hockham and Ritchie 2000). Nevertheless, the example serves as a striking illustration of the fact that the roles of "exploiter" and "exploited" are in no way affixed to the genders.

2.10 Empirical Approaches to the Study of Sexual Conflict

In this chapter we have emphasized that the ability to recognize sexually antagonistic coevolution and distinguish it from other forms of male-female coevolution rests upon assessing the signs of direct and indirect forces of selection acting on coevolving male and female traits. The bulk of this book will deal with more or less anecdotal accounts of male-female interactions, the types of traits involved in these interactions, and the forces of selection acting upon them. We will argue that seemingly manipulative or persistence traits in males are very common, as are female resistance traits. There is a huge variety of interactions between males and females that suggest conflict. Although many would argue that an understanding of the current function of a trait will, to some extent, provide insights into the evolutionary processes responsible for its origin and/or maintenance (see Reeve and Sherman 1993), this is certainly not always the case. For example, one might conclude that overt aggression between parent and offspring indicates a conflict of interest between the two, but this is not necessarily so (Mock and Parker 1997). Aggression could in theory be an outcome of an honest signaling system, negotiating need and supply (Godfray 1995). Similarly, overt premating chases, struggles, or fights between males and females may be indicative of sexual conflict, yet this is also not necessarily so (Parker 1979, Crump 1988). The converse also holds true: the lack of overt manipulation or aggression in male-female interactions does not in itself demonstrate that sexual conflict has not been involved in the evolution of that interaction (Holland and Rice 1998, Gavrilets et al. 2001). In order to determine whether sexual conflict is responsible for male-female interactions, and the traits involved, we need empirical data on the forces of selection acting on each sex, past and present (see chapter 7; and Thornhill 1980, Rowe et al. 1994, Chapman et al. 2003a,b, Cameron et al. 2003, Cordero and Eberhard 2003).

Yet unambiguous experimental data on sexual conflict are really quite scarce. No doubt, this results in part because the field is relatively young, but perhaps more because empirical studies aimed at distinguishing among forms of male/female coevolution are notoriously difficult, as indicated by the lack of consensus on the relative importance of the various hypotheses for female

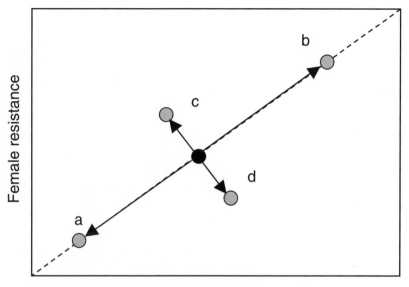

Male persistence

Figure 2.5. Much like the Red Queen, an arms race over mating rate (or other antagonistic interactions) may have little impact on the interaction itself. For this reason, sexually antagonistic coevolution may be hidden, with adaptation in one sex offset by counteradaptation in the other. If escalation (or deescalation) in armaments in one sex are matched by the other (a and b), there will be little or no change in the outcome of the interaction. Only when adaptation in one sex is not matched by the other (c and d) will we see measurable changes in the outcome of interactions between the sexes.

preference, despite 25 years of research effort (e.g., Borgia 1987, Kirkpatrick 1987b, Kirkpatrick and Ryan 1991, Andersson 1994, Kokko et al. 2003). The empirical study of sexual conflict (and sexual selection) suffers from three problems. First, as pointed out by Parker (1979), it is difficult to gain a detailed understanding of the "economy" of these interactions simply because it is hard to measure every conceivable cost and benefit to both sexes under natural conditions (see also Thornhill 1980). As a result, ideal studies that measure the effects of potentially sexually antagonistic traits in one sex on lifetime reproductive success in both bearers of those traits and their mates are exceedingly rare (Zeh and Zeh 2003). Second, the field is still suffering from conceptual ambiguities, and there is debate over precisely which kind of data we need (Chapman et al. 2003a, Cordero and Eberhard 2003, Pizzari and Snook 2003, Arnqvist 2004). Third, because of continual adaptation and counteradaptation, sexual conflict may often be "hidden" from our eyes (see figure 2.5, and chapter 7; and Chapman and Partridge 1996a, Rice 1996b, 2000, Härdling et al. 2001, Arnqvist and Rowe 2002a).

In covering those experimental data that are available, we will refer to several distinct empirical approaches that have been used to document the existence of sexual conflict over the outcome of interactions or to elucidate sexually antagonistic coevolution of traits used in these interactions (see also Chapman et al. 2003a). As in any evolutionary study, any single approach is unlikely to be sufficient. Therefore, for any particular question or species, a combination of as many approaches as possible is likely to be most helpful. Several of the currently employed approaches are listed below.

(1) A large number of *economic studies* have measured various costs and benefits to both sexes from the outcomes of sexual interactions. These studies are a first step in providing evidence that selection operates in opposing directions in the two sexes. For example, one might assess the costs and benefits to each sex of a range of mating rates. This economic approach has been particularly useful in identifying conflict in water striders (e.g., Rowe et al. 1994, Arnqvist 1997a), bumble bees (e.g., Baer et al. 2000, 2001), and fruit flies (e.g., Chapman 2001, Friberg and Arnqvist 2004). Although required in determining whether there is currently sexual conflict over the outcome of any one interaction, the economic approach suffers from two limitations. It is often very difficult indeed to firmly conclude that all possible costs and benefits to both sexes have been accurately quantified. Second, the forces that have shaped these interactions may be hidden because of past adaptation and counteradaptation.

(2) An important addition to economic studies is the use of *phenotypic manipulations* of presumed resistance or persistence traits, as a means of determining their role in the outcome and economy of sexual interactions. In addition to identifying sexually antagonistic traits, this approach reduces the second limitation of economic studies, in that we can study the economy of sexual interactions outside the current range of phenotypic adaptation/counteradaptation. Eliminating or exaggerating sexually antagonistic adaptations by phenotypic manipulation is, needless to say, not always feasible. Attempts to apply this empirical strategy include experimental manipulation of morphological male grasping (Thornhill 1984, Arnqvist 1989b, Thornhill and Sauer 1991, Sakaluk et al. 1995) and female antigrasping traits (Arnqvist and Rowe 1995).

(3) When persistence and resistance are behavioral in origin, studies of *adaptive phenotypic plasticity* can contribute to distinguishing among the forces shaping these traits. For example, we can use theory to derive predictions of optimal resistance under various environmental conditions for competing explanatory models, and then test these with experimental data. This approach has been widely employed in studies of overt premating resistance behavior in female invertebrates, such as water striders (reviewed by Rowe et al. 1994, Arnqvist 1997a) seaweed flies (Shuker and Day 2001, 2002), and crustaceans (Jormalainen et al. 2000).

(4) A powerful approach to the study of sexual conflict is the use of *experimental evolution* techniques. In these experiments, the coevolutionary effects of altered persistence and/or resistance, or of removal of sexual conflict, can be studied directly. Examples include selection experiments in fruit flies (e.g., Rice 1996b, Holland and Rice 1999, Holland 2002) and dung flies (Hosken et al. 2001, Martin and Hosken 2003a,b). Importantly, these methods also reduce the second limitation of economic studies, in that we can study the economics of male-female interactions in systems that have been manipulated to evolve away from current equilibria.

(5) A variety of different *genetic experiments* can shed light on sexual conflict. These include the use of genetic mutants that lack certain sexually antagonistic traits (e.g., Chapman et al. 1995), other forms of genetic engineering (e.g., Lung et al. 2002), and experimental comparisons of distinct genotypes that differ with regards to the expression of sexually antagonistic traits (e.g., Civetta and Clark 2000). Ingenious and creative genetic experiments can be highly informative, but few empirical systems are amenable to these types of experiments.

(6) Provided there is sufficient understanding of both the traits under study and the economy of the sexual interactions within which they are involved, *comparative studies* can be highly informative. For example, one can directly assess whether antagonistic traits in females and males have coevolved among species (e.g., Bergsten et al. 2001, Arnqvist and Rowe 2002b, Rowe and Arnqvist 2002, Koene and Schulenburg 2003) or whether evolutionary changes in the relative degree of adaptation/counteradaptation are associated with the predicted evolutionary changes in the outcome of antagonistic interactions between the sexes (e.g., Arnqvist and Rowe 2002a).

3 Sexual Conflict Prior to Mating

In some species with sexual reproduction there is no direct interaction between individuals of each sex (e.g., some salamanders, most plants, and a variety of marine broadcast spawners). For these, the union of gametes does not require direct interactions between individual males and females. These species, though sexual, do not mate. More common and familiar to biologists are those species where direct interaction (mating) is required. Here mating may be the foremost and in many cases the only interaction between the two sexes. This temporary union has often been described as a harmonious affair, where males and females, after a period of courtship, join in the shared goal of producing offspring that carry their genes.

Observations of male courtship and the discriminating nature of females prior to copulation have inspired some fine prose. The "pairing of birds is not left to chance; but that those males, which are best able by their vigorous charms to please or excite the female, are under ordinary circumstances accepted. If this be admitted, there is not much difficulty in understanding how male birds have gradually acquired their ornamental characters" (Darwin 1871). In stark contrast is the emerging view that interactions between the sexes prior to mating are often antagonistic. Consider, for example, a recent description of the "courtship" behavior of male garter snakes while entwined with the female; "rhythmic caudocephalic waves by courting male garter snakes push anoxic air from the saccular lung forward and across the respiratory surfaces such that females cannot obtain oxygen. Their stress response involves cloacal gaping" which thereby permits male intromission (Shine et al. 2003).

In those species that include mating in their reproductive strategies, at least one mating is required to attain any fitness and is in this sense "beneficial." The interesting questions here concern how costs or benefits to an individual vary with the environmental setting and the state of its partner, and how these may scale with the number of matings (or partners) undertaken. There are potential costs and benefits for both sexes of choosing among mates in any encounter. The evolution of mate choice (particularly by females) and resulting sexual selection (particularly on males) has spawned a vast literature. It is not our purpose to review this literature here (there are reviews by Thornhill and Alcock 1983, Bradbury and Andersson 1987, Andersson 1994). We will instead focus on how intersexual differences in optimal mating strategies result in conflict between the sexes, and specifically those occurring prior to copulation. This necessarily leads us to emphasize mating frequency, where the interests of the sexes typically diverge most dramatically. In this chapter, we briefly review the costs of mating to females, the resulting conflicts between males and females over decisions to mate, and the evolutionary consequences for the tactics employed by each sex. We then explore the relationship between conflict over mating, sexual selection, and the evolution of sexual dimorphism.

3.1 The Economy of Mating and the Evolution of Resistance

The benefits of multiple mating to males are typically obvious: more mating, more offspring. It is certainly not always a simple relationship, for example because of postcopulatory processes (see chapter 4). Nevertheless, the number of fertilized gametes obtained is very likely to be an increasing function of the number of matings the male achieves. This view is predicted by theory concerning the implications of anisogamy on parental investment and the sex roles (see chapter 2). We expect male fitness to be limited by access to female gametes, and access to females is clearly a necessary step. Empirical support comes from the common observation that there is much variance among males in fitness, and a major source of this variance is the mating rate (Arnold and Duvall 1994, Kingsolver et al. 2001).

Mating carries costs and benefits to both sexes, yet the balance of these is likely to often differ between the sexes. If this is so, then conflicts between individual males and females over mating will occur. It is this fact that leads to the typical pattern of competition among males for access to females. If it were the case that females had nothing to lose from additional mating, we would predict that females would simply mate whenever approached by males. Instead, females often resist mating, and sometimes at great expense to themselves. Resistance by females suggests that mating carries significant costs to females that are not compensated for by any benefits that come with mating.

3.1.1 DIRECT COSTS OF MATING

Mating can be a costly activity for both sexes. Some of these costs would not come as a surprise to us, while others would. For example, since the emergence of AIDS in the early 1980s, no one would be surprised to learn that the risk of sexually transmitted disease is an increasing function of the rate of mating. On the other hand, it may come as a surprise to many to learn that in some species the male intromittent organ can damage the female reproductive tract during mating, to the point where it reduces female longevity as a consequence (Crudgington and Siva-Jothy 2000, Blanckenhorn et al. 2002), or that copulation can corrupt immune function (Rolff and Siva-Jothy 2002). The most general mating costs, however, include energy expenditure, increased susceptibility to predation, and lost opportunities (e.g., reduced time for feeding). More recent evidence is accumulating that suggests that substances carried in ejaculates transferred during mating can themselves be costly to females (see chapter 4). Finally, resisting mating may have similar, and at times, even greater costs (see sections 3.1.3 and 3.2.1). Despite the fact that mating in most species is likely to have at least some costs, there are relatively few cases where these have been carefully quantified. Nevertheless, there is widespread evidence of costs to mating, and we refer readers to reviews of such costs (see Daly 1978, Walker 1980, Thornhill and Alcock 1983, Gwynne 1989, Choe and Crespi 1997a, Arnqvist and Nilsson 2000). Here we only briefly describe one well-studied example.

In water striders, females carry males during copulation and postcopulatory guarding, which results in a variety of direct costs to females (reviews by Rowe et al. 1994, Arnqvist 1997a). Carrying males increases the energy expended during skating on the water surface by some 20%, and skating speed is much reduced (Arnqvist 1989a, Fairbairn 1993, Watson et al. 1998). The cost of carrying males may explain why females of one species allow smaller males to guard for longer (Sih et al. 2002). Reduced speed and agility of mating females leads to increased predation rates on pairs and reduced foraging rates (Wilcox 1984, Arnqvist 1989a, Fairbairn 1993, Rowe 1992, 1994). These costs of mating have dramatic effects on the water strider mating system. For example, mating rate is reduced in the presence of predators (Sih et al. 1990), and females often stop foraging and leave the water surface when mating (Rowe 1992, Rowe et al. 1996). Similar predation costs in other species are reflected in several studies demonstrating that mating frequency declines in the presence of predators (e.g., Magnhagen 1991, Fuller and Berglund 1996, Maier et al. 2000).

3.1.2 Costs of Low Mate Quality

Both males and females may in a sense lose out by copulating with a partner of low quality. Low quality may in this context refer to individuals with high parasite loads, incompatible genotypes, low genetic fitness, or lacking in direct resources to offer. The costs of mating with these individuals are exacerbated when there are significant costs to females associated with mate searching or mating, or when there is last male sperm precedence (Parker 1983b). This fact is thought to be the major force leading to mate choice by both females and males (see chapter 2) and is the subject of much literature on sexual selection by mate choice (e.g., Bradbury and Andersen 1987, Kirkpatrick and Ryan 1991, Andersen 1994). The mating biases that result from mate choice are usually discussed as resulting in preferences for certain phenotypes (high quality). But one could just as easily view them as resulting from discrimination or resistance against certain phenotypes (of low quality). Although nothing would be gained by recasting all traditional discussions of mate choice in terms of resistance, it is important to note that choice entails resistance and that resistance may result in the evolution of persistence traits in the other sex (see also section 7.7).

3.1.3 COSTS OF RESISTING MATING

Given that mating itself is often costly to females, we would expect female resistance to mating attempts by males to be a common feature of mating systems. In fact, some form of overt and potentially costly female resistance is common among those animal species where mating occurs. Often this takes the form of accepting/favoring preferred males and avoiding/resisting others. In these cases, resistance has been termed "mate choice" or "preference" and this is, again, the subject of most studies of sexual selection (see chapter 2 and sections 3.1.2 and 3.3). Costs of such discriminating behavior are referred to as "costs of choice" (reviewed by Pomiankowski 1988), but such costs usually do not result from a physical interaction with the male attempting to mate. In other cases, the act of resisting mating attempts can be very vigorous and costly (reviews by Smuts and Smuts 1993, Clutton-Brock and Parker 1995, Gowaty and Buschhaus 1998).

During struggles between persistent males and reluctant females, injuries or even death may occur. Female dungflies (*Sepsis cynipsea*) try to dislodge males attempting to copulate by shaking vigorously and studies suggest that the male clasping forelegs cause injuries to female wings during the struggles (Blankenhorn et al. 2002, Muhlhauser and Blanckenhorn 2002; see also Leong et al. 1993). Injuries inflicted on reluctant female elephant seals by males attempting to mate them are more dramatic. Over a 20-year study of one breeding area, researchers found 17 dead females, their bodies usually located at harem peripheries or shorelines. Death was most often due to a blow or bite to the head or spine, or massive internal bleeding due to crushing of ribs and organs (Le Boeuf and Mesnick 1991). From the pattern and severity of injuries, the authors concluded that harassment by male elephant seals was the most likely explanation for death.

In many water striders, females resist mating with males by somersaulting and rolling on the water surface (Rowe et al. 1994, Arnqvist 1997a). These struggles last from a few seconds up to several minutes, and females in natural populations can be harassed every few minutes throughout the day (Rowe 1992). The struggles increase energy expenditure in females (Watson et al. 1998) and increase predation rates (nearly fivefold) on both sexes (Rowe 1994). Studies of crustaceans likewise suggest energetic costs to females of resisting male mating attempts (Jormalainen 1998). Experiments have also shown that active resistance behavior may impose a cost to females in the form of decreased egg volume (Jormalainen et al. 2001). When compared to individuals isolated during their resistance stage, female *Idotea baltica* housed with males for the entirety of the mating sequence (last intermoult to oviposition) suffered decreased glycogen levels and subsequently smaller egg size.

Male harassment of females occurs frequently in many species of all major groups. One well-studied case, where harassment rates are very high, is the solitary bee *Anthophora plumipes* (Alcock et al. 1977, Stone 1995). These bees forage on the nectar of a small range of flowers. Females divide their time and resources between foraging, constructing nests, and producing and feeding offspring. Males harass females on their foraging flights by chasing and pouncing on them. These events are vigorous, with females often being knocked from flight to the ground. Harassment peaks in midday with males harassing females at a rate of 11 pounces per minute. These mating attempts rarely result in mating, but nonetheless result in high costs to females, decreasing foraging returns by up to 50%. When males were present, females reduced their use of the most profitable flowers on the outsides of plants to avoid male harassment. When males were experimentally removed, females returned to favoring these outer flowers. Similar avoidance behavior has also been documented in other insect species (e.g., Krupa et al. 1990).

3.1.4 COSTS TO FEMALES AS A SIDE EFFECT OF MALE-MALE COMPETITION

Sexual selection through male-male competition is most often thought to be congruent with sexual selection by female choice, and therefore to reflect female interests (Darwin 1871, Berglund et al. 1996, Wiley and Poston 1996, Bonduriansky and Rowe 2003). The viewpoint that all forces of selection are aligned is reflected in Darwin's writings. In the *Origin* (Darwin 1859) he argued that in many cases "sexual selection will give its aid to ordinary [natural] selection, by assuring to the most vigorous and best adapted males the greatest number of offspring," and in the *Descent* (Darwin 1871) he argued that female choice favors the most "vigorous and well armed males." These arguments are based on the assumption that the winners of male competition are likely to be the most vigorous subset of males, and it is these males that females prefer to mate with because of the direct or indirect benefits they offer. In essence, male competition reinforces female choice. Indeed, there is evidence to support the reinforcing nature of male competition and female choice (Berglund et al. 1996, Wiley and Poston 1996). However, this alignment of the forces of sexual selection is not universal. In some species, those traits favored in male competition conflict with those favored by female choice and in others, competition among males may incidentally harm females.

In the red collared widowbird *Euplectes ardens*, males have two ornaments, a red collar and an elongated tail. Red collars appear to be favored by male competition for territories, and the long tail by female choice. Intriguingly, there appears to be a trade-off between these two ornaments (Andersson et al. 2002). Therefore, it is expected that male competition tends to favor different individuals than female choice. In the cockroach *Nauphoeta cinerea*, pheromones play a role in the establishment of dominance hierarchies in males as well as attracting females. Moore and Moore (1999) describe a case of conflicting sexual selection in this species, with male competition favoring pheromone compositions that were discriminated against by females. It was suggested that females discriminate against these males to avoid aggression.

Another case of conflicting selection has been described in the carrion fly *Prochyliza xanthostoma* (Bonduriansky and Rowe 2003). This species has a complex set of premating behaviors that involve competition among males and courtship of females by males (Bonduriansky 2003). Males have an elongated head (and antennae) that is used in both combat with other males and courtship with females. Laboratory studies demonstrated that males with elongated heads were at a disadvantage in male-male combat over territories, but these same males were preferred by females (Bonduriansky and Rowe 2003). Incidentally, female preference for males with elongated heads in this species may result from direct benefits, as females have the interesting habit of expelling some ejaculate after mating and eating it. Apparently, consuming the ejaculate has beneficial effects on females, and males with elongated heads transfer larger ejaculates (Bonduriansky et al. 2005). Finally, Sih et al. (2002) suggested that male competition favors large size in the water strider *Aquarius remigis*, whereas females prefer smaller males, possibly because they are less energetically expensive to carry (Watson et al. 1998).

In general, selection resulting from male competition may tend to bias the subset of males from which females can choose. In the four cases mentioned above, the bias is away from the phenotypes that females prefer, and in at least two cases this appears to have direct costs for females. These studies suggest that the common view that the selective forces of male-male competition and female choice are reinforcing is not always true, and that this may result in costs of male competition to females.

Male competition for access to females may have much more direct costs. In Manitoba, Canada, red-sided garter snakes overwinter communally in underground limestone dens. Groups can number in the thousands. During the spring emergence period, the sex ratio outside the den becomes strongly male biased, as females tend to disperse while males tend to lurk at the den. Late-emerging females are therefore exposed to intense "courtship" by many males—so intense, in fact, that one can see writhing balls of hundreds of males with a single female inside (figure 3.1; Shine et al. 2000, 2003). The consequences for females range from delayed dispersal to increased risk of predation and death by crushing (Gibson and Falls 1975, Shine et al. 2000, 2003).

A variety of direct costs to females of male competitive traits result from male guarding behavior (see chapter 4 and table 4.1). One associated behavior

Figure 3.1. A ball of male garter snakes attempting to mate a single female (center, head up) as she leaves the overwintering den. (Reprinted from Shine et al. 2003, with permission from the University of Chicago Press.)

is takeover, where other males seek to replace guarding males, with expensive consequences for females. Such take-over attempts are common in some explosive breeding amphibians (Arak 1983). For example, mating in the common toad (*Bufo bufo*) is restricted to a brief period of the season, and competition for access is therefore intense. Males climb on top of females and grasp them firmly with their forearms in a position called amplexus. Although the grip is strong, males are susceptible to takeovers by competitors or even multiple competitors. These take-over attempts can be vigorous enough that the female, for whom they are competing, drowns in the process (Davies and Halliday 1979). Other cases of females drowned by males include ducks (section 3.2.1) and otters (Mestel 1994) in water, and dungflies in liquid dung (Parker 1970b).

3.1.5 Sexual Conflict and the Evolution
of Sexual Cannibalism by Females

One of the most notorious cases of costly mating occurs when it pays one partner to mate but the other to kill and consume the suitor. Sexual cannibalism has long been a matter of some controversy, and various hypotheses have been proposed to account for the evolution of this remarkable behavior (figure 3.2). Defined as cases where the female kills and consumes her mate during some stage of courtship and/or mating, it occurs frequently in spiders and several other

Figure 3.2. Females of the golden orb-web spider *Nephila plumipes* frequently cannibalize males before, but also during, copulation. Males weigh less than 5% of females, presumably a result of selection for small males: large males are more likely to be attacked and killed by females prior to copulation (Elgar and Fahey 1996). In this picture, a mature female is in the process of consuming a male suitor. (Photo by D. Paul.)

groups of invertebrates (Elgar 1992) and is thought to play a significant role in the evolution of sexual traits in both sexes (Robinson 1982, Elgar et al. 1990, Barth and Schmitt 1991, Elgar 1991, Prenter et al. 1994). It was originally suggested that sexual cannibalism after sperm transfer may have evolved as an extreme form of paternal investment (Thornhill 1976, Parker 1979, Buskirk et al. 1984) or mating effort (Andrade 1996), such that males offer their soma as a blood meal to the female, thereby increasing the quantity or quality of offspring sired.

Although the paternal investment hypothesis has received a lot of attention in popular articles, it has been rejected as a general explanation of the evolution of sexual cannibalism for three different reasons. First, males are potentially able to mate with multiple females even in some species classically characterized as sexually cannibalistic such as the black widow spider, which lowers the potential benefits of self-sacrifice (Buskirk et al. 1984, Breene and Sweet 1985, Birkhead et al. 1988). Second, male mating behavior seems poorly designed for sexual cannibalism in most sexually cannibalistic species—they typically try to escape cannibalism (Polis and Farley 1979, Gould 1984, Breene and Sweet 1985, Birkhead et al. 1988, Elgar 1992, Lawrence 1992; but see Forster 1992, Andrade 1996, Andrade and Banta 2002). Third, and most importantly, sexual cannibalism most frequently occurs before copulation, at least in spiders. Elgar (1992) found that sexual cannibalism primarily occurred before mating in 40 out of the 49 species of spiders where sexual cannibalism had been reported. The paternal investment hypothesis is clearly not applicable in these cases, since the male is consumed prior to sperm transfer.

Parker (1979) noted that whenever the potential benefits males may gain from attempting to remate with other females exceed those gained by "self-sacrifice," there will be sexual conflict over the act of mate consumption. This may generally be the case when sexual cannibalism occurs before sperm transfer, and Newman and Elgar (1991) presented a model to account for premating sexual cannibalism that dealt exclusively with female foraging considerations. They showed that sexual cannibalism prior to copulation even by unmated (virgin) females can, in theory, evolve solely as a result of female foraging considerations. The act of sexual cannibalism may be said to be beneficial for females but not for males, and the model thus predicts sexually antagonistic coevolution between the ability to kill and consume mates in females and the ability to escape female attacks in males. In contrast, Arnqvist and Henriksson (1997) suggested that the act of premating sexual cannibalism may be nonadaptive and represent a "spillover" effect of aggressive behavior which is adaptive in earlier ontogenetic stages. If this is the case, neither sex can be said to benefit directly from sexual cannibalism and there will be no sexual conflict over mate killing. There is currently mixed experimental support for the assumptions and predictions of both of these models, and thus our understanding of the evolution of premating sexual cannibalism remains incomplete (see, e.g., Andrade 1998, Maxwell 2000, Johnson 2001, Kreiter and Wise 2001, Schneider and Elgar 2001, 2002, Herberstein et al. 2002).

The fact that females seem to benefit from mate consumption in many sexually cannibalistic species in terms of fecundity advantages does, however, suggest that premating sexual conflict often underlies the maintenance of this behavior. Male adaptations to avoid being cannibalized, such as long legs, agility, and vigilance, would then at least to some extent represent sexually antagonistic adaptations (see Elgar 1991, 1992, Prenter et al. 1994).

3.1.6 SEXUAL CONFLICT AND THE EVOLUTION
OF INFANTICIDE BY MALES

In taxa with extended periods of maternal care of offspring, females are often unreceptive to further mating/reproduction until their current offspring either become independent or die. This fact has set the scene for one of the most extreme costs of sexual interactions: the killing of offspring by males. Hrdy (1979) originally suggested that such infanticidal behavior may be beneficial for a killer in reproductive competition with other males, provided that (i) the male is not closely related to the offspring he kills, (ii) the premature death of the offspring renders the female receptive to mating/reproduction sooner, and (iii) the killer has a certain probability of fathering the future offspring produced by the female. Infanticide by males is a remarkably widespread phenomenon among social mammals, including primates, and although its causes and consequences are still matters of controversy, there is now strong empirical support for Hrdy's original hypothesis (see, for example, Borries et al. 1999, van Schaik and Janson 2000, Soltis et al. 2000). Thus, there is little doubt that in some groups infanticide by males is favored in males. Of more immediate interest here is the fact that this male behavior is very costly to the mothers of the killed offspring and thus represents intense sexual conflict.

Given the obvious and dramatic costs to females, who have made a substantial investment in the killed offspring, researchers in this field have suggested a suite of female counteradaptations to infanticidal behavior by males. Such counteradaptations include (see Hrdy 1979, Hiraiwa-Hasegawa 1988, Smuts and Smuts 1993, Sterck et al. 1997, Schneider 1999, Wolff and Macdonald 2004) (i) aggression toward males, (ii) alliances with noninfanticidal males, (iii) timing of reproduction to periods when fewer potentially infanticidal males are around, and (iv) strategies to cloak paternity, such as multiple mating with many males and/or concealed ovulation. Although there is evidence suggesting that several of these potential counteradaptations are associated with reduced risk of infanticide, this is generally correlational rather than experimental. Needless to say, this partly reflects the practical limitations involved when conducting experimental research on social mammals in their natural environment. Rather than discussing this large and somewhat controversial field here, we refer the interested reader to the recent monograph edited by van Schaik and Janson (2000).

Infanticide by males is, however, not restricted to social mammals (Hausfater and Hrdy 1984, Parmigiani and vom Saal 1994, van Schaik and Janson 2000). One of the best-studied systems, and perhaps the only one where substantial data exist on rates of infanticide and its consequences for lifetime fitness in natural populations, is actually a spider: the eresid *Stegodyphus lin-*

Figure 3.3. In the spider *Stegodyphus lineatus*, infanticidal males steal and dispose of female egg sacs to gain mating opportunities. This is illustrated here, where a male is emerging from a female nest carrying her egg sac, about to literally push the eggs over the edge of the nest opening. (Photo by J. M. Schneider.)

eatus (figure 3.3. and see Schneider and Lubin 1996, 1997, Schneider 1999). Females of this remarkable species exhibit suicidal maternal care. Females lay a single clutch of some 80 eggs, and then guard the egg sac. The female opens the egg sac to release the spiderlings, feeds them by regurgitation, and is herself then consumed by the young within 14 days. Males do the mate searching in this species, and since the mating and egg-laying seasons overlap extensively in the field, males often encounter females guarding egg sacs. The following four facts strongly suggest that infanticide is adaptive for males: (i) females will produce a second clutch if the first is lost, (ii) females will remate prior to laying a new clutch, (iii) there is complete mixing of sperm, and thus shared paternity, when females have mated with two males, and (iv) a male can expect to encounter on average only 1–2 females in his lifetime, so a female guarding an egg sac represents a valuable opportunity for reproductive success for a male under natural conditions. This is especially true for males emerging late in the season. It therefore should come as no surprise that most males that encounter a female guarding an egg sac literally try to steal and dispose of the egg sac. The rate of infanticide in the field is surprisingly high: some 8% of all clutches are lost to infanticidal males. Infanticide by male *S. lineatus* is clearly very costly for the female, as (i) the estimated probability of surviving until the time when offspring hatch from the egg sac is only about half that of females who were not forced to lay a second clutch, because of the time and energy required to produce a second clutch, (ii) second clutches are significantly smaller than first clutches, and (iii) offspring of second clutches emerge later in the season and thus have less time to feed prior to the winter. Females exhibit at least two strategies to reduce the risk of infanticide by males. First, females with egg sacs are very aggressive toward intruding males. Intrusions invariably result in aggressive fights, during which males sometimes get injured or even killed and cannibalized. The relative body size of the two contestants has a large effect on the outcome of these fights, and it has been suggested that body size evolution has been affected by these conflicts in *S. lineatus*. Second, females maturing early in the season, when the risk of infanticide is highest, have a longer interval between maturation and egg laying than have females maturing later. This suggests that females may be modulating the timing of egg laying to reduce the risk of infanticide.

3.2 Adaptations for Persistence and Resistance

If encounters between males and females commonly involve conflict over whether to mate or not, male strategies that allow them to circumvent female premating resistance behavior altogether could, under certain circumstances, be beneficial and thus become established. For example, in some natural populations, male fruit flies (*Drosophila melanogaster and D. simulans*) patrol sites where adult females emerge from their pupae, and copulate with teneral females while their bodies are still soft and transparent and before their wings have unfolded (Markow 2000). Such females may have very little, if any, possibility to resist copulations. A similar evolutionary scenario may have been involved in the evolution of pupal mating in insects (Deinert et al. 1994, Brower 1997). For example, mating in the mosquito *Opifex fuscus* takes place on the water surface from which the adults emerge (Slooten and Lambert 1983). Adult males search for, grasp, and guard female pupae which are close to emergence, and copulate with females just before they emerge from the pupae. The sex ratio at the time of mating is extremely male biased (some 260 males to each female). Another intriguing example occurs in some bats that gather at hibernation sites to spend the winter in a state of torpor. While females quickly enter into a winter-long dormancy, males choose warmer niches and stay active for longer or even "wake up" periodically during the winter. Males use this mobility to copulate with dormant females that can offer no resistance. Females subsequently store sperm throughout the winter and fertilize their eggs in early spring. The habit of copulating with torpid females is widespread among bats (e.g., Thomas and Fenton 1979, Tidemann 1982, Phillips and Inwards 1985, Gebhard 1995).

Sidestepping of female resistance is, however, rarely possible, and it is not surprising that persistent mating attempts (sexual harassment) are frequent among animals. We can view this harassment as a "persistence" trait that evolves in response to resistance. Clashes between persistent and resistant individuals can be costly and sometimes gruesome. Given this high-stakes battle, one might expect that any trait that increases the effectiveness of persistence will be favored. One dramatic example is the funnel-web spider *Agelenopsis aperta*, where males have evolved a strategy to literally anesthetize reluctant females (see Singer et al. 2000). Approaching the female, males spray a "toxin" at the female that causes her (and occasionally the male himself) to enter an "unconscious" state. The male then hauls the female around on the web, repositions her, and finally inseminates her while she is still "knocked out." Although costs to females are unknown, they seem likely in this system. A similar, and equally remarkable, example is the infamous "sexual sting" common in some groups of scorpions. During courtship, the male uses his pedipalps to grasp those of the female and then pulls and pushes her until he finds a substrate suitable for spermatophore deposition. This walk, aptly named "promenade à deux," lasts from a few minutes to many hours (see Polis and Sissom 1990 and Benton 2001 for reviews). The male then deposits a spermatophore and maneuvers the female so the spermatophore contacts her genital opening. In some species, the male routinely stings the female during or prior to the promenade by puncturing the female's body. Although the sting typically remains within the female for many minutes and males may administer multiple stings during a single mating interaction (Tallarovic et al. 2000), it is not known whether the male actually envenomates the female. Polis and Sissom (1990) suggested that venom is transferred and that the sexual sting may literally drug the female, to subdue her aggressive resistance behavior (males of some species of scorpions suffer a risk of being cannibalized by females during sexual encounters). Interestingly enough, Inceoglu et al. (2003) recently reported that the scorpion *Parabuthus transvaalicus* produces two distinct types of venom. They hypothesized that males may inject females with the less toxic type, which may be more efficient at paralyzing females. This dual venom system might allow males to effectively make their mate more sluggish while avoiding the deleterious effects of the more toxic components they are capable of producing, although this remains to be demonstrated.

A much more common persistence trait, however, is grasping structures that are employed during intersexual struggles. Likewise, any trait arising, including antigrasping traits, that increases the ability of females to resist or avoid mating attempts will be favored. In the next section we discuss some well-studied examples of harassment and resistance (see Clutton-Brock and Parker 1995), and some traits that enhance persistence and resistance.

3.2.1 HARASSMENT AND RESISTANCE

In several species of ducks, males frequently harass females on land, water, and in the sky (McKinney et al. 1983). Males grasp at females with their beaks and attempt to mount them. It is not unusual for the antics of one pair to attract the attention of many more males, resulting in a mass of struggling males enveloping a single female in the middle. Struggles can be violent enough that females may die. Following a study of mallards at a pond in England, J. S. Huxley attributed 7–10% of annual female mortality to the mating attempts of these "unsatisfied" males (McKinney et al. 1983).

The mating system of several pinnipeds (seals, sea lions, and walruses) is characterized by frequent male harassment of resistant females. Northern elephant seals (*Mirounga angustirostris*) are dramatically sexually size dimorphic, with males weighing several times more than females (Le Boeuf and Mesnick, 1991). These seals are polygynous, gathering in large harems controlled by a single dominant male or bull. Females arrive at the rookery pregnant each year, give birth to their offspring and then nurse for approximately one month, never once leaving the breeding ground. During the last days of nursing, females will mate 1–5 times with the dominant male before weaning pups and returning to the sea. Although courtship is direct and aggressive, there is usually little resistance to mating attempts by dominant males (Mesnick and Le Boeuf 1991).

Peripheral or subdominant males surround harems throughout the breeding season waiting for opportunities to mate. These males gain mating success either by sneaking into harems or by attempting to copulate with females as they are exiting harems (Le Boeuf 1972). The presence of bull males reduces harassment and copulation attempts within the harem, but females are relatively undefended while traveling along the shore and back to sea, and most departing females are intercepted and mounted (Mesnick and Le Boeuf 1991). Although females often strongly resist these attempts, a male can physically overpower a resistant female by slamming down the full weight of his head on her back and biting her neck vigorously. Injuries sustained during these interactions can result in the death of the female (see section 3.1.3).

Females may attempt to mitigate the high price of resistance by allowing peripheral males to mate when interception is unavoidable. Mesnick and Le Boeuf (1991) showed that females who were receptive to mating suffered fewer blows and mounts while exiting harems, although their number of copulations was indeed higher. Departing females were more receptive than they were when in harems, but were much more likely to terminate copulations, compared with bull matings.

Similar patterns have been documented in southern elephant seals where females prefer to join larger harems, as harem females experience reduced levels of harassment. Females of this species also experience heightened harassment levels when leaving breeding grounds and often accept secondary male copulations to reduce the possibly dangerous effects of resistance (Galimberti et al. 2000a,b,c). Southern elephant seal males also show a unique behavior called group raids, where several secondary males will invade territories and attempt to forcibly establish residence or abduct females from harems (Campagna et al. 1988). Males, females, and pups had severe and obvious wounds attributed to raiding activities. Several pup deaths were also observed as a result of group raids.

Although the severity of costs for lethal interactions is obvious, quantifying the costs of nonlethal male harassment is more difficult. Boness et al. (1995), however, were successful in showing that increased levels of male harassment on lactating gray seals (*Halichoerus grypsus*) resulted in reduced suckling time and slower-growing pups. Females who pupped later in the season experienced higher levels of harassment, due to difference in operational sex ratio, and weaned significantly lighter offspring. The authors thus suggest that fitness costs of male harassment may have contributed to reproductive synchrony in this species.

When female receptivity for copulation is short and there is no potential for sperm storage, theory predicts that selection will favor precopulatory guarding (Parker 1974). Female crustaceans are receptive for a very short time, immediately following molt into the reproductive form, when the exoskeleton is soft enough to allow fertilization. Precopulatory guarding by males is accordingly widespread throughout this taxon. In several aquatic species, guarding is accomplished by males pairing with and carrying females until molt and copulation, a strategy that almost ensures the guarding male's paternity of the following brood (Jormalainen 1998). Although females must also ensure fertilization of their eggs and may benefit from the insurance gained by such precopulatory guarding, it is likely that the optimal guarding duration will differ between the sexes. Female resistance to mating attempts has been observed in more than 15 species by over 20 authors (reviewed by Jormalainen 1998). Such widespread occurrence of female resistance in study species strongly suggests intersexual conflict over the duration of precopulation.

Early resistance by females could indicate the costly nature of lengthy precopulas. Mathis and Hoback (1997) showed that females were more likely to be eaten in precopulatory pairs of the amphipod *Gammarus pseudolimnaeus*. Males of this species started guarding earlier with increasing threat of predation, indicating that search time, not guarding time, may be costly for males. However, in an interpopulation study of *Hyalella azteca*, Strong (1973) showed that populations who experience higher predation have shorter precopulation duration as determined by female resistance. Ward (1986) also showed that precopulatory pairs of *Gammarus pulex* are more susceptible to predation than solitary individuals.

In response to unwanted precopula attempts, females will kick violently and arch their bodies, making pair formation almost impossible. Several authors have thus concluded that it is females who control the start of precopulatory pairing (Jormalainen 1998). Resistance is generally strongest in females who are farthest away from their next sexual molt (Strong 1973). As they approach molting, resistance generally diminishes, and females allow pairing more readily (Jormalainen et al. 1994a).

Male harassment of females is also common to some species of the Poeciliid fish, including guppies and some mollies. Guppies have a promiscuous mating system, where two classes of male-female interactions occur. Under some conditions, males display to females with a conspicuous sigmoid display, which females appear to assess and then choose to copulate or not. A great deal of study has been directed toward this type of mating interaction, and consequently guppies have become a model system for studies of female preference and sexual selection (Houde 1997, Magurran 2001). However, under other conditions, males will frequently harass females that are reluctant to mate, and attempt to sneak copulations with thrusts of their gonopodium (intromittent organ). Thrusting is a less efficient means of sperm transfer, but is successful to some degree (Evans et al. 2003). Harassment can be costly to females, with foraging rates in the laboratory being decreased up to 25% when pursuing males are present (Magurran and Seghers 1994b). Similar indications of female resistance, costly harassment, and the use of thrusting mating attempts have been reported in other poeciliids (e.g., Bisazza et al. 2001, Schlupp et al. 2001, Pilastro et al. 2003, Plath et al. 2003)

In Trinidadian guppies variation among populations in predation risk has played a major role in shaping a variety of traits, including life history, social behavior, and sexual ornaments (e.g., Endler 1980, Reznick and Endler 1982, Reznick et al. 1990, Rodd and Sokolowski 1995). Interestingly, the reliance of males on the thrusting tactic is thought to be greater in sites of high predation (Endler 1987, Magurran 2001, but see also Farr 1975), perhaps because the conspicuous sigmoid display attracts predators. If we view thrusting as a persistence trait in males, one might expect adaptation to facilitate the success of thrusting in high-predation populations. Recently, Kelly et al. (2000) have reported that the gonopodium tends to be larger in high- than low-predation sites. This is expected if sperm transfer during thrusting with reluctant females is more efficient with elongated gonopodia. More recent evidence suggests that the gonopodia may be adorned with grasping hooks that facilitate sperm transfer during gonopodial thrusts (Yun Cheng, personal communication). As in other taxa (Eberhard 1985), gonopodia among species in the family Poeciliidae are highly diverse in shape as well as length (Constantz 1989). There is also much interspecific variation in the degree to which the thrusting tactic is used by males. It would make an interesting study to determine whether some of this interspecific diversity in gonopodia is attributable to variation in mating systems.

Male harassment against females is common also in many other groups of fish with internal fertilization (e.g., Bisazza et al. 2000), but in no case are the associated adaptations as remarkable as in the Malabar ricefish *Horaichthys setnai* (Beloniformes) (see Kulkarni 1940). Males of this little species (2 cm) are unique among teleosts in producing true spermatophores. The Malabar ricefish spermatophore is club shaped and contains large numbers of sperm, and each male can produce several hundred spermatophores. The tapering end is pointed, sharp, and armed with several rows of barbs (see figure 3.4). The mating behavior is similar to that of many Poecilids. During the breeding season, males follow females, swimming below and behind them at a distance of a few centimeters. The male then makes a sudden dash toward the female at very high speed, and lashes out the gonopodium (his modified anal fin) with which the spermatophore is delivered. As the gonopodium strikes against the female genital opening, the pointed end of the spermatophore is literally stabbed into the female skin and remains very firmly attached. After a while, the spermatophore bursts open at the base of the armed end and the sperm migrates into the genital opening of the female. After about an hour, the spermatophore is empty. Adult females can carry a large number of attached spermatophores.

The structure of the male Malabar ricefish gonopodium is very elaborate and complex, and although it is clearly used to stab females during the transfer of spermatophores, there are no detailed studies of its functional morphology. It is perhaps worth noting that the Malabar ricefish is the only known species in its genus. Somewhat puzzled by the reproductive biology of the Malabar ricefish, Kulkarni (1940) suggested that mating may carry substantial costs to females. First, he noted that females are aggressive toward harassing/stalking males. Second, the area near the genital opening of females is covered by "genital pads," where the skin is markedly thickened and hardened (see also section 3.2.3). In what must be one of the first clear discussions of a female resistance adaptation, he suggested that these pads function to reduce the cost of mating to females as they may become injured not only by the insertion of spermatophores but also by the "powerful strokes" of the gonopodium (the terminal portion of which carries various hooks and pointed processes). Incidentally, the exact location of the genital opening and the genital pads is strikingly variable across females.

3.2.2 GRASPING TRAITS

Darwin (1871) noted that there is an infinite diversity of "organs of prehension" in males that clearly function to grasp the female prior to or during mating. In some species, these seem to be absolutely required for mating; in others, they

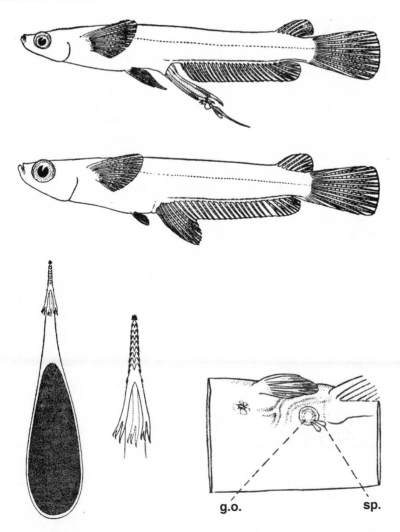

Figure 3.4. Male Malabar ricefish (top) stab females (second row) with "darts" loaded with sperm. The barbed spermatophores (bottom left), that are filled with sperm (dark mass in the center), are deposited in the area of the genital opening (bottom right). Note the genital pads—areas of thickened skin in (bottom right picture). g.o., genital opening; sp., spermatophore. (From Kulkarni 1940.)

are not, but instead give males some advantage over others. That advantage, Darwin believed, came from preventing the female's escape, where she could be mated by another male, or in preventing takeovers of the female by other aggressive males (see sections 3.4.1 and 3.4.2). As early as 1907, female resistance to mating was recognized as a potential force in shaping the evolution of grasping devices (Edwards 1907, cited by Richards 1927). Female resistance is common even in those cases that were used by Darwin as examples of male competition, such as the clasping devices of crustacea (section 3.1.3). In the face of female resistance, any trait that enhances a male's success in overcoming this resistance will be favored, and grasping devices are an obvious candidate. Consider, for example, the antennal claspers of the water strider *Rheumatobates rileyi* (figure 3.5), which are employed during the precopulatory struggle. The antennal grasp is released as soon as male genitalia are inserted in the female; therefore, they can play no role in guarding the female from other males, or in preventing takeovers. In these systems where claspers are employed only against resisting females, intersexual conflict is particularly likely to be involved in their evolution. Finally, if claspers are favored in mating by some female trait (resistance in this case), then we may view them as selected by female choice. This argument is developed in more detail in section 3.4. Here we discuss some of the diversity that exists in grasping devices in animals.

Modifications of the male pregenital segments for clasping females are very common among the insects (Darwin 1871, Eberhard 1985, 1996, Arnqvist 1997b), and sexual conflict may have played a role in many or all of these devices. In some cases, these have been demonstrated to function as persistence traits, but there is some controversy over the generality of this conclusion (see section 3.4 and Eberhard 2002). The earliest studies on these traits centered on the scorpionflies of the genus *Panorpa* (Thornhill 1980, 1984, Thornhill and Sauer 1991). The notal organ is a large muscular set of claspers on the abdominal segments of this insect. Males attempt to grasp a female's forewings with these claspers prior to mating, and females vigorously resist. When Thornhill (1980, 1984) disabled the claspers by covering them in beeswax, males were unable to overcome female resistance and consequently could not mate. Interestingly, when males had a food gift to offer females and females were therefore not resistant, the claspers were not required for mating.

A similar clasping structure occurs on the terminal abdominal segments of male sagebrush crickets (*Cyphoderris strepitans*), called the "gin trap" (Sakaluk et al. 1995). To copulate, a female climbs on top of the male and begins feeding on the male's fleshy wings, which "preoccupies" her for a period while the male transfers his spermatophore. During copulation, the female is secured to the male with his gin trap. Sakaluk et al. (1995) reasoned that females may attempt to terminate matings prior to complete transfer of the spermatophore if males have little remaining wing to feed on during mating. If so, then the gin trap may function to retain females in cases where they attempted to dismount prior to spermatophore transfer. To conduct the experiment, both wing availability and functioning of the gin trap were manipulated. As expected, females tended to terminate matings early when the male's wings were par-

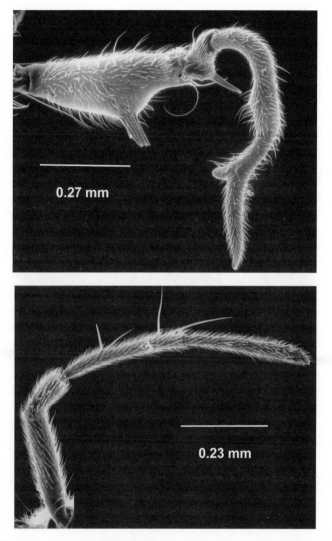

Figure 3.5. Scanning electron micrograph of the antennae of male (top) and female (bottom) *Rheumatobates rileyi*. In this water strider males grasp resistant females with their antennae. (Reprinted from Westlake et al. 2000, with permission from the National Research Council, Canada.)

tially removed, and males with the gin trap enabled were able to prolong the mating longer than those with it disabled.

In one water strider species, *Gerris odontogaster*, males have a pair of abdominal clasping processes that are employed to grasp females during the vigorous premating struggles. Arnqvist (1989b, 1992) found that males with

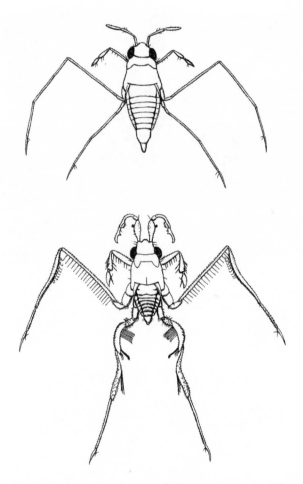

Figure 3.6. Adult female (top) and male (bottom) *Rheumatobates rileyi*. Males of this species, and some others in the genus, have all three pairs of legs and the antennae (figure 3.5) modified for grasping resistant females during premating struggles. (Reprinted from Westlake et al. 2000, with permission from the National Research Council, Canada.)

longer claspers could withstand more somersaults by struggling females and therefore achieved greater mating success. In fact, there is a variety of modifications of male water striders that appear to function as grasping apparati (Arnqvist 1997b). Perhaps the most extreme cases occur in the genus *Rheumatobates* (Westlake et al. 2000). In this genus, species exist where every leg and even the antennae are apparently modified for grasping (figures 3.5 and 3.6). Even water strider genitalia appear to be shaped by female resistance (see section 3.4 and reviews by Arnqvist 1997, Fairbairn et al. 2003). More generally, animal genitalia show the hallmark of sexual selection, with rapid evolution creating such diversity that, in many closely related species, species can be distinguished only by their genitalia (Eberhard 1985, 1996, 2004, Arnqvist 1997, 1998). Clasping structures on genitalia are common, and this suggests that conflict may be involved in their remarkable diversification.

Males and females of crustaceans where males carry females during the precopulation period are significantly dimorphic in a number of morphological traits. In addition to large body size, male crustaceans often have highly modified walking legs used to grasp females in precopulation and in male-male agonistic interactions. Resistance to these mating attempts is not uncommon, but we are unaware of any strong evidence for a resulting selection on these clasping devices. Conlan (1991) found significant sexual dimorphism in legs of guarding species, where no such dimorphism was found in nonguarders. However, the most significantly enlarged leg, the second gnathopod, is not used to carry females, but is held free and used in male-male agonistic interactions and takeovers (Borowsky 1980, Conlan 1991). Borowsky (1980) also found sexual dimorphism of lateral plates in three crustacean genera, where the plate is modified in females to show a distinct process. Males grasp onto this process and the gnathopod and the female's point of attachment fit together like two puzzle pieces. Vueille (1980) also described the modified peraeopods of two isopod species as "sexual hooks" which grasp the female tightly during precopulation and are developed only in older males. However, clipping the nails from male *Idotea baltica* had no effect on the precopulatory duration, though the nail is used as the point of attachment in pairs (Jormalainen and Merilaita 1993, 1995).

Biting or holding is an almost universal mating behavior of elasmobranchs (sharks, rays, and skates). While the intensity, severity, and location of bite scars vary across species and seasons, the majority of documented bite scars are found on females (Pratt and Carrier 2001, Kajiura et al. 2000). Though males too may exhibit scars, these can be attributed to either inter- or intrasexual aggression (Kajiura et al. 2000, Gordon 1993). Pre-, post-, and copulatory biting have been observed in elasmobranchs (Pratt and Carrier 2001, Yano et al. 1999). Copulatory biting most likely serves as a holdfast during copulation, especially in larger species where mating is nonswimming (Kajiura et al. 2000, Pratt and Carrier 2001, Tricas and Le Feuvre, 1985). In such cases, the male bites either the anterior or posterior surface of the pectoral fin to gain leverage for mating. Precopulatory biting, however, is less clear in its significance. Alternatively, the precopulatory releasing theory, as suggested by Springer (1960), hypothesizes that such biting facilitates female cooperation and receptivity. Maruska et al. (1996) suggested that precopulatory biting in the Atlantic stingray, *Daysatis sabina*, a species with an extremely protracted mating season, functions to induce female reproductive cycles, mediated through skin receptors.

It is, however, possible that such biting, both pre- and copulatory, serves to overcome female resistance, serving as an aggressive form of harassment. The intensity of mating scars in some species, such as the blue shark, supports this hypothesis as precopulatory courtship bites are severe enough to pose a serious threat of blood loss and infection (Pratt and Carrier 2001). In addition to the visible costs of precopulatory biting, Klimley (1980) also observed female avoidance behaviors in response to male aggression, including pivoting and rolling when males attempted to bite the pectoral fins. Castro et al. (1988) also witnessed female resistance in the chain dogfish as the female struggled violently against a sturdy grip on her tail. When she finally became motionless, the male then moved up to grasp her pectoral fin and copulation ensued. Females may also bite back in response to a persistent male (Gordon 1993).

Nordell (1994) observed the mating behavior of the round stingray, *Urolophus halleri*, classifying biting into "noncopulatory" and "copulatory." Noncopulatory bites were directed toward the posterior or lateral portion of the disk, while copulatory bites, which lead to copulation, were targeted at the anterior part of the disk. Females often escaped from noncopulatory nips and buried themselves in the sand. Males would then search out females, uncover them, and proceed with precopulatory nips which may or may not lead to copulation. Although Nordell suggests several alternative reasons for precopulatory nips, it is possible that this is yet another example of male harassment in elasmobranchs.

There is significant sexual dimorphism of many elasmobranchs, which may be directly related to courtship biting, and represents a grasping trait. Male sharks often have longer teeth with fewer cusps. Similar dimorphic dentition is also widespread among rays (Pratt and Carrier 2001). One particularly interesting example of sexually dimorphic dentition is found in the Atlantic stingray, where the male's teeth become sharper and recurved during mating season, returning to the femalelike molariform dentition in the nonmating months. The grip tenacity of the sharper teeth was shown to be significantly greater (Kajiura and Tricas 1996). This temporal plasticity could increase male mating success and increase the possibility of overcoming female resistance.

Biting as a grasping strategy is not limited to sharks and rays. For example, male tiger beetles secure females by biting with their mandibles, and mandibles are sexually dimorphic in size and shape (Fielding and Knisley 1995, Kritsky and Simon 1995). In seals, males tend to possess large canines that are used to bite the female's neck before and during copulation (see section 3.2.1). Males of the Lake Eyre dragon, *Ctenophorus maculosus*, similarly bite at the neck and head of resistant females (Olsson 1995). In the salamander, *Desmognath wrighti*, males tenaciously bite females with premaxillary teeth that have been highly modified for the breeding season (Arnold and Houck 1982). Apparently the bite is meant to deliver hormones into the female circulatory system from glands in the male chin. Such glandular secretions are often employed before and during mating in amphibia, and they appear to play a role in persuading females to mate (Arnold and Houck 1982). Male *Notophthalmus viridescens* employ these substances most vigorously when courting a resistant female (Arnold 1977). The male will chase a retreating female and attempt to capture her with his hind legs. If successful, the male will hold the female in amplexus for an hour or more while delivering secretions from courtship glands. The glands are usually located on the chin, head, and cloaca. The product of these glands is sulfated mucin, which has been experimentally shown to influence male mating success (Arnold and Houck 1982).

Amphibians also commonly possess grasping structures and these may function in male combat or to reduce takeovers of females while in amplexus, but may also have a role as persistence traits (reviewed by Shine 1979). Nuptial pads, found almost exclusively on males, are rough patches of skin on the thumb, digits, inner arms, and/or chest of many anuran species. They vary in keratinization and spinousity, becoming more pronounced during the mating season due to increased androgen levels (e.g., Epstein and Blackburn 1997, Jungfer and Hodl 2002). In some frog species, nuptial pads are so well developed as to become discrete spines. Spines can be found on the thumb, digits, chest, and humeral bone. Taylor (1949, as cited in Flores 1985) reported a sharp, prepollical spine (located by the thumb) on *Centronella spinosa*. He presumed a grasping function for these traits, as they would be capable of piercing the female's skin during amplexus. Although females of this species also possess a prepollical spine, it is completely covered by skin. Other species in this family (Centrolenidae) have prepollical spines; however, spines are present and equally developed in both sexes in some species, being completely covered with skin in males and females (Flores 1985). Well-developed and sexually dimorphic prepollical spines also occur in the Hylidae (Shine 1979). Patches of chest spines have been documented in several toads and frogs, notably in the genus *Bufo* and the genus *Rana* (e.g., Dubois and Matsui 1983, Duellman and Ochoa 1991). The spines of *Rana blanfordii* are distributed along the chest, inner forearm, and the first three digits. These spines may assist in grasping females but correlations between male aggression and the presence of nuptial spines have led other authors to conclude that spines evolved in the context of male-male competition (Shine 1979).

Humeral spines may also assist in the grasping and holding of females during amplexus. These large, strong, bony projections of the upper arm are present in the genus *Telmatobius*. One species is fully aquatic, found in fast flowing streams of Bolivia (De La Riva 1994). The presence of such humeral spines may be highly advantageous for amplexus maintenance in such an unstable mating environment. Similar traits on the digits, forearms, and chest also occur in this genus, regardless of humeral spine presence (e.g., Benavides et al. 2002). Male northern leopard frogs, *Rana pipiens*, have disproportionately larger forelimb muscles in addition to nuptial pads (Yekta and Blackburn 1992). These muscles are used in grasping females and may represent another possible adaptation for overcoming female resistance or increasing mating duration.

3.2.3 ANTIGRASPING TRAITS AND OTHER FORMS OF RESISTANCE

It is expected that the evolution of grasping traits that increase the efficiency of persistence in one sex will be matched by the counteradaptation of resistance traits in the other sex. Although grasping traits are very common (section 3.2.2), evidence for antigrasping traits is much slimmer, having received very little study. A common response is to simply avoid areas where harassment is common (reviews by Smuts and Smuts 1933, Clutton-Brock and Parker 1995). In other cases, females possess traits that simply reduce the costs of grasping traits in males. For example, skin thickness is commonly sexually dimorphic in those elasmobranches where biting by courting males occurs. Kajiura et al. (2000) found that the dermal layer on female pectoral fins was 50% thicker than that of males in the Atlantic stingray. Pratt (1979) also found that, in those areas in which bite marks from harassing males were most often found, the dermal layer of females was twice the thickness of that of males. Likewise, the "genital pads" of Malabar ricefish (see section 3.2.1) may represent an adaptation to minimize the cost of spermatophore darts and strokes of the gonopodium. In these preceding examples, female traits possibly function to reduce the damage inflicted by male grasping devices.

In two groups of insects, diving beetles and water striders, females possess traits that apparently function to reduce the efficiency of male grasping devices (figure 3.7). Experiments have been conducted to verify the efficacy of antigrasping traits of water striders. In the genus *Gerris*, females of some species have modified pre- and genital segments that function to increase their ability to resist mating attempts by males (see section 3.4 and Arnqvist and Rowe 1995, 2002a, Andersen 1997, Arnqvist 1997a). These include elongated and upturned connexival spines that flank the genitalia, and a downward bending of the genital opening. During premating struggles, males attempt to grasp the female by hooking their inflated genitalia around the female's abdominal tip (sections 3.2.2 and 3.4). In the field, Arnqvist et al. (1997) found that females of *Gerris incognitus* with more elongated spines tended to mate at lower rates than those with less elongated spines. An experimental manipulation of spine length in this species demonstrated that females with larger spines could repel unwanted mating attempts more quickly, and consequently mated at a lower rate than did those with smaller spines (Arnqvist and Rowe 1995).

An unusual form of response to male harassment has evolved in certain robber flies of the genus *Efferia* (figure 3.8; Diptera, Asilidae). Males of this genus fly actively through the vegetation in search of females (see Dennis et al. 1986). When approached by a male, females generally take to the wing and

Figure 3.7. Females of some species of diving beetles are polymorphic for the surface structure of their pronotum and hemilytra (see section 3.5.1). In the top panel, the rough granulate morph (left) and smooth morph (right) of *Graphoderus zonatus verrucifer*. Males in these species tend to have more developed adhesive structures in populations where the rough morph is more common. Suckers on the mesotarsus (mid-page) and protarsus (bottom of page) in male *Graphoderus zonatus verrucifer* are shown. (Reprinted from Bergsten et al. 2001, with permission from the Linnean Society of London.)

Figure 3.8. In many robber flies (Diptera, Asilidae), males capture females in flight or when perching. A short but very animated "struggle" follows which may end with the female fleeing the male, with copulation, or in some species with the male being killed and consumed by the female. In other species, females use an unusual but efficient resistance strategy: they sometimes play dead when caught by the male. Picture shows a mating pair of *Efferia apicalis*. (Photo by G. Beaton.)

may then be pursued by the male. If the male succeeds in grabbing the potential mate in flight, the female struggles violently and often successfully to dislodge the male. In a few species, however, females sometimes use another strategy to achieve this goal: if grasped by a male they exhibit thanatosis (playing dead) (Dennis and Lavigne 1976). Once the female ceases to move, the male apparently no longer recognizes the lifeless female as a potential partner, loses interest, and so releases the female. A reasonable hypothesis is that this death-feigning behavior represents a female resistance strategy. The fact that thanatosis is a widespread antipredator behavior in insects suggests that females may have been preadapted to evolve this particular form of resistance.

In sexually dimorphic taxa where males search for and harass females using visual cues to distinguish potential mates from other males, a female mutant that looks like and/or behaves like a male should enjoy reduced harassment. However, this benefit may be frequency dependent: as the frequency of male look-alikes increases in the population, males should alter their search image, thus diminishing the rewards of looking like males (Robertson 1985, Miller and Fincke 1999, Sherratt 2001). Interestingly enough, this process seems largely responsible for a remarkable phenomenon among insects: many damselflies, and a few butterflies, have a sex-limited female di- or polymorphism, where one morph closely resembles the male (andromorph) and the other/s is/ are more cryptically colored (heteromorph/s) (e.g., Robertson 1985, Cook et al. 1994). This has been well studied in Coenagrionidae damselflies. Here, body coloration is determined by a single bi- or triallelic autosomal locus with female-limited expression (Johnson 1964, 1966, Cordero 1990, Andrés and Cordero 1999) and andromorphs not only look like males but may also mimic male behavior in interactions with males (Robertson 1985). Several studies have shown that andromorphs indeed experience reduced harassment by males compared to hetromorphs (e.g., Robertson 1985, Hinnekint 1987, Forbes 1991, Cordero 1992, Cordero et al. 1998, Andrés et al. 2002), but this benefit comes at two costs: predation rates may be higher for the more conspicuous andromorph (e.g., Johnson 1975, Robertson 1985) and they may suffer some risk of remaining unmated (e.g., Cordero et al. 1998). Because males learn to identify andromorphs if present in high enough frequencies (Van Gossum et al. 1999, 2001), frequency-dependent selection should tend to maintain a balanced polymorphism among females (Andrés and Rivera 2001, Sirot and Brockmann 2001). As might be expected, the frequency of female morphs within populations is often remarkably stable over time (Hinnekint 1987, Andrés et al. 2002). However, selection should also be density dependent: the benefits of looking like a male increase (higher rates of harassment) and the costs decrease (lower risk of remaining unmated) with increasing male density. The frequency of andromorphs does indeed increase with population density in several species (Hinnekint 1987, Forbes et al. 1995, Andrés et al. 2002) Thus, it seems clear that this female-limited dimorphism, or polymorphism, is maintained by a combination of frequency- and density-dependent selection (see Andrés et al. 2000, 2002). At the heart of this is the fact that females can resist costly sexual harassment and superfluous matings by mimicking males.

Male mimicry as a female strategy to reduce the costs of male harassment is, however, not limited to insects. For example, gravid female lizards may develop male body coloration (Galan 2000) and female fiddler crabs sometimes mimic male behavior (Burford et al. 2001). More generally, Burley (1981) suggested that, under certain circumstances, individuals of both sexes might benefit from concealing their sexual identity in order to avoid costly intra- and intersexual interactions.

3.2.4 EXPLOITATION OF SENSORY BIASES

As discussed earlier (section 2.6), direct selection on the female (or male) sensory system, morphology, or behavior, in some context other than mating, may have the side effect of biasing mating success in the other sex (e.g., West-Eberhard 1979, 1983, 1984, Kirkpatrick 1987a,b, Ryan 1990, Basolo and Endler 1995, Endler and Basolo 1998). Given the complexity of sensory systems, there is potential for preexisting biases toward a great variety of colors, sounds, and even behaviors and morphologies. Where conflicts over mating occur, these preexisting biases in one sex provide a target for "exploitation" or "manipulation" by the other (West-Eberhard 1983, Arak and Enquist 1993, 1995, Enquist and Arak 1993, Holland and Rice 1998). Consider a case where females are resistant to costly multiple mating or to mating with certain types of males (small, drab, etc.). If a mutation arises in males which results in some trait that exploits a sensory bias in females, those possessing the mutation may be able to sidestep resistance, and the trait will spread (section 2.7). The evolution of such an exploitative trait sets in motion the potential for either sexually antagonistic coeveolution (see sections 2.7.1 and 2.7.1.4 and Parker 1983a,b, Holland and Rice 1998, Gavrilets et al. 2001) or coevolution toward indirect benefits (see section 2.7). There is growing empirical evidence that preexisting biases play a central role in the evolution of mating biases (reviews by Ryan 1990, Endler and Basolo 1998, Boughman 2002) and sexual conflict may in theory have been important for the origin, elaboration, and maintenance of many male traits traditionally believed to have evolved by other forms of male-female coevolution. Unfortunately, the consequences of the exploitation of preexisting biases for female fitness have not been documented, and therefore the extent to which sexually antagonistic coevolution has actually played a role is not known.

There are some intriguing possibilities. Consider, for example, the description of courtship in garter snakes that was quoted at the beginning of this chapter. Females have a stress response, clearly evolved under natural selection, which includes cloacal opening so that feces and musk can be extruded (Shine et al. 2003). Males apparently exploit this during courtship, by inducing stress, cloacal opening, and thus the opportunity to copulate. Similarly, the males of neotropical orchid bees (Apidae, Euglossini) have the peculiar habit of collecting fragrances from flowers (Dressler 1982, Eltz et al. 1999). The fragrances are absorbed with the forelegs and transferred into specialized, hair-filled cavities in the strongly enlarged male hind legs. Males also "process" the fragrances by applying labial gland lipids to the sites of storage. They then release fragrance bouquets, presumably to attract females that use fragrances to locate orchid flowers. Although the evolutionary origin of this remarkable behavior and the associated structures almost certainly must have involved sensory exploitation, it is not clear whether male fragrance signaling in any way interferes with female foraging.

In a small group of fish, consisting of three monophyletic genera within the Glandulocaudinae tetras (Weitzman and Menezes 1998), another most remarkable form of sexual dimorphism has evolved. In the genus *Corynopoma* (one species), males are provided with a long and slender extension of the gill cover, or operculum, the tip of which is enlarged, flattened, and provided with a dark

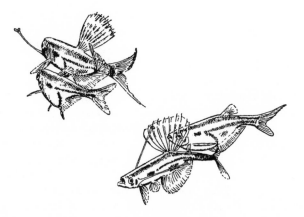

Figure 3.9. During courtship, the male *Corynopoma riisei* swims in front of the female with extended paddle. As the male twitches and jerks the tip of the paddle, the female is clearly attracted to this secondary sexual trait: she follows and nips at the paddle tip. Since the shaft of the paddle is transparent but the tip is pigmented, the resemblance to a moving prey organism is striking. (Reprinted from Nelson 1964, with permission from University of California Press.)

chromatophore (figure 3.9). This "paddle" is normally inconspicuously placed along the body, but is erected at a right angle from the body during parts of the complex courtship (see Nelson 1964, Richter 1986). The male orients parallel with, and somewhat in front of, the female with the paddle extended at a right angle from the body, then twitches and shakes the tip of the paddle in front of the female. The female responds to this by slowly turning toward and following the tip of the paddle, typically snapping, nipping, or biting at the tip of the paddle (figure 3.9). This is then followed by a lightning-fast movement involving both the male and the female, during which sperm is somehow transferred from the male to the female genital pore. During this exercise, the female and the intertwined male often break the water surface and literally jump up into the air (Nelson 1964). These fish feed primarily on various insects and crustaceans, and several authors have suggested that the paddle may function to "lure" females into proximity. The dark tip of the *Corynopoma* paddle both looks like and is moved like a small prey organism, and the fact that females are attracted to, attack, and bite the tip of the paddle does indeed strongly suggest that this peculiar male trait has evolved to exploit female foraging behavior (Nelson 1964, Wickler 1968, Bussing and Roberts 1971, Bussing 1974). Biting by females can be quite vigorous, and can result in male paddles being damaged or even lost (Nelson 1964, Bussing 1974). Males of the closely related genus *Pterobrycon* (two species) also bear very similar paddles, which are used by males during courtship in the same way and are attacked by females (see Bussing 1974, Weitzman 1975). Amazingly enough, however, the paddles of male *Pterobrycon* are not gill cover extensions, but are instead formed by greatly enlarged and extended scales on the mid flank. The paddles of both Pterobrycon species are pigmented at their distal ends, but male *P. landoni* carry a single slender paddle on each side (very similar indeed to that in *Corynopoma*) while *P. myrnae* males have two wider paddles on each side (see figure 3.10). In a third and newly described genus (two species), the male pectoral fin is instead extended and modified into a paddle. Although nothing is known about the reproductive behaviour in this new genus, it seems very likely that the paddles are used in much the same way as in the other two genera (Weitzman, personal communication). No species in the closest relative, the genus *Gephyocharax*, appear to have any similar modification (Burns and Weitzman 2005).

This little group of tetras not only offers one of the clearest examples of a male secondary sexual trait that has evolved to exploit female foraging behavior (cf. Wickler 1968, Proctor 1992, Rodd et al. 2002), but is also perhaps the best example of parallel evolution of male sensory exploitation. Nonhomologous "lures" have clearly evolved independently at least three times in this group, as modifications of either gill covers, humeral scales, or pectoral fins. Further, it seems as though all three forms of male "lure" (i) are engaged by males in similar ways and in similar stages of the courtship sequence, (ii) elicit female attacks and biting, (iii) are slender and elongated with a wider distal end, and (iv) are transparent apart from the pigmented distal end at which (v) all are armed with dark chromatophores that "flash" a silvery yellow or copper color in front of the females during courtship (Weitzman personal communication). Thus, the similarity is not restricted to functional morphology and behavior, but also includes structure and possibly even innervation of these derived secondary sexual traits. It is truly remarkable that no detailed experimental and/or comparative work on male-female interactions has yet been conducted in the group. Although both Wickler (1968) and Nelson (1964) suggested that male secondary sexual traits function to "deceive" "unresponsive females" to mate, the extent to which sexual conflict over mating has influenced male-female coevolution in this group therefore remains unknown.

The ultimate reason why we generally expect males to evolve an ability to exploit and manipulate females, rather than the opposite, is of course that females typically invest more heavily in offspring than do males and thus have a lower potential reproductive rate (see chapter 2). Actually, cases where the sex roles are reversed, such that males instead limit the reproductive success of females, offer a few very striking illustrations of the fact that the roles of "exploiter" and "exploited" are in no way inherent in the sexes but simply reflect the relative reproductive investment made by males and females in any given species and situation. For example, in the dance fly *Rhamphomyia longicauda* (Empididae), males carrying arthropod prey items enter a swarm of females. The male then appears to choose a female and donates his prey item,

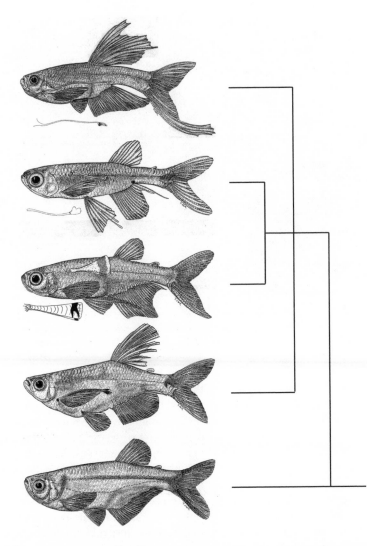

Figure 3.10. Phylogeny of the tribe Corynopomini within the glandulocaudine fishes, a group of primarily neotropical tetras. From the top are illustrations of males of the following taxa: *Corynopoma riisei*, *Pterobrycon landoni*, *Pterobrycon myrnae*, a new species of a new genus currently under description, and *Gephyrocharax atricaudata*. In this group, there have been at least three independent origins of male "paddles" which are used to attract females during courtship by mimicking prey items. In *Corynopoma*, the paddle is formed by the gill covers, in *Pterobrycon* by humeral scales, and in the new genus by the pectoral fins. Males of the most basal taxa, *Gephyrocharax,* lack paddles altogether. (Reprinted from Burns and Weitzman 2005, with permission from the Smithsonian Institution Division of Fishes; drawings by Tamara Clark.)

Figure 3.11. In the dance fly *Rhamphomyia tarsata*, the hind femur of females bear rows of flattened leg scales. The legs are wrapped around the body when swarming, creating the impression of an enlarged abdomen (see also figure 2.4). (Reprinted from LeBas et al. 2003, with permission from the Royal Society of London; photo by N. R. LeBas.)

and the pair fly off to copulate while the female feeds on the "nuptial gift." In this species, females are equipped with inflatable abdominal sacs and conspicuous leg scales that greatly exaggerate the apparent size of the female abdomen (see figure 3.11). Male *R. longicauda* prefer to pair with large females, presumably because they are more visible and/or because large females tend to be more fecund, and Funk and Tallamy (2000) suggested that females have evolved deceptive secondary sexual traits which exploit the male preference for large females and thus manipulate males into allocating their paternal effort in a suboptimal manner. Such seemingly "deceptive" female ornaments are also found in other dance flies with a similar mating system (e.g., *R. marginata*, *R. tarsata,* and *Empis borealis*) (Svensson and Peterson 1987, Cumming 1994, Svensson 1997, LeBas et al. 2003). Whether males have responded at all by evolving resistance to such displays is not known (see Hockham and Ritchie 2000). Although it seems that males must have suffered from these deceptive signals at least during their origin, this need not be true in extant populations. LeBas et al. (2003) found that the size of female ornaments was related to their fecundity, and argued that female ornaments may have evolved to "honestly" reflect female size/fecundity. A male preference for giving gifts to females with large ornaments may then in fact be maintained by direct benefits to males (LeBas et al. 2003).

Female dance flies do not, however, have exclusive right to deception in these intriguing insects (see Cumming 1994). Males of many species offer prey items as nuptial gifts to females, which incidentally may be reused by the male for several consecutive matings (Preston-Mafham 1999). Some species wrap the gift in secreted silk before carrying it into the swarm, and a subset of these are known to insert their mouthparts into the gift and consume the prey item prior to entering swarms, leaving little but an empty wrapping for the females. In yet other species, males construct altogether empty silk "balloons" which are handed over to females at mating (Cumming 1994, Sadowski et al. 1999). In *Empis opaca*, males sometimes carefully prepare balloons of willow-seed fluff which they then use as nuptial "gifts" with some success (Figure 3.12.) (Preston-Mafham 1999). Why females are attracted to and accept non-nutritional gifts such as these is a bit of a puzzle. Although balloons may be a "handicap" which signals male genetic quality (Cumming 1994), in which case a preference for males with large balloons may be maintained in females due to selection to secure indirect benefits, this preference may also simply represent a sensory bias with little or no current adaptive value to females (see Sadowski et al. 1999). Again, it does seem that costly deception must have been involved at least at some phase during the evolution of non-nutritional gifts from nutritional gifts. A formal comparative analysis of the evolution of mating systems and nuptial gifts in Empidid dance flies would be very interesting indeed, but has not yet been conducted.

3.2.5 CONVENIENCE POLYANDRY

A final response of females or males to persistence traits in the other sex may simply be to give up and mate, and thereby avoid further costly harassment, a behavior that has been referred to as "convenience polyandry" (Thornhill and Alcock 1983). There is a variety of evidence suggesting that this is just what some females do in insects (Thornhill and Alcock 1983), crustaceans (Jorma-lainen 1998, Thiel and Hinojosa 2003), birds (McKinney et al. 1983, Westneat and Stewart 2003), nonhuman primates (Smuts and Smuts 1993), and other mammals (Clutton-Brock and Parker 1995).

A well-studied example occurs in water striders. When harassment rates are experimentally elevated or the costs of mating are reduced, females in the genera of *Aquarius* and *Gerris* tend to reduce their resistance to male mating attempts and therefore increase mating rates (e.g., Arnqvist 1992, Rowe 1992, Weigensberg and Fairbairn 1994, Lauer 1996, Ortigosa and Rowe 2003). The fact that female resistance behavior is sensitive to both the costs of resistance and mating itself can lead to insights into the forces that shape optimal resistance level. Manipulations of female resistance, however, can be more informative about the forces of selection that are operating on resistance. For example, if one hypothesizes that one cost of mating is reduced foraging efficiency, then

Figure 3.12. Instead of offering females prey items, male *Empis opaca* sometimes prepare balloons of willow-seed fluff and use these to attract females. In this picture, a male is mating a female who is fruitlessly trying to feed on such a "fake" nuptial gift. (Photo by K. Preston-Mafham, Premaphotos Wildlife.)

it may be expected that a manipulation of female energy states would affect the level of resistance employed—hungry females would be particularly resistant to mating. This approach has been employed extensively in water striders (e.g., Rowe et al. 1994, Arnqvist 1997a).

3.3 Sexual Conflict and Sexual Selection

In sections 3.1 and 3.2 we have made the case that the costs of mating frequently lead to resistance by females (and sometimes males) to mating attempts by the other sex. The fact that resistance itself is often costly (section 3.2) reinforces the view that females are attempting to avoid matings that are more costly than resistance. Resistance has two potential results: one is to reduce mating rate, and the second is to bias mating success of males toward those certain phenotypes that can overcome resistance. The first seems obvious (and see section 3.2.5): every male spurned is potentially one less mating for the female. The second can be a simple side effect of reducing the mating rate and is, by definition, sexual selection by female mate choice (section 2.1, Halliday 1983, Maynard Smith 1987, Pomiankowski 1988). Consider a scenario where the female's degree of resistance to mating is independent of the phenotype of the male that is attempting to mate. We have argued that this is often the case because, once fertilized, mating per se is frequently costly. If some male phenotypes are better able to overcome this resistance, they will achieve higher mating success. By the logic above, these males are preferred, even if females resist them to exactly the extent that they resist males of the nonpreferred phenotype. In this way, costly mating, and the resulting resistance of females to mate, can be a powerful force of sexual selection. The scope of its action is a function of the extent of resistance (either male or female) across taxa. We have argued above (sections 3.1 and 3.2) that sexual conflict over mating, leading to resistance in either sex, is a common feature. If so, then sexual conflict will commonly underlie sexual selection in nature. Therefore, the variety of persistence traits that we have discussed thus far are, in fact, sexually selected by female resistance.

We have argued that the source of this resistance is natural selection, and that sexual selection can be viewed as a side or pleiotropic effect of resistance. Natural selection arises from the fact that mating can be costly. This scenario has a direct analogy with preexisting bias models of sexual selection (Rowe et al. 1994, Holland and Rice 1998). In these models some sensory trait of females is shaped by natural selection to be particularly sensitive to some signal. For example, female eyes may have evolved to be sensitive to the color red, because their preferred food is red. As a result females are attracted to red, and this preference resulted from natural selection. Males may then exploit this sensory bias by evolving red signaling traits, which attract females to them, and thereby increase their mating success. Now, red signaling traits in males are favored by female preference, which is a pleiotropic effect of natural selection on the female sensory system (see Rodd et al. 2002). In the examples discussed in sections 3.1 and 3.2, resistance to mating likewise appears to be under natural selection, and it sets up a bias for traits in the other sex that enhance their ability to overcome resistance. These sexually selected traits are, for example, the inflated abdomens of some dance flies, the gin trap of the sagebrush cricket, the recurved teeth of the Atlantic stingray, and even the suffocating caudocephalic waves employed by courting male garter snakes.

3.4 Mate "Screening" and Other Alternative Explanations for Resistance Traits

One alternative for a subset of persistence traits (grasping devices) is male-male competition. Darwin believed that the only force of sexual selection favoring grasping devices was in male-male competition. In one case grasping traits, employed by a male in copula, may serve to prevent takeovers of that female by another male. In another, grasping traits prolong association with an otherwise passive female and thereby limit access to her by other males. This may now be described as mate guarding and could occur prior to copulation (precopulatory guarding, see section 3.2) or after (postcopulatory guarding, see section 4.3). Therefore, any trait that allows males to continue association with females in the face of intruders would be favored. Some grasping traits surely fit this function (section 3.2). For example, the second gnathopod of gammarid crustaceans is highly modified in males but is held free of females during mate guarding and employed only in male-male interactions (Borowsky 1980). Two types of data to confirm the role of grasping traits in preventing takeovers would suffice. First, the hypothesis predicts a correlation between expression of the grasping trait and the efficiency of that male in fending off intruding males. Second, the efficiency with which males repel intruders could be compared between males with natural and experimentally disabled graspers.

One example of the former occurs in the explosive breeding crayfish *Orconectes rusticus* (Berrill and Arsenault 1984). In this species, males in breeding form have enlarged chelae (the familiar fore pincers, common to some crabs, crayfish, and lobster). Over a brief period in spring, males wander the substrate in search of females. When a male locates a female, he attempts to grasp her with his chelae. Females typically respond by tail flipping away. If a male successfully grasps a female, she continues resistance for a brief period before appearing passive. During copulation, other males often attempt to interrupt the pair. Berrill and Arsenault (1984) observed 128 such attempts. Of these, about one-third resulted in interruption, and in each of these cases, the interrupting male had chelae as large as or larger than those of the interrupted male. These data provide good evidence that male competition plays an important role in the evolutionary enlargement of male chelae in this species. In another crustacean, *Asellus aquaticus*, male size has similarly been shown to be a determinant of the success of takeovers. However, the only manipulative data that we are aware of are not particularly convincing. Thornhill (1984) compared takeovers of female scorpionflies copulating with males possessing unmanipulated notal claspers to those with disabled claspers. Three of 20 attempts on unmanipulated males were successful, and two of 17 attempts on disabled males were successful. More experimental data on the selective value of clasping devices for repelling intruding males, and other forms of male-male competition, would clearly be informative.

An alternative to the view of sexual selection as an incidental byproduct of female resistance is the view that females resist males as a means of "screening" out poor-quality males (e.g., West-Eberhard 1983, Crump 1988, Eberhard 1996, 2002, Wiley and Poston 1996, Bisazza et al. 2000, Kokko et al. 2003, Pizzari and Snook 2003). Eberhard (2002) has also referred to this process as "selective resistance." The essence of this argument is that female resistance does not result from some general cost of mating, but from the costs of mating with poor-quality males (section 3.1.2). Those males that females favor are either resisted to a lesser extent, or by virtue of their high quality are able to overcome the hurdle of resistance. High-quality males are either those that have "good genes," expressed as vigor, or those likely to produce manipulative (sexy) sons because manipulative traits such as grasping devices are inherited in their sons. We think that there are both empirical and theoretical data to suggest that this reasoning is unlikely to explain the generalized resistance to frequent mating that seems so common in nature (for discussion, see Arnqvist 1992, Cameron et al. 2003, Chapman et al. 2003a,b, Cordero and Eberhard 2003, Córdoba-Aguilar and Contreras-Garduno 2003, Eberhard and Cordero 2003).

It is certainly true that we generally expect sexual selection to lead to both sexy (manipulative) sons and good-gene effects (section 2.4). Yet, the costs of resistance that females pay are a form of direct selection (section 3.2), whereas the benefits that they gain from screening males are indirect (see sections 2.2, 2.3, and 2.4). Theory suggests that indirect benefits are a weak force in the face of direct selection (section 2.8, Kirkpatrick 1996, Kirkpatrick and Barton 1997, Cameron et al. 2003). If correct, the benefits of mating high-quality males, even if screening is effective, are unlikely to substantially affect the level of resistance that females show, when resistance is costly. The answer to this question, however, lies in experiments. Those experiments conducted so far do not support the view that the costs of female resistance are more than balanced by the indirect benefits to resistant males, or that indirect benefits are required to explain female resistance to mating.

Seaweed flies are a good example of generalized female resistance that results in sexual selection for persistence traits in males (Day and Gilburn 1997). When males and females of the seaweed fly *Coelopa frigida* encounter one another in mats of seaweed, males attempt to mate and females resist with a premating struggle. About half of these struggles lead to mating (Crean and Gilburn 1998). Large males tend to be more successful in achieving mating, and this is especially so when struggling with small females. The large-male advantage is compatible with both "screening" for high-quality males and the sexual conflict hypothesis. However, large males are not favored in these struggles because females chose to selectively resist these males less than smaller males. Instead, large males are simply able to endure *longer* struggles and thereby achieve more mating (Crean and Gilburn 1998). One could argue that although resistance is generalized, its "aim" is still to favor the high-quality males. Two lines of evidence argue against this view. First, large-male advantage was particularly strong with small females (Crean and Gilburn 1998). There is no reason to expect that the larger, perhaps higher-quality, females are less interested in obtaining a high-quality male than small females, as would be required under the "screening" hypothesis. But these data are exactly what is expected under the sexual conflict hypothesis—large males are simply able to overpower females and this is particularly true when their size advantage is exaggerated by struggling with a small female.

The second line of evidence comes from a study by Shuker and Day (2001). These authors reasoned that if females were struggling for the good genes of high-quality males, then females that had already mated a large (high-quality) male and stored his sperm would be more resistant to mate again. They tested this hypothesis by examining the repeatability of female resistance over multiple encounters with males. Their data demonstrated that female resistance was highly repeatable, as was the selection for large males, and that females were no more or less likely to resist when they did or did not have stored sperm from high-quality males. Their conclusion, based on these studies on *C. frigida* (Crean and Gilburn 1998, Shuker and Day 2001) and a variety of other seaweed flies (Day and Gilburn 1997, Crean et al. 2000), is that sexual conflict over mating frequency underlies sexual selection for large size in this group.

Very similar studies, with very similar results for the cause of sexual selection on male persistence traits, were previously conducted on a variety of water strider species (reviews by Rowe et al. 1994, Arnqvist 1997a). One further technique that has been employed in water striders is to experimentally alter the ecological costs of either resistance or mating, and then determine the selective outcome for male persistence traits (section 3.2.5). These experiments focus attention on resistance as a naturally selected biasing mechanism. For example, under the sexual conflict hypothesis, one might expect that if costs of mating were reduced for females, conflict over mating would be reduced. If so, we would expect females to be less resistant to mate, with the consequence that sexual selection on persistence traits would be reduced. Data have supported this view. One cost of mating in *Gerris buenoi* is reduced female foraging rates. Ortigosa and Rowe (2002) reasoned that this cost would be reduced in satiated females, and therefore selection on male size would be reduced. Their results showed that satiated females were less resistant to mating, and consequently there was no mating advantage to large males. In contrast, hungry females were resistant to mating, and this favored large males that could withstand resistance. Another approach would be to elevate the costs to females of resistance, with the expectation that females would reduce their level of resistance, with the consequence that selection on male persistence traits would be reduced. Experiments of this sort on two species (*Gerris odontogaster* and *Aquarius remigis*) and various persistence traits (abdominal claspers, body size) have supported this prediction (Arnqvist 1992, Weigensberg and Fairbairn 1996).

In these experiments, with seaweed flies and water striders, indirect benefits were not assessed. Therefore, these data cannot rule out the possibility that some indirect selection has shaped female resistance to some extent. This is a difficult question, though we do expect that there will be at least some effect of "good genes": males that overcome resistance are likely to be relatively vigorous (section 2.8). Yet these data do support the view that the costs of mating alone can account for female resistance and that sexual selection on male persistence traits is simply a side effect of this resistance. Therefore, in these species indirect selection is not necessary to account for female resistance.

3.5 Case Studies in Sexually Antagonistic Coevolution

In section 2.7. we outlined the theory of sexually antagonistic coevolution. The theory suggests that arms races between persistence and resistance traits may occur (figure 2.5). Our exploration of persistence and resistance traits in this chapter suggests that they are common to a variety of taxa (sections 3.2 and 3.3). Yet we currently have relatively little evidence from natural populations of coevolutionary arms races in action. This may in part be due to the fact that interacting persistence and resistance traits are often difficult to identify and quantify (Arnqvist and Rowe 2002b). In this section, we highlight three systems where comparative studies of the coevolution of persistence and resistance traits have been performed.

3.5.1 DIVING BEETLES

In some diving beetles (Dytiscidae), females resist mating attempts of males by repeatedly diving when males attempt to grasp them (Aiken 1992, Bergsten et al. 2001). We do not have direct evidence that mating is costly to females, although their vigorous resistance to mating suggests that this is so. Males have evolved adhesive structures on their tarsae that appear to facilitate grasping females by the pronotum and elytra along her dorsal surface (Aiken and Khan 1992). Interestingly, females in some species have likewise evolved conspicuously altered dorsal surfaces that appear designed to be more difficult to grasp (figure 3.7 and see section 3.4). Although these surfaces appear more difficult to grasp, no experiments on their utility as antigrasping traits have been conducted.

Females of some species are polymorphic for this proposed antigrasping trait: discrete female morphs with smooth or rough dorsal surfaces exists in the same species. As one moves among populations of the species (*Graphoderus zonatus verrucifer*), the frequency of the rough morph changes noticeably. Johannes Bergsten and colleagues reasoned that if a coevolutionary arms race were occurring in this species, one would expect a positive association between the degree of expression of male grasping traits and the proportion of females in the population with the antigrasping rough morph (Bergsten et al. 2001). Eight populations were sampled to determine the frequency of rough female morphs and the expression of grasping traits (size and number of adhesive suckers) in males. Their results were exactly as predicted from an arms race: there was a close correspondence between the degree of armaments expressed in females (proportion of the rough morph) and the expression of armaments in males (size and number of adhesive suckers). This was interpreted as the first comparative evidence of an ongoing arms race resulting from sexual conflict over mating.

In a more recent and taxonomically broad study, Miller (2003) described the coevolution of these structures in 52 species of diving beetles. Her main aim was to determine the order of evolution of persistence and resistance traits to determine if it fit the expectations of an arms race driven by sexual conflict over mating. By mapping traits of interest on the phylogenetic tree of the group she was able to draw a picture of an arms race over evolutionary time. Within the group she found strong evidence that the adhesive grasping structures of males evolved very early. It was after this event that females evolved antigrasping traits, presumably to counter the grasping traits possessed by males.

3.5.2 WATER STRIDERS

A recent comparative study of the water strider genus *Gerris* suggests a similar arms race fueled by sexual conflict over mating (Arnqvist and Rowe 2002a,b, Rowe and Arnqvist 2002). In this group, mating is costly to females, male harassment is frequent, and resistance by females can be intense (sections 3.2 and 3.3, reviewed by Rowe et al. 1994, Arnqvist 1997a). Both males and females have evolved grasping and antigrasping structures that increase the efficiency of persistence and resistance during premating conflicts (sections 3.2 and 3.3, and figure 3.2 and 3.13).

Under the arms race hypothesis, one would expect correlated evolution of these armaments, where exaggeration of grasping traits is met by exaggeration of antigrasping traits. In their study of 16 species of the genus, Arnqvist and Rowe (2002b) found exactly this pattern (figure 3.14). They went on to determine whether this arms race had the predicted effect on the mating rates and behaviors of the species. If persistence and resistance traits involved in determining mating rates are really matched over evolutionary time, then we may

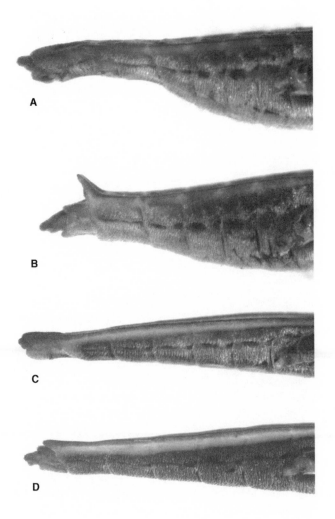

Figure 3.13. The degree of antagonistic armaments varies extensively in the water strider genus *Gerris*. More armed species such as *G. incognitus* have males (A) that have evolved grasping (e.g., prolonged genital and pregenital segments in conjunction with a relatively flattened distal part of the abdomen) and females (B) that have simultaneously evolved antigrasping traits which aid in obstructing the male grip during premating struggles (e.g., markedly prolonged/erect abdominal spines and a less accessible genital tip). On the other extreme are less armed and less dimorphic species such as *G. thoracicus*, where these adaptations are much less pronounced in both males (C) and females (D). (From Arnqvist and Rowe 2002b.)

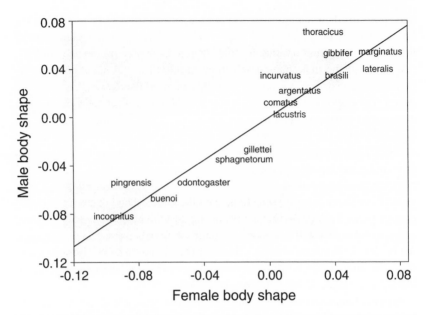

Figure 3.14. When species of *Gerris* are plotted by indices of male and female armament level, species tend to cluster tightly to a line. This indicates that male and female armaments are coevolving as one would expect in sexually antagonistic coevolution. The figure shows this relationship, with the most armed species in the bottom left corner of the plot, represented by *G. incognitus* (figures 3.13A and B). The least armed species, represented by *G. thoracicus* (figures 3.13C and D), appear in the top right corner of the plot. Remarkably, the variable over which they are arming themselves, mating rate, changes little as you move along this line. This observation suggests that the evolutionary arms race represented in this plot is balanced. In general, adaptation in one sex it met by countereradaptation in the other. (Data from Arnqvist and Rowe 2002b.)

see little variation in mating rates, despite an ongoing arms race. This appears to be the case in this genus of water striders (Arnqvist and Rowe 2002a, Rowe and Arnqvist 2002). Despite a large range among species in the degree to which males and females are armed, and in mating rates, there is very little correspondence between the two. The result matches a prediction from theory (Parker 1979, 1983a; see figure 2.5); much like the red queen, the two sexes are racing over evolutionary time, yet staying abreast all the while (Rice and Holland 1997). If this pattern is common, then arms races may often be hidden if only current interactions between the sexes are observed (Chapman and Partridge 1996a, Rice 1996b, 2000, Härdling et al. 2001, Arnqvist and Rowe 2002a,b). To unmask the conflict, Arnqvist and Rowe (2002a) looked at those species that deviated slightly from this pattern of balanced coevolution (figure 2.5). The expectation was that in those species where males were more armed than females, mating rates would be high. Likewise, where females were more armed then males, mating rates would be low. This pattern was supported, and the effects on male success in harassment and on mating rates were dramatic.

3.5.3 BEDBUGS

In these last two examples male traits are aimed at physically overcoming female resistance to copulation. Another way of evading female premating resistance might be to bypass the normal "mechanics" involved in copulation and sperm transport by, for example, piercing the female through the skin and ejaculating into the body of the female. Needless to say, the circumstances in which such a system could originate seem rather restrictive. For example, male genitalia must allow such transfer, thus inseminated sperm must be more successful at actually fertilizing eggs, and the potential trauma must not be too costly to the partner. It is therefore perhaps a bit surprising that different forms of traumatic, or hypodermic, insemination have evolved independently many times in invertebrates. It occurs, for example, in rotifers (Aloia and Moretti 1973), Acanthocephalans (Doyle and Gleason 1991), pinworms (Hugot 1984), free-living flatworms (Michiels 1998; see chapter 6), gastropod snails (Trowbridge 1995), and Strepsiptera (Chapman 1998). It is, however, in heteropteran insects of the infraorder Cimicomorpha that traumatic insemination is most widely spread (see Schuh and Stys 1991). In some basal taxa within this group (e.g., Prostemmatinae), the male genitalia enters the female reproductive tract during copulation in a seemingly normal manner. However, a spine at the genital tip perforates the wall of the bursa copulatrix and the male ejaculates into the female hemocoel. The sperm then migrates through the hemolymph to the ovaries, where the eggs are fertilized. In the more derived families, traumatic insemination instead takes place by males puncturing the female body wall.

There is no doubt that within the bedbug family (Cimicidae) traumatic insemination has been most thoroughly studied, and that male and female adaptations to this mode of mating are most diverse (see the review by Carayon 1966). Male bedbugs initiate matings by simply climbing onto and mounting females. They pierce the female abdominal wall with their sharp intromittent organ and inject sperm and accessory gland fluids directly into the blood, leaving visible melanized scars in females (figure 3.15). Although males of most species are strikingly indiscriminate, and frequently mount other males as well as nymphs, piercing and actual sperm transfer only seem to occur in matings with adult females (Stutt 1999 but see below).

As discussed elsewhere, mating carries general costs to females and elevated mating rates can compromise female fitness (see section 3.1). In taxa with traumatic insemination, matings are clearly associated with additional costs. These include (i) repair of the wound, (ii) leakage of blood, (iii) increased risk

Figure 3.15. The male bedbug (*Cimex lectularius*) mounts the female sideways, while grasping her with his legs. He then curves his abdomen under the female and punctures the ventral side of her abdomen with his daggerlike intromittent organ (seen here in the lower right part). Sperm and seminal fluid are then inseminated into the abdominal cavity of the female during the "copulation," which typically lasts a few minutes. The male in this picture (top) has just terminated copulation and is dismounting. (Photo by A. Syred, Microsopix.)

of infection through the puncture wound, and (iv) immune defence against sperm or accessory gland fluids that are introduced directly into the blood. Evidence that matings are indeed costly to female bedbugs comes from three different sources. First, dense laboratory colonies of several genera (e.g., *Caminicimex, Ornithocoris*, and *Haematosiphon*) often dwindle and go extinct as a result of a much higher mortality rate in adult females compared to males (Usinger 1966, personal observation). When kept in close quarters, these bedbug populations are rapidly reduced to contain males only. This is consistent with females suffering significant costs from repeated traumatic insemination. Second, interspecific matings are sometimes lethal to female bedbugs. For example, female *Hesperocimex sonorensis* die 24–48 hrs after mating with male *H. cochimiensis*, their abdomens being swollen and blackened (i.e., melanized) as a result of a massive immunoreaction (Ryckman and Ueshima 1964). Similarily, mating with *Cimex hemipterus* males reduces the life span of female *C. lectularius* (Newberry 1989) and results in an obvious melanization at the site of intromission (Walpole 1988). Artificial insemination experiments have shown that this effect is due to a toxic effect of heterospecific ejaculates (Davis 1965a). It has even been suggested that such deleterious interspecific mating is responsible for the displacement of *C. lectularius* by *C. hemipterus* in human habitations (the "satyr effect"; Ribeiro and Spielman 1986, Newberry 1989). At the very least, these instances show that traumatic insemination provides a challenge to the female immune system. Third, and most importantly, by performing controlled experiments where the rate of traumatic insemination was varied, Stutt and Siva-Jothy (2001) were able to directly demonstrate that elevated mating was indeed associated with reduced longevity and reproductive success in female *C. lectularius.*

Considering the spectacular mode of copulation in bedbugs, it is not surprising that females have evolved behavioral, physiological, and morphological resistance adaptations. Female resistance behavior varies widely across different genera. Although females of some genera (e.g., *Cimex*) largely remain passive prior to and during mating (Stutt 1999, personal observation) females in many other genera shake vigorously and often successfully escape harassing males by running away (e.g., *Leptocimex* and *Haematosiphon* [Usinger 1966], *Ornithocoris* [personal observation]). The existence of some form of physiological resistance adaptations to the male ejaculate is evident from the observations of lethal interspecific matings discussed above, even if these adaptations are currently poorly understood.

The most striking aspect of female resistance in bedbugs is, however, the remarkable evolution of a "secondary" reproductive system termed the paragenital system (see Carayon 1966). It typically consists of (i) a modification of the abdominal wall (swollen and often folded) in the region where males pierce females, called the ectospermalege, and (ii) a distinct internal pocket or sac attached to the inner surface of the abdominal wall, called the mesospermalege, that is located under the ectospermalege, is filled with haemocytoid cells, and receives the inseminated ejaculate. The morphology, elaboration, and location of these structures, together forming the spermalege, vary markedly across genera (figure 3.16). In what is considered the most basal taxa, the genus *Primicimex*, females lack these adaptations altogether. Copulations take place in a rather large area of the abdomen, as is evident from the distribution of mating scars in females. In other genera, such as *Afrocimex* and *Bucicimex*, females have evolved an ectospermalege, but the mesospermalege is restricted to a diffuse mass of cells (amoebocytes). In yet others, for example *Cimex* and *Ornithocoris*, the mesospermalege is a well-defined organ delimited by membranous tissue. Finally, in *Stricticimex* and *Crassicimex* the mesospermalege is attached by a conductor cord to the genital tract of the female. In these latter species, the sperm and seminal fluid is contained in the paragenital system and never actually enter the female hemolymph. Three independent experiments performed on the human bedbug, *C. lectularius*, collectively demonstrate that the spermalege indeed functions to reduce the direct costs of traumatic insemination in this species. First, Davis (1965b) found that while females reproduced normally after artificial inseminations into the mesospermalege, such inseminations into the abdominal cavity adjacent to the mesospermalege were usually

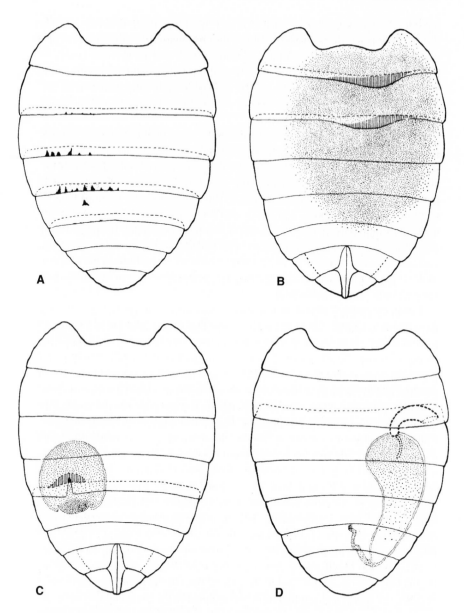

Figure 3.16. Examples of different paragenital systems in bedbugs (Cimicidae). (A) *Primicimex*, (B) *Afrocimex*, (C) *Cimex*, and (D) *Stricticimex*. Black spots represent mating scars in A only. See text for further explanation. (Reprinted from Usinger 1966, with permission from the Entomological Society of America.)

fatal to females. Second, Morrow and Arnqvist (2003) assessed mating costs by varying the rate of insemination on one hand and the rate and mode of piercing trauma to females on the other. They showed that elevated mating rate shortened female life span but did not find any significant effects on lifetime egg production. More importantly, while additional experimental abdominal piercing in the ectospermalege had no effect on females, even a very low rate of such piercing outside the spermalege reduced female lifetime egg production by some 50%. This experiment thus demonstrated that the spermalege efficiently reduces the direct costs of piercing trauma to females. Third, Reinhardt et al. (2003) found that the costs of additional piercing outside the ectospermalege, compared to within, were further elevated when piercing females with needles which had been contaminated with a bacterial solution cultured from bedbug colonies. They concluded that the spermalege reduces the risk of costly infections with pathogens that are introduced into females during traumatic insemination.

In general, the spermalege is an adaptation found only in females. There is, however, one noteworthy exception. In the genus *Afrocimex*, both sexes have a well-developed ectospermalege (but only females have a mesospermalege). The male ectospermalege is slightly different from that found in females and, amazingly enough, Carayon (1966) showed that male *Afrocimex* suffer actual homosexual traumatic inseminations. He found that the male ectospermalege often showed characteristic mating scars, and histological studies showed that "foreign" sperm were widely dispersed in the bodies of homosexually mated males. Sperm cells of other males were, however, never found in or near the male reproductive tract. It therefore seems unlikely that sperm from other males could be inseminated when a male that has himself suffered traumatic insemination mates with a female. The costs and benefits, in any, of homosexual traumatic insemination in *Afrocimex* remain unknown.

The bedbug system without doubt provides one of the most interesting and spectacular systems for studying mating costs, sexual conflict, and sexually antagonistic coevolution. At this point, there is little doubt that this is indeed a system in which sexual conflict over mating has had a major effect on the coevolution of male and female reproductive behavior, physiology, and morphology. There is, however, much that remains to be understood about the evolution of traumatic insemination in this group, and formal comparative studies of the coevolution of male and female traits across various Cimicid taxa should perhaps be especially rewarding. It is, for example, not clear whether traumatic insemination has originally evolved as a male strategy to bypass female premating resistance, via intragenital traumatic insemination (as in Prostemmatinae; see above) to bypass female postmating resistance to the sperm and/or seminal fluid, or perhaps for some other reason.

4 Sexual Conflict after Mating

Darwin's (1871) definition of sexual selection ("the advantage which certain individuals have over other individuals . . . in exclusive relation to reproduction") was not limited to processes occurring prior to pairing, but his subsequent discussion was. The literature on sexual selection has ever since been preoccupied with male-male competition over mates on one hand and female mate choice on the other (Andersson 1994). It is only about 30 years since the concept of sexual selection was first expanded to include processes that generate variation in reproductive success among males after mating has taken place; namely, sperm competition (Parker 1970c, Birkhead and Møller 1992, 1998, Simmons 2001a) and cryptic female choice (Thornhill 1983, Eberhard 1996). Sperm competition is typically defined as the competition between sperm of two or more males over the fertilization of eggs. In line with our definition of female choice (chapter 2), we define cryptic female choice broadly as resulting from any process occurring after intromission by which female traits bias fertilization or offspring production toward certain male phenotypes (see also Eberhard 1996). One of the insights resulting from attention to postmating events is the realization that conflicts of interest between males and females do not end once mating has taken place. There are at least two conspicuous issues over which the interests of the sexes may differ. First, there may be conflict over female reproductive behaviors after mating, if these behaviors affect which male's sperm will fertilize the eggs of a given female (e.g., reproductive rate, remating rate, and sperm utilization pattern). Second, in taxa with parental care, there may be conflict over the relative contributions of the sexes to care. This chapter is devoted to the first of these conflicts, and the next chapter will focus on the second.

Before considering these conflicts in nature, it might be interesting to examine the events taking place after mating in our own species. Although there is an almost complete lack of relevant studies of the "economics" of reproduction in humans, we do understand some of the proximate functions of male and female traits affecting postmating interactions. The male accessory reproductive glands, including the prostate gland, the seminal vesicles, and Cowper's gland, are productive chemical factories. They produce a complex cocktail, including a large number of enzymes, fibrinolysins, and hormones (including many sex hormones, such as the pituitary gland hormones LH and FSH), prostaglandins, protease inhibitors, high concentrations of many other polyamines and peptides, free amino acids, citric acid, sialic acid, fructose, cholesterol, carotenoids, and large amounts of zinc (Mann 1964, Mann and Lutwak-Mann 1981, Blandy and Lytton 1986, Aumüller and Krause 1990). The seminal fluid forms about 99% of the ejaculate transferred to females during copulation. Although it is clear that a normal secretory activity of these glands is central for male fertility, the specific functions of most of the seminal substances is not well understood (Mann and Lutwak-Mann 1981, Franks 1983, Aumüller and Krause 1990). Some may simply be by-products of the active metabolism and rapid turnover of proteins and peptides in these glands (Mann 1964), but it seems highly unlikely that such a complex and presumably costly underlying machinery would have evolved had it not served vital reproductive functions (Blandy and Lytton 1986, Clavert et al. 1990).

Several seminal substances are known to enhance sperm viability inside the female reproductive tract, by providing energy and osmotic buffer (Mann and Lutwak-Mann 1981), and thus act in the direct reproductive interest of males. However, this is apparently not the entire story (Mann and Lutwak-Mann 1981, Clavert et al. 1990). The fact that many substances are known to have hormonal and immunoregulatory functions (Aumüller and Krause 1990, Kurpisz and Fernandez 1995) suggests that at least some are targeted more directly toward females (Eberhard 1996). One possible function of male seminal substances is to provide a reliable signal to females of a successful copulation, thus allowing the synchronization of reproductive events within the female to the availability of viable sperm (cf. Gillott 1988). This synchronization would obviously benefit both sexes. Nevertheless, the sheer complexity of the seminal fluid seems redundant if the function is to simply signal that sperm has been transferred, and our limited knowledge of how some of these seminal substances suggests a more varied role.

The human vaginal wall is known to be permeable to many substances, including hormones, and radio-labeling experiments have shown that seminal proteins indeed pass through this wall and into the bloodstream (Albone and Shirley 1984). Although few direct experimental studies exist, there is at least some evidence suggesting that substances in the seminal fluid may directly affect female reproductive events. For example, Habib et al. (1980) showed that certain concentrations of zinc induce a stimulatory affect on the binding of hormones in the human endometrium, which is likely to affect the probability of pregnancy. Human seminal fluid also contains a number of substances in concentrations high enough to elicit smooth-muscle contraction within the female (Mann 1964, Mann and Lutwak-Mann 1981), which obviously might affect the transport of both sperm and eggs in the female reproductive tract (Clavert et al. 1990, Eberhard 1996). In support of this possibility, proteins and hormones transported to females via the seminal fluid are known to affect pregnancy rates in several other mammal species (Ericsson and Baker 1966, Banerjee 1968, Mann and Lutwak-Mann 1981, Eberhard 1996) and experimental blocking of secretion from male accessory glands results in markedly lowered fertility (Clavert et al. 1990). These facts are not in themselves informative about the type of selection involved in the evolutionary origin and maintenance of these substances, but they do show that at least some seminal substances have direct effects on female reproduction (Aumüller and Riva 1992).

Surprisingly little is known about what happens to the various components of the ejaculate after deposition in the vagina (Mann and Lutwak-Mann 1981, Aumüller and Riva 1992). It is clear, however, that the ejaculate is not particularly welcome in the female reproductive tract. The mammalian vagina is a generally hostile and acidic environment, which is no doubt essential to avoid invasion and subsequent infection by pathogens (Austin 1965). But this hostility is not limited to a general unfriendliness to foreign bodies, as females also produce a series of substances that appear targeted specifically toward components of the ejaculate. For example, the presence of sperm in the reproductive tract triggers a massive release of phagocytotic leukocytes (Kurpisz and Fernandez 1995). Females produce a series of specific antibodies against sperm which effectively incapacitate spermatozoa, and elevated concentrations of such antibodies can even cause female infertility (Kurpisz and Fernandez 1995). The reason for targeting sperm is still unclear, but could involve benefits to females from selecting healthy or more resistant sperm to fertilize their eggs (see Curtsinger 1991, Birkhead et al. 1993, Keller and Reeve 1995, Eberhard 1996), or could represent defence against overly aggressive or numerous sperm (see section 4.2.2.1.1. below). The female reproductive tract also appears hostile to seminal proteins and peptides, most of which are broken down and disappear rapidly (Mann and Lutwak-Mann 1981, Szecsi and Lilja 1993). This break-down is partly due to the acid environment and partly to substances in the ejaculate itself (Szecsi and Lilja 1993), but also to the presence of specific female antibodies against seminal fluid proteins, and to various proteolytic enzymes in the cervicovaginal secretion (Hocini et al 1995, Kurpisz and Fernandez 1995).

A striking aspect of human seminal fluid is that it contains a series of defences against the hostile environment of the female reproductive tract. For example, human semen is particularly rich in zinc. The concentrations (4–500 µg/ml) are actually higher than in any other part of the body (Blandy and Lytton 1986). Although zinc affects reproductive physiology in both sexes in several ways (Bedwal and Bahuguna 1994), high concentrations of zinc are known to impair human immune function (Rink and Kirchner 2000), and zinc in mammalian semen could thus suppress female immunoresponses (Leonhard-Marek 2000). Concordantly, zinc concentration in human semen is correlated with male fertility, and zinc is believed to prolong the lifetime of the sperm within the female reproductive tract (see Bedwal and Bahuguna 1994). Likewise, human semen contains many proteins and peptides known to prevent or suppress immunological reactions inside the female reproductive tract (Clavert et al. 1990). Finally, the seminal fluid contains various polyamines, several of which are known to act as protease inhibitors (Mann and Lutwak-Mann 1981). These polyamines may interfere with proteolytic enzymes within the female reproductive tract.

The traditional view of the seminal substances in humans is that they function only to help sperm survive in the female genital tract (Mann 1964), and that the hostile environment provided by females serves to prevent infection by pathogens and to select certain sperm over others to fertilize their eggs (Austin 1965). The short, and admittedly very incomplete, review of human seminal fluid and its fate inside the female presented here indicates that this may not be the entire story. It bears at least a superficial resemblance to an "arms race" between the sexes. At least some of the substances that human males transfer to females within the seminal fluid actually seem to affect reproductive processes inside the female after copulation has taken place. Females, on the other hand, seem to sometimes resist the presence and actions of these products, and produce substances which cause rapid breakdown of ejaculate components. Males in turn appear to resist this attack with the addition of immunosuppressants and proteolytic inhibitors to the seminal fluid. Could it be that male and female humans are involved in chemical warfare?

At present, these questions remain little more than intriguing, yet highly speculative, hypotheses. There are certainly alternative explanations for those observed proximate functions of the male and female traits discussed above (Aumüller and Riva 1992, Eberhard 1996). For example, many of the substances produced by the accessory reproductive glands of humans may be byproducts of the active metabolism in these glands or may simply be waste products without selected function in the ejaculate (Mann 1964). Others are known to serve a role in metabolic cascades which activate and/or nourish sperm (Mann and Lutwak-Mann 1981, Aumüller and Krause 1990). Similarly, seminal substances which target females more directly could represent a complex ejaculate "signal" which is the result of male-female coevolution that has not been antagonistic (Eberhard 1996). Unfortunately, these questions are difficult to resolve without careful experiments of the sort that are perhaps too difficult or unethical to conduct in humans.

In the following sections, we first look more closely at some general issues regarding the reproductive interests of males and females after mating and then turn to other animal taxa, where various types of experimental studies have shed important light on these postmating sexual conflicts.

4.1 Female Reproductive Effort and the Conflicting Interests of the Sexes

Reproduction is associated with a series of costs. In species that are strictly semelparous, such as certain mayflies, which lay a single batch of eggs and then die, the optimal female reproductive effort is simply to allocate all available resources to offspring production. In all other organisms, however, current reproductive effort will be traded against the potential for future reproduction, and we expect reproductive effort to balance current benefits with these future costs. The study of these costs of reproduction is at the heart of life history research (see Stearns 1992 and Roff 1992 for reviews). Needless to say, the optimal balance between current and future reproduction may be affected by various factors, such as environmental conditions or female age and state, but typically we expect reproductive effort to evolve to some intermediate level (see figure 4.1). This basic logic has been used successfully to understand the evolution of egg production rate in insects, clutch size in birds, and litter size in mammals (Stearns 1992, Roff 1992).

Under true genetic monogamy, where neither sex mates outside of a lifelong pair bond, the relative interests of the partners coincide (Partridge and Hurst 1998, Holland and Rice 1999, Arnqvist et al. 2000). Since a male will father all the offspring of his mate, a reproductive rate that optimizes female fitness will also optimize male fitness: what is best for the fitness of the female is also best for the fitness of the male. This harmonious situation changes dramatically as soon as the male risks not siring all of the future offspring of his mate. Whenever the female is likely to remate with additional males, at any point in her life, the focal male's interest in the future offspring of his mate diminishes or disappears altogether. For example, pair bonds are rarely maintained between seasons in many "monogamous" birds and mammals. Similarly, most insect females remate at regular intervals with different males during their reproductive life (e.g., Thornhill and Alcock 1983, Choe and Crespi 1997a). In both of these cases, males have a limited interest in the future performance of their mates. Under these circumstances, a higher than optimal offspring production rate for his mate will benefit the male (he will father them), even if producing these offspring comes at a substantial future cost to females (figure 4.1) (Eberhard 1996). Sexual conflicts over female reproductive effort stemming from such asymmetries may be one of the most common forms of sexual antagonism (Trivers 1972, Parker and Simmons 1989, Arak and Enquist 1995, Rice 1998a, Chapman et al. 1998, Arnqvist and Nilsson 2000).

Considering this conflict over female reproductive rate, we expect the evolution of male traits aimed at inducing, stimulating, or manipulating females to produce offspring at a rate that is higher than that favored by natural selection in females. Mutations in males that cause their mates to reproduce at a higher

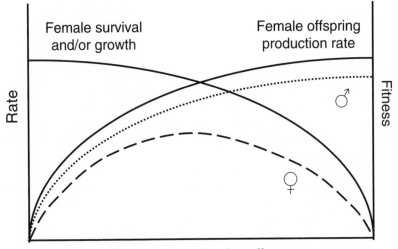

Figure 4.1. Graphical illustration of sexual conflict over female reproductive effort. In itero-parous species, individuals trade the increase in offspring production that results from an increased current reproductive effort against an accompanying decrease in future repro-ductive success due to the costs of reproduction. For females, there will therefore be an optimal intermediate reproductive effort at any point in life which maximizes female life-time fitness (dashed line). In contrast, unless there is true genetic monogamy, a given male does not benefit from the future performance of his mate. The male fitness derived from reproducing with a particular female (dotted line) therefore increases monotonically with the reproductive effort of his mate, and males that induce higher female reproductive effort will be favored, even if this is costly to females.

than optimal rate would in many cases spread in a population. Nevertheless, the potential of males to elevate female reproductive effort beyond female optima may be limited in many species. There are, however, some important exceptions to this conjecture.

4.1.1 SEMINAL SUBSTANCES WITH GONADOTROPIC EFFECTS

When males of internally fertilizing animals mate with females, they transfer much more than spermatozoa. The larger part of the ejaculate is typically made up of seminal fluid, containing various accessory substances produced by males. From a biochemical point of view, these seminal fluids are exceptionally complex. As discussed in the first part of this chapter, mammalian seminal fluid contains an array of proteins, peptides, and other substances produced in the male accessory glands. The seminal fluid of most other animals is not essentially different. Most male birds lack specialized accessory sex glands and the seminal fluid is instead produced mainly in the reproductive ducts (Fujihara 1992). Although the accessory substances of birds have received little study, avian seminal fluids are often similarly rich in various peptides, proteins, and other substances (Alaghbari et al. 1992, Fujihara 1992, Cerolini et al. 1997). Males of many invertebrates, including, for example, ticks (Oliver 1986), crustaceans (Diesel 1989), and insects (Chen 1984, Raabe 1986, Gillott 1988, 2003), are also known to add a multicomponent seminal fluid to their ejaculate. Many substances in the seminal fluid are known to provide nourishment or otherwise assist sperm, but others directly affect females and are thus candidates as adaptations employed in sexual conflict over female reproductive rate.

Although it is clear that mating stimulates female egg maturation, ovulation, and egg laying in most animals (see Eberhard 1996), it is only in insects that this effect has been unequivocally attributed to the transfer of seminal substances that directly stimulate female gonads (i.e., gonadotropins) and thus elevate reproductive activity (but see references above for mammals; Barnes et al. 1977 and Klepal et al. 1977 for barnacles; Ward and Carrel 1979 for nematodes). A variety of experimental techniques have been used to study the effects of seminal substances on female reproductive processes in insects, including implantation of male accessory glands and injection of either extracts of male accessory glands, or the purified/synthesized substances themselves into females. These studies demonstrate that the most general effect of seminal substances on female insects is to elevate reproductive rates (see Chen 1984, Raabe 1986, Gillott 1988, 1996, 2003, Miller et al. 1994, Eberhard 1996, Chapman et al. 1998, Klowden 1999 for detailed reviews). The effects of male seminal substances have been particularly well studied in *Drosophila* fruit flies, and this system will be discussed in detail later in this chapter (section 4.7). However, a few other examples deserve special attention.

First, we need to briefly review the regulation of reproductive rate in female insects (see figure 4.2). Egg maturation and oviposition of these eggs is primarily regulated by juvenile hormone (JH), typically through the following pathway (see Nijhout 1994). Juvenile hormone is produced by the corpus allatum (and in some species possibly in the male accessory glands; see Borovsky et al. 1994). The production rate of JH in the corpus allatum is stimulated by neuropeptides (i.e., allatotropins) acting in a dose-dependent manner on receptors in the neural system. Juvenile hormone then regulates the rate of vitellogenin (yolk protein) synthesis in the fat body, and the rate of vitellogenin synthesis is proportional to the concentration of JH in the blood (i.e., the hemolymph). Vitellogenins, large lipoproteins, make up the bulk of the volume of eggs. During the process of egg maturation (vitellogenesis), the eggs accumulate vitellogenin from the hemolymph. The reproductive rate of female insects has been shown to be proportional to the titers of both JH and vitellogenin in the hemolymph (Nijhout 1994, Ramaswamy et al. 1997).

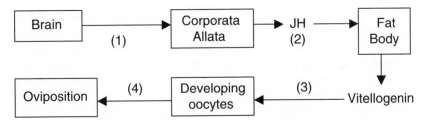

Figure 4.2. A simplified illustration of the main regulatory pathway of female reproduction in insects (see Nijhout 1994). Juvenile hormone (JH) is produced by the corpus allatum, and production of JH is under neural control. The titer of JH in the hemolymph then directly affects the rate of vitellogenin (yolk protein) synthesis in the fat body. During the process of egg maturation (vitellogenesis), the eggs then accumulate vitellogenin from the hemolymph. Insect males are known to transfer both (1) neuropeptides and (2) JH to females, both of which increase the rate of vitellogenin synthesis and thus the rate of egg maturation. It has also been suggested that males may transfer substances (3) that directly increase the rate of vitellogenin accumulation in the developing eggs. Males of several insects are also known to transfer myotropins, which cause muscle contractions in the oviduct (4) of females.

Butterflies are of special interest here, partly because they are known to transfer enormous quantities of seminal fluid in the form of a spermatophore. Males frequently transfer substances corresponding to 10–15%, sometimes as much as 25%, of their body weight in a single mating (Forsberg and Wiklund 1989, Svärd and Wiklund 1989). Many of the substances in butterfly spermatophores are involved in a complex chain of biochemical reactions that lead to the release of energy and respiratory substrates to the sperm (e.g., Osanai and Chen 1993) (see also section 4.5) and should thus be beneficial to males. Although oviposition in butterflies is triggered in part by the presence of viable sperm, mechanical stimulation during mating, or the mere presence of a spermatophore (Raina et al. 1994), seminal substances do appear in the female hemolymph after mating, and some of these are known to elevate female egg production rates. Male reproductive glands and spermatophores of many butterflies are rich in JH (Ramaswamy et al. 1997, Park et al. 1998; but see Cusson et al. 1999). For example, male *Cecropia* silk moths produce large amounts of JH in their corpus allatum, possibly even more than the female, which is accumulated in their accessory sex gland (Peter et al. 1981, Shirk et al. 1983). Stored JH is then transferred into the spermatophore, and thus to the female during mating (Shirk et al. 1980). In Heliothinae moths, substances produced by the male accessory glands have been shown to be at least partly responsible for the stimulatory effect of mating on female oviposition (Bali et al. 1996, Park et al. 1998). This effect is most likely the consequence of JH transferred in the seminal fluid, since females injected with JH respond similarly with elevated oviposition rates (Ramaswamy et al. 1997). Moreover, endogenous female JH synthesis also increases after mating (Nijhout 1994), and there is some evidence that this results from the stimulatory effect of seminal substances on the female corpus allatum. For example, Bali et al. (1996) showed that substances in male *Helicoverpa zea* ejaculates contribute significantly to the elevation in female reproductive rate observed after mating. Virgin females injected with extracts from the male accessory glands greatly increased rates of egg maturation and oviposition, relative to control females that were injected with either extracts of other male tissues or saline. The fact that gland-injected and mated females actually laid a much larger proportion of the eggs they matured during the experiment (39% and 69%, respectively), compared to control females (< 5%), further suggests that more than one ejaculate component independently affect egg maturation and oviposition. In support of this suggestion, Bali et al. (1996) also found that topical application of a JH analogue to the abdomen of virgin females almost doubled the rate of egg maturation compared to control females, but that it did not affect oviposition rates.

In another intriguing study, Fan et al. (1999) showed that a synthetic form of a seminal fluid protein from *Drosophila* fruit flies, the so-called sex peptide, had a gonadotropic effect in the moth *Helicoverpa armigera*. The *Drosophila* sex peptide stimulated JH production in the corpus allatum of female moths, in vitro as well as in vivo. The authors concluded that a substance similar to the *Drosophila* sex peptide may be present in the seminal fluid of male moths, and that this substance is partly responsible for the elevated reproductive rates seen after mating. Similarly, Park et al. 1998 reported that the oviposition rate of female *Heliothis virescens* mated with males that had their accessory glands surgically removed was much lower than that of females mated with either sham-operated or normal males, and that it did not differ from that of virgin females. In summary, males of some species of butterflies apparently use at least two different routes to increase female postmating reproductive rate (see figure 4.2). First, they produce and transfer a female gonadotropin (JH) during mating. Second, they also seem to transfer neuropeptides that stimulate the endogenous production of this gonadotropin in corpus allatum of females. Both of these routes will increase JH titers in females, which in turn increases production of vitellogenin, resulting in elevated rates of egg maturation. Both types of substances are commonly found in the ejaculate of other insects as well (e.g., Eberhard 1996, Klowden 1999).

A quite different example of male seminal substance function occurs in the Colorado potato beetle (Chrysomelidae, *Leptinotarsa decemlineata*). Smid and Schooneveld (1992) discovered that a monoclonal antibody they believed to be specific for a putative neuropeptide in the part of the brain that innervates the corpus allatum also reacted to a peptide produced in the male accessory glands. This peptide, transferred to the female during mating (Smid 1998a), was subsequently characterized and designated Led-MAGP (Smid et al. 1997).

The closest homologue of Led-MAGP is a prion protein described from chickens. The infamous prion proteins are best known for causing neuro-degenerative brain diseases in mammals, and their only known nonpathogenic function is as a signal, which induces transport of proteins across cell mem-branes by endocytosis (ingestion by a cell) (Brockes 1999). However, further similarities exist. Prion proteins are known to bind with other proteins, forming an aggregate (Brockes 1999), and Led-MAGP similarly tends to establish aggregates (Smid 1998b). This binding ability of Led-MAGP, in combination with its structural similarity to prion proteins, led Smid (1997) to suggest the following function for this peptide. When in the female hemolymph, Led-MAGP binds to the most abundant protein in a reproductively active fe-male: vitellogenin. The uptake of vitellogenin by the egg normally involves specific egg membrane receptors (Nijhout 1994), but Led-MAGP could cir-cumvent this route by inducing endocytosis of "tagged" vitellogenin. Vitello-genin would then be internalized by the egg, together with Led-MAGP pro-vided by males (Smid, personal communication), leading either to an increase in the net rate of egg maturation or an increase in egg size. This fascinating scenario has, however, still not been tested, and alternative explanations exist (Smid 1998a). It is nevertheless notable that similar "sticky" peptides have been described in the male accessory glands of several other insects (Miller et al. 1994). If Led-MAGP indeed functions the way Smid (1997) suggested, this might compromise female interests. Unfortunately, little is known about the effects of receiving large doses of seminal fluid on female reproduction in the Colorado potato beetle. Boiteau (1988) reported a slight increase in fecundity from multiple matings, but this result was questioned by Orsetti and Rutowski (2001), who found no effects on fecundity but a dramatic decrease in fertility both among females mated multiply and among females mated repeatedly to the same male. The latter result is, however, unusual among insects (cf. Arn-qvist and Nilsson 2000).

Summing up, males of a wide range of organisms can elevate female post-mating reproductive rate by transferring ejaculatory substances which have various gonadotropic effects to females. These seminal signals may often come under sexual selection by cryptic female choice (reviewed by Eberhard 1996), simply because males who transfer relatively more efficient, compati-ble, persuasive, and/or larger amounts of these signals will elicit a stronger reproductive response in females. This is supported, for example, by the fact that females often tend to exhibit increased rate of offspring production when mated with large males (e.g., McLain et al. 1990, McLain 1998, Savalli and Fox 1998a,b), males that transfer large ejaculates (see Vahed 1998 for a review), males that copulate for a long time (e.g., Thornhill 1983), or males that otherwise achieve high relative fertilization success (i.e., $P2$; Clark et al. 1995, Clark and Begun 1998, Arnqvist and Danielsson 1999b). If these substances tend to elevate female reproductive rate beyond its optimum, we would expect to see females counteradapt by evolving resistance to this male "manipulation" of their reproductive machinery. We will return to this discussion in section 4.6.

4.1.2 NUPTIAL FEEDING

Males in a variety of species, within insects, birds, and mammals, provide females with food items prior to or during mating (we do not regard seminal products as being food items here; see section 4.5) (Wickler 1994, Vahed 1998). Males may benefit from providing their mate with food if it increases the number or the quality of surviving offspring fathered (paternal investment) or if it increases the number of eggs fertilized (mating effort) (Trivers 1972, Gwynne 1984, Simmons and Parker 1989, Vahed 1998). Female reproductive rate does indeed increase as a result of such food donations in many species (Vahed 1998), but it is not known whether this stems from male attempts at elevating female reproductive rate beyond its optimum. The following suggests that this is not the case. When an increase in reproductive effort is observed as a result of intake of food items provided by the male, it is perhaps more likely to reflect a shift in optimal female reproductive effort due to an enhanced nutritional state of the female and does not necessarily compromise female interests in any way (Parker and Simmons 1989, Karlsson 1998, Stjernholm and Karlsson 2000, Wedell and Karlsson 2003) (cf. figure 4.1). In support of this view, elevation of reproductive rates with nuptial feeding does not appear to increase female mortality rates in some insects (see Arnqvist and Nilsson 2000 for a review). The fact that female fecundity can be elevated without an associated decrease in life span is at least indicative of a shift in optimal female reproductive effort.

Nevertheless, conflict over female reproductive effort could have contributed to the evolution of nuptial feeding in animals with extended parental care, such as mammals and many birds. If females use their own nutritional status as a cue when making decisions about the number of offspring to produce, males could exploit this sensory route by providing females with extra food during the oviposition/implantation period. Under this scenario, females could be manipulated into overestimating food abundance and/or paternal effort, and thus suffer from a larger than optimal investment during subsequent parental care. Indeed, male courtship feeding does increase both clutch size and egg size in several bird species (see Wickler 1994). However, again, it is not clear whether this reflects a shift in optimal female reproductive effort or male manipulation of female reproduction. The latter would be contingent not only upon female resistance to such manipulation but also upon the partitioning of parental care (see chapter 5), as well as the relationship between parental care and offspring quality. We know of no data directly supporting the hypothesis.

4.1.3 MALE DISPLAY TRAITS

Males of many species perform elaborate courtship displays prior to mating and/or are ornamented in various ways (Darwin 1871) and such traits are traditionally believed to have evolved by female mate choice (Andersson 1994; see chapter 3). However, Nancy Burley (1986, 1988) suggested an additional selective advantage of "aesthetic" traits. She suggested these traits might also benefit males because they elevate the reproductive or parental effort of their mates, allowing attractive males to reduce their own relative parental investment. Given that offspring quality and/or attractiveness may be inherited by offspring, Burley (1986) proposed that it may pay one partner to elevate their reproductive/parental effort when paired to an attractive mate if the costs of doing so are more than offset by an increase in offspring quality. She termed this the differential allocation hypothesis, and pointed out that it predicts that individuals should vary their reproductive effort in response to the attractiveness of their mate. The differential allocation hypothesis is thus related to the "sexy sons" hypothesis for the evolution of polygyny in socially monogamous birds (Weatherhead and Robertson 1979; see chapter 5), in that both postulate that a female reduces the number of offspring produced in her lifetime in order to achieve a higher offspring quality.

Female reproductive output is an increasing function of male "attractiveness" in many species (Sheldon 2000). For example, both female zebra finches (*Poephila guttata*) and hoopoes (*Upupa epops*) lay more eggs when mated to males with more intense songs (Balzer and Williams 1998, Martin-Vivaldi et al. 1999), and clutch size is positively correlated with the number of "decorative" stones carried to nest cavities in the black wheatear (*Oenanthe leucura*) (Moreno et al. 1994). Similarly, Norris (1990a,b) found that female clutch size in great tits (*Parus major*) was positively related to the size of the black breast stripe of their mate. However, since attractive males may provide their mates with superior resources (e.g., territories) and/or may be mating assortatively with more fecund females, correlational data does not in itself confirm that female reproductive effort is influenced by stimuli provided by her mate (Rintamäki et al. 1998, Hirschenhauser et al. 1999, Sheldon 2000). Studies where females show increased reproductive effort when randomly assigned to more attractive males, such as those of Petrie and Williams (1993, *Pavo cristatus*) and Massa et al. (1996, *Melopsittacus undulatus*), partly solve this problem. The ideal experiment, however, would involve independent manipulation of the male stimuli itself. Three studies have provided such evidence (figure 4.3). First, Burley (1986, 1988) showed that female zebra finches made a greater reproductive effort when paired with males made more attractive by colored leg bands. Second, de Lope and Møller (1993) showed that female barn swallows (*Hirundo rustica*) started breeding earlier and produce larger broods when the tail feathers of their mate had been elongated. Third, Swaddle

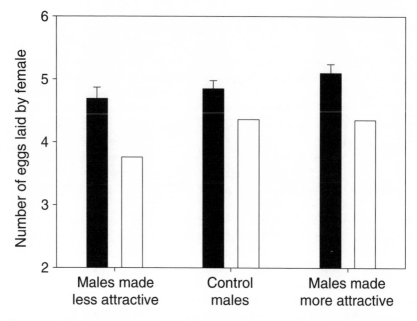

Figure 4.3. Experimental evidence suggesting that female reproductive effort is elevated as a direct result of male ornamentation. Both barn swallow (solid bars) and zebra finch (open bars) females mated with males made more attractive tend to lay more eggs per clutch compared to females mated with males made less attractive. In barn swallows, manipulation of male attractiveness involved shortening or elongating the tail feathers and in zebra finches attaching color bands of various color to males (red, attractive; green, unattractive). In both cases, the results indicate an increased female reproductive effort. (Data from Burley 1986 and de Lope and Møller 1993; see also Swaddle 1996.)

(1996) found that female zebra finches laid more eggs when paired with males provided with a symmetric arrangement of colored leg bands. These data provide strong evidence that male display traits, independent of their bearer, elevate female reproductive effort in at least some cases. How can this be, given that these females may bear costs to their future reproductive output?

Burley's original discussion of differential allocation argued that these costs were offset by indirect genetic benefits to offspring. Yet is certainly possible that increased female reproduction with attractive males may reflect a shift in optimal female reproductive effort because these males provide superior direct benefits (Wedell 1996). Since the experimental studies mentioned above all involve species with biparental care, it has been suggested that females may be responding adaptively to honest signals revealing male ability to provide care for offspring. However, three facts suggest that this may not generally be the case. First, we may not generally expect the evolution of honest signals for ability to provide paternal care, simply because honest signals often are costly and thereby decrease the care-giving ability of their bearers (Kirkpatrick et al.

1990, Westneat and Sargent 1996; but see Price et al. 1993, Wolf et al. 1997, and Kokko 1998). Second, several experimental studies have shown that attractive males provide less paternal care, indicating that they may not generally be "good fathers" (Burley 1986, 1988, de Lope and Møller 1993, Møller and Thornhill 1998, Mazuc et al. 2003). Third, and most importantly, females exhibit elevated reproductive effort when mated with attractive males also in a number of species where males do not provide any paternal care, such as peacocks (*Pavo cristatus*, Petrie and Williams 1993), black grouse (*Tetrao tetrix*, Rintamäki et al. 1998), mallards (*Anas platyrhynchos*; Cunningham and Russel 2000) and several insects (see Eberhard 1996, Møller and Thornhill 1998, and Sheldon 2000 for references).

Thus, it seems that the "good father" hypothesis cannot always explain why females elevate their reproductive effort when paired with more attractive males. The idea that elevated reproductive effort may nevertheless be beneficial for females can be rescued, however, if indirect benefits are considered (Kokko 1998, Møller and Thornhill 1998). Burley (1986, 1988) stressed that an elevated reproductive effort may be costly for females, but suggested that females mated to attractive males may enjoy indirect genetic benefits that more than outweigh these direct costs. Such indirect benefits would be manifested as an increased survival or reproductive success of offspring. Yet studies showing that attractive males tend to father offspring of better condition or higher survival potentially confound the causes and consequences of the very effect they are trying to explain: that females tend to invest more in offspring production with attractive males (Rintamäki et al. 1998, Gil et al. 1999). This issue was highlighted in an experimental study by Cunningham and Russel (2000). They showed that, while offspring of attractive male mallards were indeed in better condition, this was caused by a nongenetic maternal effect (females laid larger eggs with attractive males) rather than by any "good genes" provided by the male.

An alternative to these explanations based on the notion that females benefit from responding to male attractiveness with an elevated reproductive effort is that male traits evolve to exploit and manipulate female reproductive investment against the primary interests of females. For example, Wachtmeister (2001) noted that behavioral displays in socially monogamous birds which are directed toward their mates typically continue for extended periods of time (i.e., long after pair formation), and argued that sexual conflict over relative reproductive effort is the most likely explanation for the evolution of such traits. Given that females of various species use cues provided by other individuals to regulate their own reproduction, it seems likely that males could evolve to exploit such sensory routes. For example, many experimental studies on birds have demonstrated that various stimuli, such as male song, other vocalizations, and general exposure to males, directly affect female sex hormone titers as well as rates of ovarian development and oviposition (e.g., Hinde and

Steel 1978, Bluhm et al. 1984, 2000, Shields et al. 1989, Wingfield 1994, Wingfield et al. 1989, Massa et al. 1996). Using a neural network model, Wachtmeister and Enquist (2000) demonstrated that display traits in males may quite generally evolve to exploit and manipulate females into making suboptimal reproductive decisions (e.g., reproductive effort, timing of reproduction), while females will counter by evolving resistance to such traits.

Incidentally, any male trait which elevates female reproductive rate above its naturally selected optimum (seminal gonadotropins, display traits, etc.) would be predicted to cause an accelerated rate of senescence in females (Promislow 2003). The implications of this interesting link between sexual conflict theory and life-history theory have not yet been fully explored, but there is some empirical support for this suggestion. Friberg and Arnqvist (2003) showed that female fruit flies housed with attractive (large) males aged at an accelerated rate compared with those housed with less attractive (small) males. This was evident not only from an elevated female mortality but also from a more rapid decline in egg-adult survival over time of the offspring in the former females.

In summary, it is clear that male "display traits" often do elevate female reproductive rate. Whether this is generally counter to the net interest of females or not, however, is contentious (cf. Westneat and Sargent 1996, Wachtmeister 2001) and there is no conclusive evidence of either to our knowledge. There is some evidence suggesting that females indeed suffer increased direct costs of reproduction as a result of pairing with attractive males (Burley 1986, 1988, de Lope and Møller 1993, Massa et al. 1996, Mazuc et al. 2003), but it is difficult to assess whether indirect benefits counterbalance such negative effects. Sheldon (2000) argued that maternal favoritism provides strong evidence for the "good-genes" process of sexual selection by female choice. We suggest that this is not necessarily true (see also Wachtmeister and Enquist 2000, Colegrave 2001, Cunningham and Russel 2001, Gil and Graves 2001). Many male display traits may be favored in males because they tend to elevate female reproductive effort, and females may be responding by evolving resistance to such traits. Under these scenarios, attractive males would be "bad mates" without necessarily being "good fathers," and antagonistic coevolution between the sexes would contribute to the evolution of male display traits and female reproductive responses to these (cf. Wachtmeister and Enquist 2000).

4.2 Female Mating Behavior, Sperm Competition, and the Conflicting Interests of the Sexes

Whenever a female mates with more than one male during a single reproductive period, the ejaculates of these males will be present simultaneously in the female at the time of fertilization. It is now more than 30 years since Geoff

Parker (1970c) reasoned that the outcome will be sperm competition: the ejaculates of different males will compete inside the female to fertilize her eggs. Parker (1970c, 1984) also recognized that sperm competition favors two kinds of adaptations in males. On one hand, males will often be under selection to mate fairly indiscriminately and to displace the sperm of previous males. This will favor male adaptations that increase the focal male's ability to achieve matings with previously mated females, and adaptations that increase their ability to outcompete the ejaculates of previous mates. These can be referred to as offensive adaptations. On the other hand, once a male has mated he will be under selection to prevent future sperm competition. This will favor male adaptations that prevent or delay remating by their mates and those that favor their ejaculate over those deposited later. These are referred to as defensive adaptations. It is now evident that sperm competition has led to a wide variety of male adaptations of several kinds: examples include behavioral (e.g., mate guarding), morphological (e.g., male claspers, modified genitalia, mating plugs), and physiological (e.g., seminal substances which delay female remating) traits. Since our primary focus is not on sperm competition per se, we refer the interested reader to the excellent and comprehensive reviews of Smith (1984), Birkhead and Møller (1992, 1998), and Simmons (2001a) for more complete discussions of traits involved in sperm competition.

Our interest in sperm competition here stems from the fact that male adaptations for sperm competition can be costly to females (table 4.1). Parker (1979, 1984) was again first to recognize that since females are the evolving battleground upon which sperm competition games among males are played (see also Thornhill and Alcock 1983, Eberhard 1996, Stockley 1997), conflict between the sexes may play a crucial role in these evolutionary interactions. Two main types of sexual conflict are involved (figure 4.4). First, male defensive adaptations aimed at delaying or preventing remating by their mate may often be in conflict with female interests. Second, male offensive adaptations rendering their ejaculates more competitive may compromise female interests.

4.2.1 MALE DEFENSIVE ADAPTATIONS AND SEXUAL CONFLICT

In many species, females that have already mated are known to benefit from remating with other males. For example, they may require "refueling" with viable sperm, or additional mates may be able to provide females with other direct or indirect benefits (Arnqvist and Nilsson 2000, Jennions and Petrie 2000). Whenever this is the case, female interests may conflict with those of her previous mate/s, provided that former mates will suffer loss of paternity if females remate. Here, male traits that prevent or delay remating in their mates will be favored. Therefore, defensive adaptations in males may spread in a population even if they are deleterious to females. For example, imagine a promiscuous species in which females begin to suffer from a shortage of viable

Table 4.1.
Examples of male sperm competition adaptations that are, or have been suggested to be, costly to females.

Male trait	Male benefit	Costs to females	Female counter-adaptations	References
Mate guarding behavior	Prevention or delay of remating in females	Energetic costs, increased predation risk, interference with oviposition, risk of injury, restriction of female mate choice	Female aggregation, avoidance of males, dislodgement of guarding males, antimale morphologies, concealment of reproductive state, choice of males causing least costs as mates	Parker 1970b, Borgia 1981, Martens and Rehfelt 1989, Krupa et al. 1990, Davies 1992, Schroder 1993, Alcock 1994, Rowe 1994, Arnqvist 1997a, Arnqvist and Rowe 1995, Watson et al. 1998, Zeiss et al. 1999
Mating plugs	Prevention of remating or ejaculate rejection in females, serve as a donor of seminal signals	Interference with oviposition, obstruction of replenishment of sperm, restriction of polyandry and female mate choice	Removal or dissolution of plugs	Hartmann and Loher 1974, Koprowski 1992, Eberhard 1996, Baer and Schmid-Hempel 1999, Duvoisin et al. 1999, Baer et al. 2000
Seminal receptivity inhibitors	Prevention or delay of remating in females	Obstruction of replenishment of sperm, restraint of remating for other direct benefits, decrease of life span, restriction of female mate choice	Decreased sensitivity to seminal substances, breakdown of substances, ejaculate rejection	Simmons and Gwynne 1991, Civetta and Clark 2000, Eberhard 1996, Holland and Rice 1998, Arnqvist and Nilsson 2000
Transfer of anti-aphrodisiacs	Prevention or delay of remating in females	Restraint of remating to gain direct benefits	Breakdown of antiaphrodisiacs	Andersson et al. 2000
Infliction of harm upon females	Prevention or delay of remating in females	Toxic or injurious effects of male traits or behaviors		Johnstone and Keller 2000

Table 4.1. *(cont.)*

Male trait	Male benefit	Costs to females	Female counter-adaptations	References
Seminal toxins	Death or incapacitation of sperm from previous mates, protection of ejaculate from female "attack"	Decrease of life span	Decreased sensitivity to seminal substances, breakdown of substances	Harshman and Prout 1994, Chapman et al. 1995, Keller 1995, Chapman and Partridge 1996b, Lung et al. 2002
Aggressive sperm	Outcompetition of sperm of other males	Infertility due to polyspermy or immunization to sperm, restriction of cryptic female choice	Various physiological, immunological and structural internal barriers against sperm	Birkhead et al. 1993, Eberhard 1996, Howard et al. 1998, Palumbi 1998, Frank 2000
Infertile sperm	Delay of remating in females	Infertility	Reduced storage of infertile sperm	Cook and Wedell 1999, Wedell 2001
Sperm allocation	Optimal partitioning of ejaculates	Infertility due to lack of fertilization		Warner et al. 1995, Alonzo and Warner 2000a,b
Deceptive alarm calls	Prevention of remating in females	Energetic costs, increased predation risk, restraint of remating for indirect benefits		Møller 1990
Male aggression toward mates	Prevention or delay of remating in females	Energetic costs, restraint of remating, reduced foraging success	Postpone pheromone production, avoidance behavior	Alcock and Forsyth 1988, Watson 1986, 1991, 1998, personal communication
Male grasping traits, genital spines/claspers	Avoidance of takeovers, achievement of takeovers, prolongation of copulation	Energetic costs, restraint of remating for other benefits, increased infertility, risk of injury, decreased life span.	Dislodgement behaviors, anti-male morphologies	Kraus and Lederhouse 1983, Crump 1988, Thornhill and Sauer 1991, Arnqvist and Rowe 1995, Byrne and Roberts 1999, Crudgington and Siva-Jothy 2000

Note: that this is not meant to be a complete list of cases and that the degree of empirical support for the putative female costs and counteradaptations varies greatly (see references).

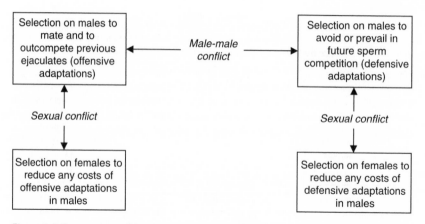

Figure 4.4. Sperm competition creates two types of selection in males. They are selected to avoid or reduce sperm competition in females they have already mated with, thus protecting their own ejaculate once deposited (sperm defence). At the same time, they are selected to displace previously deposited ejaculates (sperm offence). Both of these routes may spark sexually antagonistic coevolution. Whenever male adaptations carry direct costs to females, females are expected to evolve to depress these costs thereby potentially reducing their efficiency. (After Parker 1984.)

sperm, resulting in infertility, three days after a given mating. Next, imagine a novel mutation expressed in males which does nothing else than permanently turn off receptivity to all further matings in their mates. This will obviously come at a great cost (increased infertility) to any female mated to a male carrying the mutation. Yet this mutation could rapidly spread to fixation because of its beneficial effects in males, despite the fact that it would reduce overall population fitness: males carrying the mutation would outcompete males not carrying the mutation since they would monopolize paternity of their mates' entire offspring production (cf. Parker 1979, 1984, Johnstone and Keller 2000). Needless to say, females would then be expected to counteradapt by evolving resistance to the receptivity shutdown caused by the mutation.

It may be argued that females are particularly likely to be exposed to male manipulation of their remating rate (Eberhard 1996), since females may often benefit from turning off their receptivity in the short term (while still fertile) if mating has some cost. As the benefit of remating to females increases gradually after an initial mating, so does sexual conflict with her previous mate (see figure 4.5) (Andersson et al. 2000). This has two important implications. First, it suggests that sensory pathways in females that are open for male exploitation should be common, because they may exist for the purpose of regulating mating rate. Second, female ability to counteradapt may be constrained, since evolving overly efficient resistance to male receptivity-inhibiting cues may actually be deleterious to females. For example, reduced sensitivity to a male-derived remating inhibitor may result in females instead mating too frequently.

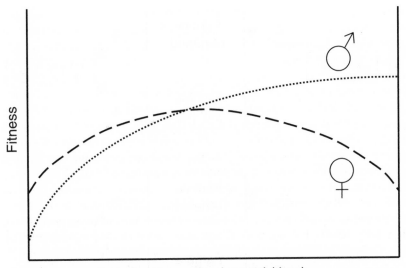

Female intermating interval (time)

Figure 4.5. Graphical illustration of sexual conflict over female remating rate in a hypothetical promiscuous species. Females exhibit an optimal intermating interval (dashed line) representing the trade-offs between various direct costs and benefits involved in mating. In contrast, male fitness (dotted line) increases monotonically with the intermating interval of his former mates, since males benefit from any postponement of sperm competition with subsequent males. Thus, the interests of the male and female may be confluent immediately after a given mating, since both partners benefit from delaying remating. However, this confluence of interest will gradually turn into conflict as time passes beyond the optimal female intermating interval. We thus expect males to evolve mechanisms that delay female remating beyond their optimal intermating interval, and females to evolve resistance to these male adaptations.

Male traits that prevent or delay remating in their mates may incur at least three types of costs to females, with quite different evolutionary implications (figure 4.6). First, there may be direct costs, given that delayed remating may reduce female fitness. Any male traits that deny their mate the opportunity of receiving material (e.g., viable sperm, nutrients) or genetic (e.g., superior or compatible genes) benefits through additional mating would fall into this category. Second, male traits that delay female remating may have additional costs to females that are incidental side effects of the trait. Mate guarding and some seminal substances may serve as examples of such defensive traits. Third, some male traits may be beneficial to males for no other reason than directly causing harm to females, provided that this delays remating in females.

4.2.1.1 COSTS OF DELAYING REMATING IN FEMALES

Despite the fact that delaying remating may commonly be costly to females, there is little direct evidence suggesting that male adaptations indeed delay

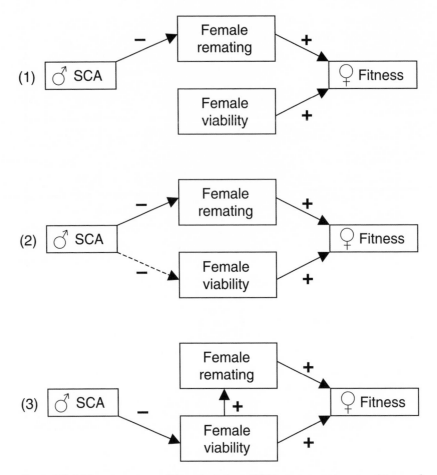

Figure 4.6. Male sperm competition adaptations (SCA) may generate several types of costs to females, indicated here with minus signs. A positive sign indicates a positive association. Male traits that deny their mate the opportunity of receiving material or genetic benefits from other males will generate direct costs to females (1). Male traits may also bring about additional costs in females as incidental side effects of their selected function (2). If a male can lower the remating rate of his mate only by directly lowering her viability, this may prove beneficial to males under some circumstances (3). However, there will typically be selection in both sexes for reduced viability costs to females, since males suffer from lowering the viability of their mates. We would thus expect the evolution of more benign males (cases 2 and 3 would tend to evolve toward case 1) and males are expected to lower female viability only in species where they are constrained to in order to gain an advantage in sperm competition. Note that the positive effects of remating indicated in this simplified graphic representation could represent any kind of net benefit, that "female viability" can also be thought of as female fecundity or fertility, and that female remating could be replaced by any trait which benefits females but is a disadvantage for their mates in sperm competition.

female remating beyond what is optimal for females. Likewise, there is little evidence suggesting that male-induced delays in remating increase female fitness. We believe that the paucity of data reflects the many difficulties involved in empirically determining the optimal female remating interval (Parker 1979, Arnqvist and Nilsson 2000). To unambiguously demonstrate that females do not remate even if it would be in their best interest to do so would require (i) accurate estimates of female remating rates in natural populations, (ii) evidence that male adaptations induce or otherwise cause depressed remating in females, (iii) accurate measures of the many potential costs of mating, and (iv) quantitative measures of the often elusive and temporally changing direct and indirect benefits of remating.

The most convincing case, to date, of indirect benefits of remating comes from a series of experiments on the common bumblebee, *Bombus terrestris*. It is often suggested that remating may benefit females because it tends to increase genetic diversity among offspring (see Jennions and Petrie 2000). Although theory suggest that the conditions under which this may be true are restricted (see Yasui 1998), it seems to be generally important among social Hymenoptera where diversity can contribute to a more efficient division of labor and a reduced rate of parasite transmission within the colony (Shykoff and Schmid-Hempel 1991, Liersch and Schmid-Hempel 1998, Schmid-Hempel and Schmid-Hempel 2000, Brown and Schmid-Hempel 2003). In support of this view, Liersch and Schmid-Hempel (1998) experimentally varied the degree of genetic variation within bumblebee colonies and showed that genetically heterogeneous colonies experienced lower degrees of parasitism. The benefits of remating in this situation were demonstrated in an experiment by Baer and Schmid-Hempel (1999). They artificially inseminated virgin *B. terrestris* queens either with sperm from four males unrelated to each other or with sperm from four males that were brothers, thus controlling for number of sperm donors and ejaculate volume. When allowed to start new colonies, those founded by "polyandrous" females exhibited markedly lower levels of parasitism and doubled reproductive success compared to the more "monandrous" females (figure 4.7). These results indicate that remating can result in substantial fitness benefits to females, and given that mating costs appear small, we would then expect *B. terrestris* females to remate with several different males. Yet studies of the intracolony genetic structure of bumblebees, using highly polymorphic microsatellite markers, and behavioral data demonstrate that females of this species typically mate only once in their lifetime in the wild (Estoup et al. 1995, Schmid-Hempel and Schmid-Hempel 2000). Further experiments have shown that this surprising fact appears to be the consequence of male sperm competition adaptations: males transfer both a mating plug (Duvoisin et al. 1999) and seminal substances (Baer et al. 2000, 2001) that delay or totally prevent female remating. Thus, it seems that male *B. terrestris* adaptations may actually delay remating beyond what is optimal for females.

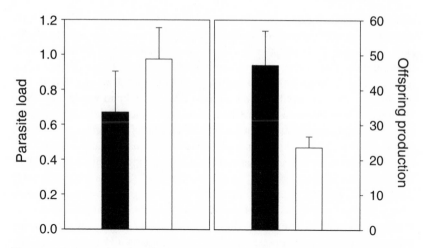

Figure 4.7. Colonies founded by bumblebee queens artificially inseminated with sperm from four unrelated males (filled bars) suffer less from parasitism and have higher reproductive success compared to colonies founded by females artificially inseminated with sperm from four brothers (open bars). This indicates that remating with several unrelated males is beneficial for females. Yet females of this species mate only once, apparently as a result of males transferring a mating plug as well as seminal receptivity inhibitors to females during mating. (Reprinted from Baer and Schmid-Hempel 1999, with permission from the Nature Publishing Group.)

It should, however, be noted that the sexually antagonistic coevolutionary dynamics should be different in social Hymenoptera compared to most other taxa, because their haplo-diploid sex-determining system changes the genetic "rules of the game."

Females of many species undoubtedly need to remate at regular intervals simply to secure an appropriate supply of viable sperm (Thornhill and Alcock 1983, Birkhead and Møller 1998, Arnqvist and Nilsson 2000). This fact is likely to conflict with male adaptations to delay or prevent female remating. Yet we are unaware of any evidence for such conflicts. The best evidence for conflicts over female acquisition of direct benefits instead comes from the study of insects where males provide another form of direct benefit: large and proteinaceous ejaculates (i.e., "nuptial gifts"; see section 4.5), such as many orthopterans and lepidopterans. Females either ingest or absorb parts of the male ejaculate and as a result of this "foraging" have increased fecundity and life span (see Vahed 1998 and Arnqvist and Nilsson 2000 for reviews). Simmons and Gwynne (1991) pointed out that sexual conflict over female remating interval should be important in such species. Selection in males will clearly favor adaptations that delay female remating, while females could benefit by remating frequently to collect and process multiple male ejaculates. In the bushcricket *Kawanaphila nartee*, the male ejaculate represents 16–20% of his body weight and female short-term fecundity is increased about 40% by the consumption of male ejaculates (Simmons 1990). Despite the massive direct benefits that females could gain from remating, Simmons and Gwynne (1991) demonstrated that normal matings cause females to become unreceptive to remating for as much as 15–20 days! They suggested that the refractory behavior of *K. nartee* females is induced by seminal substances transferred to females with the sperm. If true, it seems that male adaptations under some circumstances indeed compromise female interests in this species. Simmons and Gwynne (1991) also showed that females maintained on a low-quality diet and denied the opportunity to consume the male ejaculate tended to remate sooner. Another interesting example is the green-veined white butterfly, *Pieris napi,* where males transfer on average 15% of their body mass to females during mating (Kaitala and Wiklund 1994). Females almost double their lifetime fecundity by remating (Wiklund et al. 1993) and are thus expected to remate often. Yet females remate only once or twice during their 3–4 week life, even when given constant access to mature males (Wiklund et al. 1993, Kaitala and Wiklund 1994). This delay in remating results, at least in part, from the transfer of an efficient antiaphrodisiac during mating (Andersson et al. 2000). However, even if it seems that the optimal female remating rate should be higher than that exhibited by *P. napi* females, it is not known whether these male-derived receptivity inhibitors actually can be said to delay female remating beyond its optimum since females might be constrained in the rate at which they can process and absorb male ejaculates (Andersson et al. 2000, Wiklund et al. 2001).

Another interesting phenomenon in butterflies is the production of two distinct types of sperm: fertile eupyrene and nonfertile apyrene sperm, the latter of which lacks nuclear material (see Friedlander 1997). The majority of sperm transferred at mating is of the nonfertile type. Since female nonreceptivity to remating in butterflies is partly triggered by stretch receptors in the reproductive tract, Cook and Wedell (1999) suggested that males may have evolved apyrene sperm to exploit a sensory system in females designed to monitor the amount of sperm in storage to ensure maximum fertility. In essence, apyrene sperm could then act as "cheap fillers" which more efficiently depress female remating, with an obvious benefit to males in terms of sperm competition defense. Although not demonstrated, this could generate fertility costs to females. Wedell (2001) found that the amount of apyrene sperm stored by females is highly variable in *Pieris napi*, and reasoned that females have the ability to distinguish apyrene from eupyrene sperm, selectively storing the latter and dumping the former. If so, this may represent resistance by females to male manipulation. However, a recent experiment suggesting that apyrene sperm actually play a significant role in fertilization in *Bombyx mori* (Sahara and Takemura 2003) indicates that apyrene sperm may have alternative or multiple functions.

4.2.1.2 FEMALE COSTS AS SIDE EFFECTS

Many male adaptations impose fitness costs to females which have been interpreted as incidental side effects of their defensive function (figure 4.6 and table 4.1). The conditions for the spread of such adaptations are, however, relatively restricted. For example, imagine a male trait which delays female remating but at the same time lowers female fecundity or viability. Since a decrease in the number of offspring produced by the mate of a given male actually lowers his own reproductive success as well as that of his mate, these traits will spread in a population only if the benefit to males (reduced sperm competition) more than outweighs the cost to males (reduced fecundity of his mates). Further, such incidental costs should be quite rare, of small magnitude, and occur only when they are more or less "inevitable," simply because there will be direct selection to reduce such costs in both sexes: a male that achieves a given sperm competition advantage at the price of a certain cost to his mates will be disfavored over one that achieves the same advantage at a lower price. All else being equal, more benign males will do better (cf. section 4.2.1.3 below). To cast this simple argument in genetic terms, costs to females could be seen as antagonistic pleiotropic effects of genes which are favored by sperm competition among males. Seen this way, these genes would tend to generate a negative genetic correlation between two male fitness components: male sperm competition success and male offspring production mediated by his mates (i.e., male fecundity). In these situations, any male mutation which achieves the same defensive function with reduced negative pleiotropic effects would spread in a population, thus reducing this negative genetic correlation, and thereby resulting in the evolution of more benign males. Of course, this argument assumes that males are at least to some minor degree penalized by the lowered long-term fitness of their mates. If not, then there will be no selection for more benign males.

In some cases, however, trade-offs to males may be inevitable and the negative genetic correlation mentioned above may be constrained from evolving. This will be the case when costs to females arise as inevitable side effects of male adaptations: i.e., when those traits required to reduce female remating necessarily decrease female fitness through some route other than the remating rate. Male defensive adaptations can generally be expected to spread even under such circumstances, but only as long as there is a net benefit to males (i.e., the benefit in terms of an advantage in sperm competition is larger than the cost of the reduced fecundity of mates) (Parker 1979). Females are then expected to counteradapt by reducing costs imposed by males, thus also reducing the efficiency of male adaptations and generating interlocus antagonistic coevolution between the sexes.

For the reasons stated above, then, costs as side effects are most likely when they are an "inevitable" result of male traits that are efficient in sperm defence,

Figure 4.8. Female sierra dome spiders, *Neriene litigiosa*, that are receptive to mating "spice" their web with male-attracting pheromones. However, male spiders destroy much of the web of their mates, perhaps in an attempt to reduce pheromone transmission and, thereby, the probability of sperm competition. Picture shows a male in the process of bunching up the female's silk. (Photo by P. J. Watson.)

and they have a small cost to females or costs to females have little effect on male fitness. An example of a male adaptation that fits some of these criteria may be the destructive habits of male Sierra dome spiders, *Neriene litigiosa*. Females of this Linyphiid spider "spice" their webs with male-attracting pheromones. In an apparent attempt to interfere with the broadcast of these pheromones, and to thereby reduce the risk of attracting competing males, male spiders arriving at the web destroy much of their mate's web (figure 4.8; Watson 1986). This male strategy decreases both overt male-male competition and sperm competition but also inevitably generates a cost to females as a side effect, since the pheromone-emitting structure (the web) is also a prey-capturing device: females have to rebuild the web to regain full foraging ability. Additional costs may arise because web destruction potentially restricts female remating rate (Watson 1991, 1998). However, both costs may be relatively minor to *N. litigiosa* females and they do not show any obvious counteradaptations (Watson, personal communication).

Male mate guarding may serve as another more general example. Various forms of mate guarding are often efficient in preventing female remating, at least in the short term (Alcock 1994), but are also known to compromise female interests to some extent (e.g., by restricting female mobility and increasing their risk of predation; see the references in table 4.1). However, costs to females may often be relatively minor (Wilcox 1984, Gwynne 1989, Alcock 1994) and males may be unable to guard females without imposing such costs, which would help explain the widespread occurrence of costly male mate guarding.

Male seminal fluids of many insects are known to contain substances that reduce remating by inducing refractory behavior in females (Chen 1984, Raabe 1986, Gillott 1988, 1996, Miller et al. 1994, Eberhard 1996, Klowden 1999). In most cases, these substances are produced by male accessory glands and cause refractory behavior even when experimentally injected into virgin females. We argued above that delaying remating may often be costly per se to females, and such seminal fluid adaptations in males may therefore generally have antagonistic effects in females. Other seminal substances, aimed at elevating female reproductive rate, also sometimes compromise female interests (see section 4.1.1). In addition, seminal substances that function to delay female remating may sometimes generate additional costs in females as incidental side effects. We are, however, unaware of any direct evidence for such costs, and it is unclear whether we should expect them to be common or not, simply because we know so little about the mechanisms by which seminal substances act to delay female remating. In any case, it is clear that we would generally expect the evolution of reduced incidental costs to females of male adaptations since selection in both sexes always favors more benign males if there is any variation in the magnitude of cost to females for a given male sperm competition advantage (see above). We will discuss this at greater length below.

4.2.1.3 FEMALE COSTS AS A DIRECT TARGET OF MALE STRATEGIES

The idea that males may benefit from "intentionally" causing harm to females, provided that this reduces the risk that she will mate with other males, has a long history (see Mesnick 1997, Gowaty and Buschhaus 1998). Most of these ideas are, however, relatively limited in their applicability as they assume fairly complex and specific social interactions among individuals. A quite general problem with these ideas (cf. figure 4.6, panel 3) is that it is rarely good for males to cause harm to their mates because, in most cases, both mates will pay a fitness cost by a reduction of female fecundity, fertility, or viability (Parker 1979). Nevertheless, there are some conditions under which male strategies aimed at causing harm to their mates will be favored. These cases all assume that females exhibit a response to harm which is beneficial to their harming mate, by elevating their current reproductive effort (cf. a terminal investment) (see Michiels 1998, Lessels, unpublished) and/or by delaying remating with other males (Johnstone and Keller 2000).

Johnstone and Keller (2000) presented a game-theoretical model addressing the evolution of a male trait which causes harm to females. They assumed that the fitness cost of increasing male-inflicted harm (seminal toxin, injurious copulation, etc.) increases at an accelerating rate, and that an increased harm to females translates directly into a reduced probability of remating. When this is true, even a slight increase in harm caused by a male would require a much larger increase in female benefits from remating to make remating pay for females. Under these key assumptions, as well as a series of other conditions, Johnstone and Keller (2000) show that male traits which cause harm to females can indeed be maintained in a population. This scenario, however, cannot explain the initial origin of such harmful adaptations in males. A mutant "bad guy" will gain no reproductive advantage (inflicting harm will not deter his mates from remating, since remating with other males in a population [all good guys] will not increase the cumulative harm done to these females) but will suffer the cost of a reduced reproductive success of his mates. Johnstone and Keller (2000) suggest a way out of this dilemma. Harmful male adaptations could initially become established by other processes, either as indirect costs to females (cf. figure 4.6, panel 2) or as a result of male strategies to gain additional matings that as a side effect harm females (see chapter 2). Such initial evolutionary "hitchhiking" may then allow harmful adaptations to be further exaggerated because of their depressing impact on female remating. Another way out of this dilemma is to assume that female remating propensity is an increasing function of female condition after mating (i.e., irrespective of what costs future matings may bring about). In these cases, the benefits to a mutant "bad guy" of lowering female viability might outweigh the costs (cf. Parker 1979).

While scenarios such as that described above can certainly work in theory, key assumptions may not apply in nature (see Morrow et al. 2003a). First, male adaptations that do nothing else but cause harm to females can be beneficial to males only if female remating propensity is a decreasing function of the degree of harm she has suffered. On the contrary, female remating rates may often be higher in females in poorer condition (those who have suffered harm). In systems where mating brings any form of nutritional benefits to females, females in poor condition should accept or solicit matings more frequently. This reasoning has been used to explain the fact that female bush-crickets tend to mate more frequently if starved (Gwynne and Simmons 1990, Simmons and Gwynne 1991; but see Martin-Alganza et al. 1997). Females in poor condition may also be less able to resist male courtship or reject harassing males. For example, female *Drosophila* fruit flies suffering lowered viability as a result of inbreeding depression are known to accept matings after fewer courtships and hence remate more frequently (Bryant 1979, Van Den Berg et al. 1984, Harmsen and Clark 1987). Among crustaceans, many males mate with females directly after females molt to maturity, when the female exoskeleton is soft and females are easily injured. Yet male crabs tend to handle such females very gently despite the obvious opportunity for infliction of harm (Donaldson and Adams 1989), possibly because loss of limbs or other injuries would decrease female ability to reject harassing males later in life. Female condition may also affect remating rates indirectly through its effect on habitat use and foraging behavior. Starved water strider females, for instance, mate more frequently as a result of increased exposure to males in areas with high food abundance (Krupa et al. 1990, Rowe et al. 1994; but see Rowe 1992). Similarly, females may typically respond to harm by temporarily reallocating resources from reproduction to healing/repair, which would not be beneficial for their mates (Morrow et al. 2003a). Second, empirical tests of these scenarios have failed to find any effects of copulatory harm that may be beneficial to males. For example, dungfly females actually become less reluctant to remate as the number of harmful copulations increases (Hosken et al. 2003). Similarly, Morrow et al. (2003) experimentally inflicted various forms of harm on females immediately after copulation in three different insect species. They found that females did not delay remating or increase their reproductive rate after being harmed, but on the contrary remated sooner and laid fewer eggs in some cases.

Third, several of the assumptions of the particular model of Johnstone and Keller (2000) make it prone to predict the maintenance of harmful male traits. For instance, it does not allow evolution of the correlation between the degree of harm caused by a male and the resulting decrease in female remating propensity. Remember that selection will favor males which, for example, achieve the same delay in remating while inflicting less harm to females. More importantly, the model does not allow for the evolution of female resistance to male adaptations. Since it could be argued that selection for counteradaptations in females would be very intense (Clutton-Brock and Parker 1995), the key assumption of accelerating fitness cost to females of harm induced by males seems unrealistic.

In summary, traits in males that generate defensive sperm competition advantages by directly harming females can evolve, in theory, given certain very restricted assumptions. There are, however, a number of associated conceptual problems that collectively suggest that male traits that are designed to cause harm to their mates should be rare. In accordance with this conclusion, direct experimental tests have failed to provide support for these hypotheses and there is no evidence for male strategies or traits which are "aimed" at directly causing harm to their mates.

4.2.2 MALE OFFENSIVE ADAPTATIONS AND
SEXUAL CONFLICT

When a male mates with an already mated female, this male will benefit from any trait that increases the probability that his sperm, rather than those of the female's earlier mates, will fertilize the eggs of the female. As we might expect, males exhibit a series of offensive sperm competition adaptations aimed at displacing or otherwise gaining precedence over the sperm of previous mates (Parker 1970c, Smith 1984, Birkhead and Møller 1992, 1998, Simmons 2001a). As argued above for defensive adaptations, a male mutation that gives an offensive competitive advantage in sperm competition will spread in a population even if it compromises females' interests, provided that the male sperm competition benefits outweigh the cost to males due to the lowered fitness of his mates (Parker 1979, 1984). In these cases, we would expect male offensive adaptations to antagonistically coevolve with female counteradaptations (table 4.1). Male offensive adaptations could have at least two types of antagonistic effects in females: costs due to overly "aggressive" ejaculates and costs due to a restriction of the opportunity of cryptic female choice. Although empirical evidence for costs to females brought about by male offensive adaptations is limited, the evolution of some classes of male traits is more than likely to involve sexually antagonistic coevolution.

4.2.2.1 SPERM COMPETITION AND AGGRESSIVE EJACULATES

It is widely recognized that sperm competition selects for a suite of "aggressive" characteristics in sperm and ejaculates (cf. Parker et al. 1972): large testes and accessory glands, numerous sperm, an ability of sperm to swim fast and cost effectively, an ability to kill or incapacitate the sperm of other males, displace previous ejaculates, withstand the hostile environment of the female reproductive tract, efficiently locate the site of fertilization, and rapidly penetrate the egg (see Parker 1970c, Smith 1984, Birkhead and Møller 1992, 1998, Simmons 2001, for reviews). Accordingly, variation in attributes of the testes and sperm among species is related to the intensity of sperm competition in several groups, including insects (Gage 1994, Morrow and Gage 2000), fish (Stockley et al. 1997), birds (Møller 1988, Briskie and Montgomerie 1992), and mammals (Harcourt et al. 1981, Kenagy and Trombulak 1986, Gomendio and Roldan 1991, Hosken 1997). The role of females in sperm competition has been traditionally overlooked, although this is currently changing (see Parker 1984, Gowaty 1994, Eberhard 1996). One important insight emerging from this development is the recognition that attributes of ejaculates that increase their competitive ability may, as a side effect, bring about costs to females. If these costs do occur, then we expect antagonistic coevolution between the sexes over the aggressive characteristics of male ejaculates.

4.2.2.2 DIRECT COSTS, POLYSPERMY, AND FEMALE INFERTILITY

Direct costs to females may appear when male sperm become "too well" adapted at one of their primary functions: efficiently finding and fertilizing an egg. If male adaptations render their ejaculates overly efficient at this task, eggs suffer the risk of being fertilized by more than one sperm. This phenomenon is known as polyspermy, and results in zygote mortality and consequently reduced female fertility in many animals (Ginzburg 1972, Jaffe and Gould 1985). Braden (1953) originally suggested that the female genital tract of mammals is adapted to restrict the number of sperm reaching the site of fertilization, thereby minimizing the risk of polyspermy. Eberhard (1996) generalized this idea, and suggested that polyspermy resulting from overly aggressive sperm may be a problem for females of many taxa, and that females in response have evolved a wide range of barriers against aggressive sperm to avoid polyspermy. Such barriers may then favor (by sperm competition) even more aggressive sperm, leading to an antagonistic coevolutionary arms race between egg and sperm: eggs favored to resist penetration by sperm and sperm favored to penetrate the egg as rapidly as possible (figure 4.9) (see Birkhead et al. 1993, Rice and Holland 1997). Under this hypothesis, females would be caught in a delicate evolutionary dilemma: the evolution of overly effective barriers against aggressive sperm will prevent fertilization altogether, while too permissive barriers will result in high rates of polyspermy and associated zygote mortality. Eberhard (1996) thus suggested that female infertility might at least partially represent an evolutionary walk on a tightrope on which females in sexually reproducing species are forced to embark.

The "avoidance-of-polyspermy" hypothesis is appealing, plausible in theory, and may have important implications for the evolution of reproductive isolation (Birkhead et al. 1993, Rice and Holland 1997, Parker and Partridge 1998, Howard 1999, Frank 2000, Gavrilets 2000, Greeff and Parker 2000, Gavrilets and Waxman 2002). It also may provide an explanation for the remarkably high rates of infertility observed in many animals (mean 10–15% across 37 species in Eberhard [1996]) (Morrow et al. 2002). The empirical evidence for the avoidance-of-polyspermy hypothesis is, however, currently very limited. Nevertheless, several lines of evidence provide some support. For example, it is certainly true that females in animals with internal fertilization have evolved many general mechanisms that render fertilization exceedingly "difficult" for sperm (see Birkhead et al. 1993). Examples of female barriers against sperm include complete or partial ejection, or dumping, of ejaculates, a variety of physical barriers, such as narrow muscular ducts and passages which the sperm need to pass in order to reach the egg, chemical environments generally detrimental to sperm survival, molecular and hormonal gradients within the reproductive tract, and various immunoresponses to sperm, ranging from nonspecific phagocytosis of sperm by leucoytes to

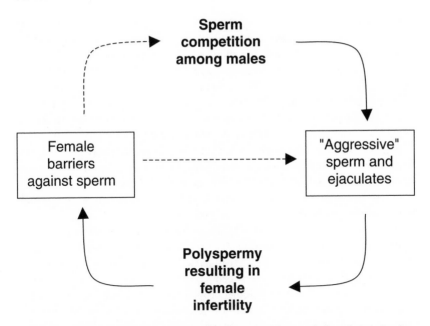

Figure 4.9. Sperm competition selects for aggressive characteristics of the ejaculate, which can result in elevated rates of polyspermy. Females are then expected to evolve barriers against these overly aggressive sperm. These barriers will affect the fitness effects in males of various sperm competition adaptations, leading to altered selection in males (both by sperm competition and by cryptic female choice) and perpetual antagonistic coevolution between the males and females. This scenario is supported by indirect experimental evidence and by the fact that polyspermy occurs naturally in a wide range of animals. The surprisingly high rates of female infertility exhibited by many organisms might in part reflect this intersexual arms race (see section 4.2.2.2)

more specific antisperm antibody responses (see Jaffe and Gould 1985, Birkhead et al. 1993, Eberhard 1996, Hunter 1996, for reviews). These female adaptations appear aimed at controlling or reducing the number of sperm that reach the mature oocytes, and may represent generalized female resistance against aggressive sperm (Braden 1953, Jaffe and Gould 1985, Hunter 1996).

However, several other hypotheses may account for these attributes of the female reproductive tract. Females may be selected to avoid infections and therefore reject non-self-elements including sperm (Austin 1965), to escape infertility resulting from immunization against sperm (Kurpisz and Fernandez 1995), or to favor certain sperm over others (Curtsinger 1991, Roldan et al. 1992, Birkhead et al. 1993, Keller and Reeve 1995, Eberhard 1996; see also section 4.2.2.2 below). The fact that polyspermy more often occurs when female barriers against sperm are experimentally circumvented provides an additional line of evidence suggesting that avoidance of polyspermy is an important function of the observed female resistance against sperm, at least among mammals. For example, the risk of polyspermy is known to increase after deep uterine artificial insemination (Thibault 1959, Viriyapanich and Bedford 1981, Hunter and Greve 1998) and polyspermy is a significant problem in in vitro fertilizations among most mammals (Nagai 1994, Asch et al. 1995, Funahashi and Day 1997, Sankai et al. 1997, Eroglu et al. 1998) where it causes zygote mortality (Han et al. 1999). The rate of polyspermy following in vitro fertilizations in mammals (up to 60–100% in some species) is also known to be repeatable across individual males (Xu et al. 1996a,b). Further, experimental studies demonstrate that the rates of polyspermy in vitro are positively related to sperm concentration around the egg, sperm mobility, and the duration of oocyte exposure to sperm (Jaffe and Gould 1985, Ho et al. 1994, Gianaroli et al. 1996, Xu et al. 1996a,b). These results are very significant as they show that female barriers do in fact limit the rates of polyspermy.

The avoidance-of-polyspermy hypothesis predicts an arms race between the sexes, such that elevated sperm persistence should be associated with an increase in female barriers to sperm (resistance). One problem with testing this prediction is that it is shared between several hypotheses (cf. Briskie et al. 1997, Presgraves et al. 1999). Another problem is, of course, that such coevolution is hard to detect: different types of traits could, for example, be involved in the two sexes (for example, longer sperm tails in males could be countered by immunological adaptations in females). No less than six different studies have provided evidence for coevolution between male sperm size on one hand and the dimensions of the female reproductive tract on the other (see section 4.2.2.3 below), but it is unclear whether this is at all related to the avoidance of polyspermy.

Few reproductive physiologists would argue with the claim that polyspermy presents a general problem for females of many organisms. The reason is simply that female gametes show refined mechanisms aimed at preventing polyspermy by rendering the egg more or less impenetrable to further sperm immediately following fertilization. The mere fact that such complex barriers have evolved and been maintained provides another observation consistent with polyspermy avoidance. The details of the sperm-egg interactions vary across taxa, but they typically involve interactions between the sperm and the egg coat and/or a cascade of biochemical pathways within the egg set off by a calcium signal, referred to as the egg activation, which causes rapid changes in the properties of the egg plasma membrane (see Raz and Shalgi 1998, Stricker 1999). Such polyspermy-blocking mechanisms have been described in a wide range of organisms, including mammals, fish, amphibians, ascidians, echinoderms, molluscs, arthropods, and annelids (see Ginzberg 1972, Jaffe and Gould 1985).

The fertilization of an egg requires precise interactions between the cell surfaces of the two gametes, to enable binding and fusion of the male and female gametes. These gamete recognition systems are known from virtually all taxa (Palumbi 1998), and pose interesting evolutionary problems (Vacquier 1998). Rice and Holland (1997) suggested that male gametes would be selected for rapid recognition and fusion, to succeed in sperm competition, whereas female gametes would be selected to slow down these processes, to allow egg activation, and thereby avoid polyspermy. The avoidance-of-polyspermy hypothesis suggests that this would generate a rapid and perpetual antagonistic arms race at a more basal level than those discussed above: between genes coding for molecules on male and female gametes which mediate gamete recognition and fusion (see also Rice 1998a, Frank 2000, Gavrilets 2000). Evidence in favor of this hypothesis comes from studies showing that such molecules indeed appear to evolve very rapidly and differ markedly among closely related taxa (see Swanson and Vacquier 2002). Most notably, the proteins that mediate these processes in certain marine gastropods and echinoderms are known to evolve at high rates and differ among closely related species (Vacquier et al. 1990, Metz et al. 1994, Vacquier 1998), and molecular studies have indicated directional selection for novel protein configurations among the coding loci (Lee et al. 1995, Metz and Palumbi 1996, Palumbi 1998, Galindo et al. 2003). The genes coding for these proteins are actually among the most rapidly evolving genes known (Evans 2000, Swanson and Vacquier 2002). Mammals seem to show a similar pattern: some of the oligosaccharide residues of the egg coat glycoproteins which are involved in mammalian gamete binding and fusion show extreme structural heterogeneity across species (Shalgi and Raz 1997) and sperm-egg fusion is generally species specific (Evans 2000). Molecular studies of mammals have also revealed directional selection for novelty in gene coding for both male (Torgerson et al. 2002) and female (Swanson et al. 2001a, 2003) reproductive proteins. Additional support for a female-resistance-driven evolution of these gamete recognition systems comes from interspecific studies of external fertilizers, showing that heterospecific sperm can actually be more efficient than conspecific sperm both in binding to (Lambert 2000) and fusing with (Gomez and Cabada 1994) eggs. These data suggest that females may have evolved resistance to conspecific sperm that is ineffective against heterospecific sperm. Recall, however, the females' evolutionary "walk on a tightrope" (see above): overly efficient barriers to sperm should be costly to females, since this would increase the risk of not achieving fertilization at all. Empirical observations suggest that sea urchins may have evolved to such an intermediate level of resistance (see Franke et al. 2002). Females of species which face relatively high sperm concentrations in their natural habitat, such as the intertidal *Strongylocentrotus pupuratus*, exhibit low levels of polyspermy even when exposed to high concentrations of sperm in the laboratory (Byrd and Collins 1975). In contrast, species which experience lower sperm concentrations in nature, such as the subtidal *Lytechinus pictus*, seem to show higher levels of polyspermy in the laboratory (David Epel, personal communication).

Not all molecules that mediate sperm-egg interactions evolve rapidly. Once the membranes of the sperm and the egg have fused, sperm introduce a protein-aceous "sperm factor" into the egg cytoplasm, which rapidly triggers egg acti-vation and subsequent polyspermy block (Raz and Shalgi 1998, Stricker 1999). Comparative studies have shown that this sperm factor is anything but species specific: it appears conserved over a wide range of taxa. Actually, many studies have now shown that sperm factors from various donor species are able to successfully elicit egg activation in eggs from closely related species, and even from organisms belonging to completely different phyla (see Stricker 1999). Such cross-reactivity between distantly related taxa strongly suggests that the molecules involved in this intracellular signaling pathway evolve very slowly. This may seem surprising, since it is in stark contrast to the intercellular signals discussed above, but is actually exactly what we would expect. As argued above, there are reasons to believe that the interests of the sperm and the egg frequently differ in gametic encounters, and this could drive rapid and diver-gent antagonistic coevolution between the molecules mediating this intercellu-lar conflict (Frank 2000, Gavrilets 2000). In contrast, once fertilization has occurred, the interests of the sexes are identical: both the paternal and the maternal genome will benefit from rapid and efficient egg activation and poly-spermy block. Because of this sexual "confluence," we would expect strong stabilizing selection on, and consequently slow evolution of, the molecules mediating egg activation.

In conclusion, a set of indirect lines of evidence collectively provides sup-port for the coevolution of offensive sperm and ejaculate traits in males and resistance to these in females. However, if we wish to invoke avoidance of polyspermy as a current mechanism of antagonistic coevolution between males and females, we need direct evidence showing that it actually occurs in nature. Unfortunately, there are very few reliable estimates of "naturally" occurring rates of polyspermy. The prime reason is simply that it is typically very difficult to retrieve all of a female's eggs and/or to confidently determine the causes of infertility in eggs with abnormal development. Various methodological prob-lems add further difficulties (e.g., Friedlander 1980). Studies in reproductive physiology are also primarily interested in describing normal fertilization events and the early developmental phases of the zygote, and most of these therefore merely note that polyspermic eggs are "rare" or that they occur "in-frequently" or "occasionally" (see Ginzburg 1972 for a review). Nevertheless, the quantitative data summarized in table 4.2 show that polyspermy undoubt-edly occurs under natural conditions in a variety of normally monospermic species. A cautious interpretation of these data suggests that average rates of polyspermy in animals with internal fertilization might range somewhere from about one to a few percent. If this has forced females to embark on the evolu-tionary "walk on a tightrope" discussed earlier, then polyspermy should be associated with a similar rate of completely unfertilized eggs such that the rate

Table 4.2.
The occurrence of polyspermy under natural conditions in various monospermic animals.

Species	Group	Estimated rate of polyspermy	References
Evechinus chloroticus	Sea urchin	0 – 77% (mean 17%)	Franke et al. 2002, 2003
Kalotermes flavicollis	Insect	~ 10%	Truckenbrodt 1964
Pseudococcus citri	Insect	3%	Schrader 1923
Pseudococcus obscurus	Insect	4%	Nur 1962
Habrobracon juglandis	Insect	1%	Speicher 1936
Drosophila melanogaster	Insect	1 – 5%	Hildreth and Lucchesi 1963, Callaini and Riparbelli 1996
Eleutherodactylus coqui	Frog	14%	Elinson 1987
Sminthopsis crassicaudata	Marsupial	6%	Breed and Leigh 1990
Sus scrofa	Mammal	1.6 – 3.8%	Hancock 1959, Thibault 1959, Hunter et al. 1999
Cavia porcellus	Mammal	3.3%	Lams 1913
Rattus rattus	Mammal	0.3 – 3.2%	Austin and Braden 1953, Austin 1956, Odor and Blandau 1956, Piko 1958
Mus musculus	Mammal	0.9 – 1.2%	Braden et al. 1954, Braden 1957
Microtus agrestis	Mammal	2%	Austin 1957
Mesocricetus auratus	Mammal	0.5 – 3.8%	Chang and Fernandez-Cano 1958, Suzuki et al. 1996, Ying et al. 1998
Homo sapiens	Mammal	0.5 – 5%	Redline et al. 1998, McFadden and Langlois 2000, Zaragoza et al. 2000, Yusuf and Naeem 2001

of hatching failure among eggs is about twice the rate of polyspermy. Thus, a tentative conclusion must be that polyspermy can be costly to females, and that it can result in infertility rates ranging from a few to several percent. If this is generally true, there is certainly plenty of fuel for antagonistic coevolution between aggressive characteristics of male ejaculates and female resistance mechanisms.

Polyspermy is not lethal to the zygote of all organisms. In some groups (notably salamanders, reptiles, birds, gastropod mollusks, some insects, and some fish; see Ginzburg 1972, Jaffe and Gould 1985) several sperm enter the egg cytoplasm, but only one fuses with the female pronucleus. The supernumerary sperm nuclei normally degenerate in such physiologically polyspermic species. Evidently, the block against functional polyspermic fertilization and resulting developmental problems of the zygote acts at a later stage in these species. The mechanism behind this polyspermy-avoidance strategy remains poorly understood (Birkhead et al. 1994), but seems to involve substances produced either by the oocyte or by the zygote nucleus (Jaffe and Gould 1985). This does not, of course, preclude the possibility that aggressive characteristics of male ejaculates and female resistance mechanisms are coevolving, but merely suggests that any antagonisms involved occur at later stages. Interestingly enough, some data suggest that triploidy in birds indeed can result from dispermic fertilizations of the egg nucleus (Lee et al. 1990, DeLasena et al. 1992; but see Thorne and Sheldon 1991).

Some theories suggest that females may be caught in an even more intricate evolutionary dilemma than that outlined above. If the problem of polyspermy is associated with sperm competition resulting from polyandry, then one might predict that females could escape the quandary by evolving to mate with one male only (i.e., monandry) (see Morrow et al. 2002). However, removing sperm competition is predicted to lead to a reduced per mating investment in ejaculates by males, which may also elevate female infertility, this time as a result of a shortage of viable sperm (see Ball and Parker 2003).

4.2.2.3 INDIRECT COSTS AND DELETERIOUS MATINGS

Some offensive sperm competition adaptations may impose fitness costs to females as incidental side effects, unrelated to their selected function, and hence be involved in sexually antagonistic coevolution. Often cited examples are components of the seminal substances in male *Drosophila* believed to incapacitate sperm stored inside the female, thus promoting sperm displacement (Harshman and Prout 1994, Price et al. 1999). Since such substances seem toxic to stored sperm it was suggested that they may also be toxic to females (Keller 1995), causing an observed reduction in female fitness which can follow the receipt of seminal fluid in *Drosophila* (e.g., Chapman et al. 1995, Chapman and Partridge 1996b). If true, the deleterious effects to females would be a side effect of a male offensive sperm competition adaptation. However, the function of most of the great number of seminal substances which male *Drosophila* transfer during mating remains undetermined (see section 4.6 below), and no substance with a dual toxic effect (to stored sperm and to females) has yet been identified (Chapman, personal communication). The results of Civetta and Clark (2000) are also not consistent with this hypothesis. Across a large number of inbred *Drosophila* lines, they found that the toxic effects to females were correlated with male sperm defensive ability but not with toxicity to other males' sperm (i.e., sperm offensive ability). Thus, while it certainly is possible that "offensive" seminal substances (Harshman and Prout 1994) could generate costs as side effects in females, we are unaware of any convincing examples.

A more likely case of indirect costs comes from studies of male genitalia in species where males short-circuit the regular pathway for sperm transport by literally penetrating the walls of the female reproductive tract during copulation. This could result in an offensive advantage by allowing sperm to migrate directly to the eggs through the hemolymph and may offer an explanation for the evolutionary origin of traumatic insemination through the female body wall as seen in some insects (see section 3.4.3). In fact the genitalia of many animals that copulate through the female reproductive tract are heavily sclerotized, and bear processes such as recurved spines and spikes (Eberhard 1985), yet the actual occurrence of "cryptic" intrusive copulation has been difficult to demonstrate. It has, nevertheless, been demonstrated in several insects. For example, males of some Nabidae bugs (i.e., Prostemmatinae) pierce the genital tract of females during copulation and ejaculate into the hemolymph (Schuh and Stys 1991), and the large, spiky, and heavily sclerotized genitalia of *Hotea* males (Scutelleridae) have also been reported to literally tear through the female vaginal wall to enable sperm deposition close to the site of fertilization (Leston; in Hinton 1964). Alternatively, male ability to penetrate female reproductive ducts may have allowed males to circumvent female barriers to secondary ejaculate substances rather than sperm. For example, in blowflies of the genus *Lucilia*, male sperm is deposited in a "normal" manner but the accessory gland material is deposited through a different genital aperture. Surrounding this aperture are genital barbs or rows of sharp teeth on the genitalia. These structures tear the cuticle of the female reproductive tract at the site of deposition of accessory gland material, which may allow it to pass more efficiently into the female hemolymph (Lewis and Pollock 1975, Merritt 1989). Interestingly enough, female blowflies rarely mate more than once, and the loss of receptivity is a direct response to substances produced in male accessory glands rather than a result of genital damage (e.g., Smith et al. 1989). Here, intrusive copulation should perhaps best be seen as male defensive, rather than offensive, sperm competition adaptation.

Although it is clear that damage to the female reproductive tract caused by male genitalia can cause significant scarring of these tissues (Leston, in Hinton 1964; Merritt 1989, Crudgington and Siva-Jothy 2000, Blanckenhorn et al. 2002), there is relatively little evidence of significant costs to female fitness (Crudgington and Siva-Jothy 2000, Blanckenhorn et al. 2002, Martin et al. 2003; figure 4.10).

4.2.2.4 CONFLICTS OVER CRYPTIC FEMALE CHOICE

Cryptic female choice refers to any process occurring after intromission by which female traits bias fertilization or offspring production toward certain male phenotypes. It is frequently suggested in the literature that, if the ability to exercise such cryptic choice is maintained because it provides females with indirect genetic benefits for their offspring, then any male adaptation which

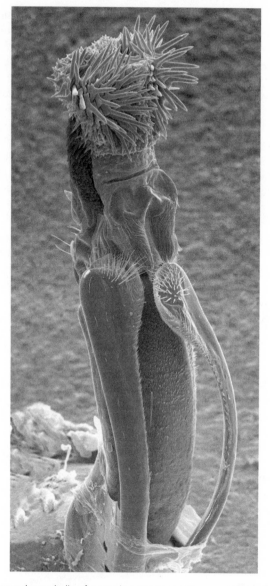

Figure 4.10. The male genitalia of many insects bear spines and barbs that sometimes are known to harm females. In the bean weevil *Callosobruchus maculatus*, the tip of the male intromittent organ is armed with sclerotized spines. Inflating within the female, these spines cause injuries and scarring in the female genital tract. (Reprinted from Crudgington and Siva-Jothy 2000, with permission from the Nature Publishing Group; photo by A. Syred, Microsopix.)

compromises female ability to freely "choose" between the ejaculates deposited by several males will reduce the benefits of cryptic female choice and will hence constitute a cost to females (e.g., Gowaty 1997b, Stockley 1997). As detailed in chapter 7, this seemingly simple argument is problematic for several reasons and it is unclear whether it is helpful to view the "denial" of achieving benefits as a cost in this context. It is also generally questionable whether females can gain appreciable indirect benefits by choosing sperm from some males over that of others (cf. Keller and Reeve 1995). Several factors suggest that the genes regulating the competitive ability of sperm are often only partially, or not at all, passed on to offspring, and that the relationship between male success in sperm competition and offspring viability or reproductive success might therefore be very weak at best (Simmons 2001b, 2003; but see Hosken et al. 2003a). It is known from *Drosophila* and mammals that the Y chromosome is enriched with genes coding for sperm characteristics (Roldan and Gomendio 1999; see also Simmons and Kotiaho 2002, Simmons 2003). There are also reasons to believe that the fertilizing and competitive ability of sperm is determined, at least in some part, by the mitochondrial genome (Ruiz-Pesisni et al. 2000, Rovio et al. 2001). Mitochondria function as "engines" in the sperm, but are not transferred by fathers to the offspring, since they are maternally inherited in most organisms (see Gemmell and Allendorf 2001). Thus, differences in sperm competitive ability stemming from mitochondrial haplotype variation have no fitness consequences for the offspring (Frank and Hurst 1996). Finally, although the amount of "heritable" (i.e., additive) genetic variation for success in sperm competition is significant in several taxa (e.g., Radwan 1998, Bernasconi and Keller 2001, Hosken et al. 2003a, Froman et al. 2002, Simmons 2003), it is limited in others (e.g., Gilchrist and Partridge 1997, Arnqvist and Danielsson 1999a). One explanation for why genetic variation for success in sperm competition may sometimes be limited was presented by Chippindale and Rice (2001), who demonstrated that Y-linked genes are highly important for male fitness in fruit flies, but that the fitness effects of these genes depends highly on the genetic background in which they were expressed (sex-linked epistatic variation, Y/X or Y/autosomal). Another contributing factor may be the male × female interactions for sperm competition success seen in both fruit flies (Clark and Begun 1998, Clark et al. 1999) and water striders (Arnqvist and Danielsson 1999a). If such interactions prove to be general, it suggests that the sperm competitive ability of a given male when mating to a particular female is a poor predictor of the sperm competitive ability of his sons, since Y-linked genes will be expressed in a different genetic background in his sons (Chippindale and Rice 2001), and since his sons will mate with females of genotypes different from their mother's.

Theory on cryptic female choice has suggested that females might be able to gain appreciable indirect benefits by choosing sperm from some males over that of others in situations where offspring viability is influenced by various genetic incompatibilities between mates (see Tregenza and Wedell 2000 and Zeh and Zeh 2001 for reviews). Even in these cases, it is important to realize that we cannot generally consider male traits/mutations which limit cryptic female choice for indirect benefits as being costly for females, even in cases where males differ considerably in viability or quality. Consider a male autosomal and sex-limited mutation that completely removes the potential for cryptic female choice (i.e., fertilizes all of the eggs of all of their mates) (Parker 1979, Andres and Morrow 2003) in a situation where offspring viability is entirely determined by alleles at another and unlinked locus or set of loci. If the "sexy sons" benefit to females from mating with "manipulative" males is larger than the lost relative viability benefits of choice, selection in both sexes will favor the initial spread of this mutation despite the fact that it removes the indirect viability benefits of cryptic female choice for females (see also chapter 2).

4.3 Conflicts over the Duration of Mating

The duration of mating varies extensively in animals with internal fertilization, both between and within species. For example, mating is often a brief matter in many insects, lasting but a few seconds. The record holder on the other extreme may be the water strider *Aquarius najas*, where pairs in the field have been reported to remain coupled for up to three months (Murray and Giller 1990)! Within species, mating and copulation duration commonly vary by one order of magnitude or more (Simmons 2001a). Red flour beetle copulations range from 30 seconds to 32 minutes (Bloch-Qazi et al. 1996) and mating duration in the water strider *Gerris lateralis* ranges from less than 4 minutes to over 7 hours (personal observation).

Once mating is initiated, the interests of the male and female, in the time to terminate the mating, will often differ. As discussed in chapter 3, mating involves a series of quite general costs which will often be an increasing function of its duration. The time-dependent costs include lost feeding and mating opportunities (Parker 1978, Rowe 1992), increased predation risk (Wing 1988, Arnqvist 1989a, Fairbairn 1993, Rowe 1994, Maier et al. 2000), energetic expenditure (Watson et al. 1998), risk of injury or infection with pathogens or parasites (Thornhill and Alcock 1983, Crudgington and Siva-Jothy 2000), and receipt of harmful accessory substances. For females, direct benefits from matings result primarily from reception of sufficient viable sperm and other resources which males may provide. Often, we would expect these benefits to accrue relatively quickly, resulting in a quickly saturating gain curve (figure 4.11). In these cases, females would benefit from terminating the mating relatively quickly.

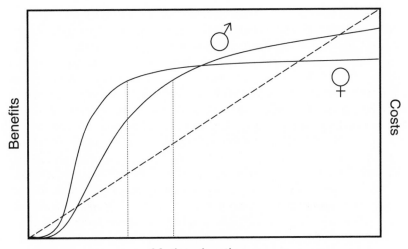

Figure 4.11. Mating involves a series of costs (dashed line) to both sexes, such as time waste, increased predation risk, and energetic expenditure. The benefits of increased mating duration to the sexes (solid lines), however, are often asymmetrical, so that males have more to gain from prolonging mating. In such cases, the optimal mating durations of the sexes (dotted lines) will differ, resulting in sexual conflict over the mating duration.

For males, the matter is a bit more complex, as mating duration can be an important determinant of sperm competition success. Many studies of invertebrates have shown male fertilization success increases with copulation duration (Parker 1978; see Simmons 2001 for a review). This effect seems to be primarily a result of more sperm being transferred with time (Simmons and Siva-Jothy 1998), but may also result from increased transfer of accessory ejaculate substances, which may induce the female to retain sperm (i.e., prevent sperm dumping) or use it to fertilize eggs (Eberhard 1996). However, prolonging copulation or mating may also serve to reduce future sperm competition from other males (Parker 1970b,c, 1974, Sillén-Tullberg 1981), by limiting access to the female by other males. These functions most likely account for the widespread occurrence of *postcopulatory* mate guarding by males (see Alcock 1994 and Birkhead and Møller 1998 for reviews). Mating duration is therefore important to males for both sperm competition offence and defence. The male gain curve (figure 4.11) is, therefore, often expected to saturate less quickly than the female's, and may even be an accelerating function (see Yamamura 1986). These different gain curves result in optimal mating durations that differ between the sexes.

There are, unfortunately, few organisms where experimental studies have directly assessed optimal mating durations for both sexes. One of these is the silkworm moth, *Bombyx mori*. Male transfer of sperm starts after 30–60 minutes of copulation, and is completed after another 15–30 minutes (Punitham et al. 1987, Suzuki et al. 1996a). Male sperm, however, does not even reach the site of storage in the female until after about 2 hours of copulation, and these sites do not begin to be filled for about 4 hours (Suzuki et al. 1996a). It is essential for males to completely fill these storage sites in order to avoid subsequent sperm displacement (Suzuki et al. 1996a), to stimulate female oviposition (Karube and Kobayashi 1999a), and to depress female production of male-attracting pheromones (Karube and Kobayashi 1999b). Thus, assuming that the ecological costs of prolonged copulations are not dramatic, it seems that optimal copulation duration is certainly longer than 4 hours in male silkworm moths. By experimentally manipulating copulation duration, Punitham et al. (1987) were able to show that female lifetime egg production peaked at a copulation duration of 3–9 hours (see figure 4.12), whereas Petkov et al. (1979) report that copulation durations of no more than 2.5–3 hours maximize female productivity. Natural copulations in *Bombyx mori* are very long, lasting some 6–12 hours (Singh 1998). It thus seems that natural copulation durations in silkworm moths may actually be closer to the male than the female optima. Notably, females have been reported to sometimes die prior to oviposition after unusually long natural copulations (Tanaka 1964).

There are many situations which differ from the, perhaps, more general one described above. For example, if mate availability is very high, and costs to males in lost mating opportunities thus accrue rapidly, optimal male mating duration may actually be shorter than that of females. In these cases, females may consequently have decreased fertility (cf. Warner et al. 1995). Similarly, if mating males protect females from predators (Walker 1980) or harassing males (Rowe 1992), optimal female mating duration may exceed that of males under some environmental conditions.

An analogous situation occurs in species where males extend mating duration by guarding females prior to, rather than after, copulation/fertilization. This precopulatory guarding occurs in various crustaceans where females are receptive to copulation for only a brief period following moult, and in many externally fertilizing frogs where males guard females prior to egg extrusion. The optimal precopulatory guarding duration in these species is often longer for males than females, and consequently conflict over mating duration is very apparent in many crustaceans (see Jormalainen 1998) and frogs (e.g., Crump 1988).

Yet another case where conflict over mating duration may occur is in those species where males first transfer ejaculate components that are beneficial for females (e.g., sperm) and then transfer components that are potentially costly to females (e.g., certain secondary substances). Sequential transfer of materials during copulation is common, at least in many insects, where transfer of sperm

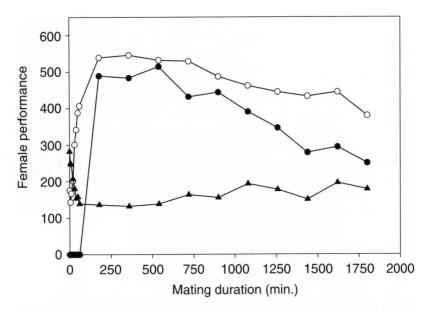

Figure 4.12. The effects of mating duration on life span (triangles; days), lifetime fecundity (hollow circles; number of eggs), and lifetime offspring production (filled circles; number of eggs hatched) in female silk worm moths (*Bombyx mori*). Female reproductive fitness increases with mating duration to a given level, and then decreases as matings are prolonged beyond this maximum (after Punitham et al. 1987). Bar at the bottom of graph indicates range of natural mating duration.

precedes that of accessory substances (e.g., Leopold et al. 1971, Merritt 1989, Duvoisin et al. 1999). If the switch point between the two materials is abrupt, then optimal female mating duration will tend toward the switch point, beyond which the net benefit of mating for females should decrease more rapidly. If transfer of substances is sequential but overlapping, female optimal mating duration will be reduced compared to the case when the transfer is not sequenced (cf. figure 4.11). Thus, in both cases, sequenced timing of the transfer of material during copulation will in theory tend to increase the difference between male and female optimal mating duration.

In summary, across the diversity of mating behaviors, it seems unlikely that optimal mating durations for males and females will often coincide. This fact leads to conflicts over mating durations, and selection for sexually antagonistic traits involved in affecting this duration.

4.3.1 MALE AND FEMALE ADAPTATIONS

One might hypothesize that many of the rich variety of grasping adaptations that occur in males of many different groups, such as claspers in invertebrates (e.g., Wing et al. 1983, Jormalainen 1998) and foreleg modifications in male

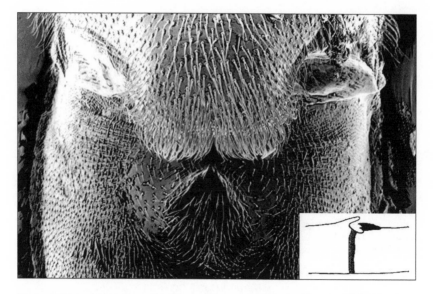

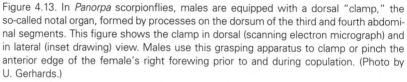

Figure 4.13. In *Panorpa* scorpionflies, males are equipped with a dorsal "clamp," the so-called notal organ, formed by processes on the dorsum of the third and fourth abdominal segments. This figure shows the clamp in dorsal (scanning electron micrograph) and in lateral (inset drawing) view. Males use this grasping apparatus to clamp or pinch the anterior edge of the female's right forewing prior to and during copulation. (Photo by U. Gerhards.)

frogs (e.g., Peters and Aulner 2000), function to prolong mating (Thornhill and Sauer 1991). However, as discussed in chapter 3, these traits may function in several other contexts; they may for example serve either to overcome female resistance to mate or to avoid takeovers by other males once mating has been achieved. Furthermore, these functions need not be mutually exclusive. In the only manipulative study that we are aware of relating to mating duration, Thornhill and Sauer (1991) manipulated the male clasping organ in a scorpion fly (figure 4.13). They found that mating duration was indeed reduced when these claspers were made inoperative, especially when males were restricted from providing salivary masses as food to their mates during mating. They concluded that the function of this male adaptation is to prolong mating duration beyond the female optimum.

Male genitalia are perhaps the most rapidly evolving morphological trait we are aware of in animals with internal fertilization (Eberhard 1985). The male intromittent organs of many organisms are variously equipped with hooks, barbs, swellings, spines, spikes, and claspers (see Eberhard 1985 for a review), and it has been suggested that these structures may sometimes function to "anchor" males to females as a means to prolong mating with resistant females (Lloyd 1979, Arnqvist 1997a, 1998, Crudgington and Siva-Jothy 2000, Schneider et al. 2001, Blanckenhorn et al. 2002; but see Eberhard 1985). Although this may be correct in many species, we are unaware of any experiments that directly test the hypothesis. Since genitalia are notoriously difficult to manipulate phenotypically, while maintaining normal function, the experiments necessary to assess current function may not be possible. One dramatic example might be the postcopulatory lock that occurs in several mammalian groups, including dogs, during which the male and female remain physically attached in a tight lock for considerable time after ejaculation (Dewsbury 1972). Dixson (1995) argued that certain features of the male genitalia cause the copulatory lock and Gomendio et al. (1998) reasoned that it may function to increase ejaculate uptake by females. However, it is not known to what extent prolongation of mating increases uptake, or whether it is at all costly to the female of the pair (Gomendio et al. 1998).

Putative female counteradaptations to male postmating adaptations may range from internal structures aimed at manipulating male genitalia to overt rejection, or resistance, behaviors during mating. Descriptive studies of functional morphology often reveal that the skeletomusculature of the female reproductive tract is complex and well developed (e.g., Heming—van Battum and Heming 1986). To our knowledge, however, it is not known in any species if this represents female counteradaptations to loosen the males' internal genital grip, or even to facilitate the grip. One way to assess whether females or males actually have the ability to influence mating duration is to manipulate the environmental conditions affecting optimal mating duration for one sex only, and determine whether observed durations change. A series of studies have attempted this, by manipulating either the benefits of mating (need for sperm, mating status) or the ecological costs of mating (starvation or predator presence) to females. Results have been mixed, sometimes in accord with simple predictions based on female economic interests (e.g., Krupa and Sih 1998, Field and Yuval 1999) and sometimes counter (e.g., Clark 1988, Cordero and Miller 1992, Wedell 1998). These experiments should be interpreted with some caution, because it can be extremely difficult to conduct a manipulation that affects the interests of only one member of the pair (Arnqvist 1997a). For example, the finding that starved females mate for shorter durations compared to well-fed females may well reflect female interests since the cost function should be steeper and the benefit function less steep (figure 4.11) in hungry females (e.g., Field and Yuval 1999, Ortigosa and Rowe 2002). However, theory on strategic male sperm allocation predicts that males should adjust copulation duration according to female condition (Galvani and Johnstone 1998), as well as a series of other factors (Parker 1998), and since starved females are likely to be less fecund (and thus yield less steep male benefit functions; figure 4.11) male optimal mating duration will also be shorter in matings involving such females. Untangling which sex is responsible for changes in mating duration is thus very difficult.

Figure 4.14. In the bean weevil *Callosobruchus maculatus*, females start vigorously kicking males with their hind legs about 3 minutes into the copulation. An average copulation lasts 4 minutes. However, if female ability to resist males by kicking is reduced, by ablation of their hind legs, matings last on average 6 minutes. (Photo by G. Arnqvist.)

Overt female resistance during mating, where females physically struggle or otherwise attempt to terminate copulation/mating, is common. One explanation for this behavior rests on indirect selection, where females struggle to extricate themselves from males as a means to sort vigorous from lower quality males, because genetically vigorous males would be better able to withstand the struggle and remain mated (Eberhard 1996). This hypothesis mirrors the "selective resistance" hypothesis discussed in section 3.4. An alternative explanation is that females struggle to terminate mating simply because prolonged mating is costly to them. Again, the best empirical evidence for this proposal comes from insects, where females of some species vigorously kick males with their hind legs at some point during copulation (figure 4.14). Female midges of the genus *Culicoides* (Diptera, Ceratopogonidae) resist mating by kicking rearward with their back legs at the male, both prior to and *during* copulation (see Linley and Adams 1975, Linley and Hinds 1975, Linley and Mook 1975, Mair and Blackwell 1996). Female kicking in these midges is correlated with a reduction in both the amount of ejaculate transferred by the male and the duration of mating, and the role of kicking has been experimentally confirmed by simply removing the female's hind legs. When female resistance is thus prevented, matings are both longer and result in larger ejaculates being transferred. Further, the hypothesis that female resistance reflects a conflict over mating duration is also supported by the fact that female resistance is condition dependent: older females and already mated females kick more vigorously at males during mating.

In the midges discussed above, it is not known whether prolonged matings are indeed harmful to females. This seems, however, to be the case in the beetle *Callosobruchus maculatus* (Coleoptera, Bruchidae). Crudgington and Siva-Jothy (2000) reported a series of experiments with this beetle, where females also use their hind legs to kick at males toward the end of copulation (Eady 1994). In this species, costs to females of prolonged mating were suspected because the "invasive" male genitalia seems to harm females (see figure 4.10). When females were prevented from kicking by ablation of legs, longer matings and more damage to the female reproductive tract was observed. Crudgington and Siva-Jothy (2000) concluded that this behavior represents a female counteradaptation in that it functions to reduce male-induced costs of prolonged mating to females, a suggestion supported by a reduction in life span experienced by females which mated more than once (but see also Fox 1993, Wilson et al. 1999, Arnqvist et al. 2004). Savalli and Fox (1998b) also demonstrated additive genetic variation for copulation duration across females, but not males, in this species. One possible interpretation of this finding is that variation in observed copulation duration is primarily determined by variation in female resistance, and that males therefore rarely reach their optimum copulation duration. Males of this species are known to transfer large numbers of sperm in each copulation, although it is uncertain whether increased copulation duration is associated with increased success in sperm competition (Eady 1995).

In summary, there are theoretical reasons to expect that the optimal mating duration of the sexes will differ, with male optima often exceeding those of females, and there is some empirical support for this assertion. There are also a series of adaptations in both males and females that appear to function to prolong or reduce the duration of mating. However, much more research is needed, addressing optimal mating durations as well as functional studies of relevant traits, before we can assess the role of sexual conflict in shaping mating duration and associated traits.

4.4 Postmating Conflicts and Male-Female Coevolution

If the examples outlined above represent sexual conflict, we would expect coevolution between resistance and persistence traits involved in postmating interactions, as have been documented for those involved in premating interactions (see chapter 3). Unfortunately, there are several reasons why it is often difficult to detect tightly correlated evolution of male and female traits in comparative studies (see Arnqvist and Rowe 2002a,b). First, even if a correlation is found, many processes other than sexually antagonistic coevolution cause correlated evolution between the sexes. Second, many of the traits involved might be difficult to detect and quantify, and this may be especially true for traits involved in postmating sexual conflict (e.g., many physiological or molecular traits). Third, it is probably rare that a given sexual conflict is entirely described by a simple coevolutionary trait-by-trait interaction. Identifying and accurately measuring all relevant traits might, needless to say, be very difficult. Fourth, there are often reasons to believe that different types of traits coevolve in the two sexes. Antagonistic physiological adaptations in males may, for example, be effectively countered by behavioral adaptations in females. In addition, traits employed may vary across species, such that an antagonistic morphological adaptation in males of, say, three related species may be matched by a behavioral female counteradaptation in one species, a physiological in another, and a morphological in a third.

For these reasons, we would expect most empirical estimates of sexually antagonistic coevolution to indicate only a weak relationship between the sexes. Nevertheless, several comparative studies have reported correlated evolution of male and female traits involved in postmating interactions. These studies have documented coevolution between various aspects of reproductive tract morphology in females and either sperm size (in insects, Dybas and Dybas 1981, Pitnick et al. 1999, Presgraves et al. 1999, Morrow and Gage 2000; and in birds, Briskie and Montgomerie 1993, Briskie et al. 1997), genital size (Ilango and Lane 2000), or other reproductive structures (Koene and Schulenburg 2003) in males. However, it should be stressed that our knowledge of how sperm size and reproductive tract morphology interact is limited, and it remains to be demonstrated whether and how sexual conflict is involved at all in their interaction (Pitnick et al. 1999, Presgraves et al. 1999). Perhaps the most convincing comparative evidence available for male-female coevolution resulting from postmating sexual conflict comes from some insects where males transfer very large ejaculates (i.e., "nuptial gifts"; see the discussion below). In any case, provided that male and female traits can be identified and sexual interactions understood, important insights into the evolutionary dynamics of sexually antagonistic coevolution will no doubt be gained by future comparative studies (cf. Arnqvist and Rowe 2002a).

4.5 Elaborated Male Ejaculates: Nuptial Gifts or Medea Gifts?

In several groups of insects, males transfer large and protein-rich ejaculates to females during mating which females either directly ingest or metabolize once deposited in the female reproductive tract (see Vahed 1998 for a review). These ejaculates can comprise 15% of male body weight in butterflies (Svärd and Wiklund 1989) and 30% in bushcrickets (Wedell 1993). It is well established that females derive direct benefits from these ejaculates in many species (see Arnqvist and Nilsson 2000 for a review), and one might well expect that females would increase mating rate as a means of accumulating these benefits (i.e., mating as "foraging"). This is, however, not generally the case. For example, females have been shown to benefit greatly from male ejaculates in a variety of butterfly genera, including *Colias* (Rutowski 1979, Rutowski et al. 1987), *Papilio* (Watanabe 1988), *Danaus* (Oberhauser 1989), and *Pieris* (Wiklund et al. 1993). Yet natural female mating frequencies in these genera average less than two per female (Svärd and Wiklund 1988, Wiklund and Forsberg 1991, Karlsson 1995). It is not immediately clear why females do not mate more frequently than they actually do in species with nuptial gifts (Arnqvist and Nilsson 2000, Wiklund et al. 2001). One possibility is that nonnutritive substances transferred in the ejaculates induce females to mate at lower than optimal frequencies.

The evolutionary origin and maintenance of nuptial gifts in insects has been the subject of many empirical studies and much debate (see Vahed 1998 for a review). The classical view is that nutritional investments by males represent either paternal investment (Thornhill 1976) or mating effort (Wickler 1985). Males of several species have been shown to benefit from producing large gifts, both in terms of increased offspring quality and quantity, and in terms of increased relative fertilization success (Vahed 1998, Arnqvist and Nilsson 2000). Although the possibility of male manipulation of females via nuptial gifts was briefly mentioned by Wickler in the 1960s (1968, pp. 218–219), Parker and Simmons (1989) first suggested that sexual conflict may be important for the evolutionary origin of nuptial feeding in insects, an idea that was further elaborated upon by Arnqvist and Nilsson (2000) (see also Simmons and Gwynne 1991, Wedell 1993, Sakaluk 2000, Vahed 2003). As mentioned earlier (see section 4.2.1.1), sexual conflict over both female remating and female reproductive rate should be common. For example, if male induction of female nonreceptivity becomes overly efficient and thus compromises female interests, we expect females to evolve resistance to the male refractory-inducing signals either by decreasing the physiological sensibility to male signals or by evolving an increased ability to neutralize the deleterious effects of male signals by metabolizing the transferred substances. In both cases, resistance would select for an increased amount of signals transferred by males, and we expect females to counteradapt with greater resistance. Actually, female ability to exploit this antagonistic interaction, by actively metabolizing (and thus neutralizing and utilizing) substances provided by males and using components of these for somatic maintenance and/or the production of eggs, may represent a key element of female resistance. This view offers an alternative explanation for the common observation that females incorporate substances provided by males into eggs or soma even in species without particularly large ejaculates (Markow and Akney 1984, Boucher and Huignard 1987, Markow et al. 1990,

Chapman et al. 1994, Eisner et al. 1996, Pitnick et al. 1997, Vahed 1998, Rooney and Lewis 1999, Markow et al. 2001). In short, females may have evolved to metabolize ejaculates as a defence against manipulative ejaculates, and may now be exploiting males. Another explanation for the appearance of male-derived substances in eggs is Smid's (1997) suggestion that certain ejaculate proteins might function to increase the rate of vitellogenesis, by binding to vitellogenin in the female hemolymph and mediating endocytosis of "tagged" vitellogenin into maturing eggs (see section 4.1.1 and Smid 1997). Such manipulative male substances would be internalized by the eggs, but can certainly not be deemed nutritive donations.

Some insight might be gained by examining the types of substances that males actually transfer, and their functions inside the female. Under the paternal investment hypothesis males would be selected to transfer maximally beneficial substances for females, such as substances directly limiting female reproductive rate (e.g., vitellins and vitellogenins or, rather, their components; Heller et al. 1998). Under the sexually antagonistic coevolutionary scenario we would instead expect to see more "manipulative" substances, such as neuropeptides which elicit responses in females that benefit males but not necessarily females. Unfortunately, relatively little detailed information is available about the precise nature of ejaculate substances in insects with elaborated ejaculates. Among butterflies, the composition and function of the ejaculate are best understood in the silkworm moth (*Bombyx mori*). The silkworm moth spermatophore is a large and complex structure (Osanai et al. 1987a), and its main function is to serve as a physiological "reactor" with a specific energy-yielding system for sperm maturation and sperm nourishment—the arginine degradation cascade (see, e.g., Osanai et al. 1986, 1987b, 1990, Kasuga et al. 1987, Osanai and Nagoka 1992, Osanai and Chen 1993). A large variety of well-characterized enzymes and their substrates are produced by male reproductive glands and are transferred to females during mating. In the spermatophore, several arginine-rich proteins and peptides are processed in a complex metabolic pathway, the end product of which is largely alanine and 2-oxoglutarate. The latter is a preferred respiratory substrate for sperm and accelerates sperm maturation. Further, several studies have shown that substances in the male silkworm moth seminal fluid cause refractiveness in females and elevate/activate female egg production (Yamaoka and Hirao 1977, Yamaja-Setty and Ramaiah 1980, Osanai et al. 1987c, 1990, Fugo and Arisawa 1992). Physiological studies of some other Lepidopterans point in the same general direction. For example, males of several species are known to transfer gonadotropins (e.g., juvenile hormone) to females during mating, and males of one species have been shown to transfer neuropeptides that stimulate the endogenous production of gonadotropins in the corpus allatum of the mated female (cf. Vahed 2003). Both types of substances elevate female egg production and thus clearly act in male interests, and perhaps female interests (see section 4.1.1). Similarly, the ejaculate of several species of butterflies contain substances which obstruct further mating of the female, either by directly and physiologically blocking the female production of male-attracting pheromones (Kingan et al. 1995, Fan et al. 1999, 2000) or by "tagging" the female with highly aromatic substances (i.e., antiaphrodisiacs; Andersson et al. 2000). These data on seminal fluid composition suggest that the ejaculates did not evolve primarily under selection for paternal investment. Substances in these large and elaborated ejaculates appear to be composed of very similar substances to those in more common smaller ejaculates. They favor males by nurturing sperm or act by eliciting various reproductive responses in females, and in no case are males known to transfer essential and maximally nutritious substances to females.

Even less is known of the exact composition of the ejaculate "gift" among insect species where females ingest the ejaculate, such as in certain Ensiferan orthopterans (Gwynne 1997), Megalopterans (Hayashi 1998, 1999), and some flies (Bonduriansky et al., 2005). The mere fact that the ejaculates in such species are "protein rich" does not offer any insights into the function of these peptides. Heller at al. (1998) studied the amino acid composition in ejaculates of five different bushcricket species, and concluded that the amino acid composition in the ejaculate differed distinctly from that in eggs. In another study, Heller et al. (2000) demonstrated the presence of carotenoids in the ejaculate of the bushcricket *Ephippiger zelleri*, and suggested that the presence of such vitamin-like substances in the ejaculate may primarily be the result of selection for paternal investment. However, this seems unlikely for two reasons. First, the carotenoids were not found in the large and protein-rich part of the spermatophore which is actually ingested by the female (i.e., the spermatophylax), which may be expected under the paternal investment hypothesis, but in the much smaller and sperm-containing part of the spermatophore which is emptied into the female reproductive tract (i.e., the ampulla). Second, carotenoids are present in high concentrations in the seminal fluid of many or even most animals where, due to their antioxidant properties, they effectively protect sperm against oxidative stress and attack. Actually, subfertility in vertebrate males is commonly associated with lowered concentrations of carotenoids in the seminal fluid (e.g., Surai et al. 1998, Omu et al. 1999).

Much more is known about the spermatophore composition in other orthopterans, particularly in locusts. In these orthopterans, a number of different neuropeptides (i.e., peptides with regulatory hormonal function) have been identified in the male accessory glands and in the spermatophore (see Schoofs et al. 1997). Among the substances that are transferred by the male to the female during mating are myotropic peptides that stimulate muscular activity in the female oviduct that is 1000 times greater than its effect in other muscular tissues (Paemen et al. 1991), specific oviposition-stimulating neuropeptides (Yi and Gillot 1999), peptides which effectively turn off female receptivity to further matings (Hartmann and Loher 1996), but also a prostaglandin-synthesizing complex (Lange 1984). As in butterflies, thus, our limited knowledge of the actual composition of orthopteran ejaculates indicates that it contains substances which favor males either directly by protecting sperm or indirectly by eliciting various reproductive responses in females. Future research may show that the ejaculates contain manipulative or regulatory substances that alter female reproduction in species where females ingest the ejaculate. Neuro-peptides that directly affect female reproduction in other insects have been found to readily penetrate the wall of the alimentary canal when ingested by females (Raina et al. 1994) and even the cuticle when applied topically (e.g., Abernathy et al. 1996).

The sexually antagonistic coevolutionary hypothesis described above is sup-ported in insects with nuptial gifts, not only by the seemingly lower than opti-mal mating rates in females (see section 4.2.1.1), but also by experimental evidence of positive dose dependency in the refractory-inducing effect of male ejaculates (e.g., Gwynne 1986, Oberhauser 1989, Wiklund and Kaitala 1995, Torres-Vila et al. 1997). The fact that the spermatophore of butterflies is gener-ally strong and has a thick and multilayered wall which is ripped open by sharp and strongly sclerotized teeth or spines in the bursa copulatrix of females (signum or lamina dentata) is also consistent with sexual conflict over the processing rate of spermatophores inside the female. These adaptations would be expected if males benefit from slowing down the processing rate of sperma-tophores and females benefit from elevating this rate. Comparative studies provide some additional support. For example, several comparative studies of butterflies have documented a positive relationship between female mating rate and the relative size of the male ejaculate (Svärd and Wiklund 1989, Gage 1994, Bissoondath and Wiklund 1995, Karlsson 1995, 1996). Comparative data also indicate that the female ejaculate processing rate, measured as the rate of decline in ejaculate weight over time since deposition in the female, may be positively related to the female remating rate (figure 4.15). This is predicted if female mating rate to some extent reflects female ability to resist male manipulation, by rapidly processing male ejaculates and thus regaining receptivity. Interestingly enough, Wiklund et al. (2001) added important new support for this prediction by showing that female reproductive output in but-terflies (measured as lifetime egg mass) coevolves closely with the rate of remating in females. If males are attempting to depress female remating rates by delivering large ejaculates that are difficult to digest and females are striving at rapidly metabolizing these in order to regain receptivity to increase remating rate, then the species-specific female remating rate reflects the balance of an-tagonistic adaptations in males and females. The current remating rate of spe-cies can then be seen as a measure of the outcome of this dynamic "arms race." The results of Wiklund et al. (2001) then indicate that as females gain a "rela-tive advantage" in this coevolutionary race, either by more rapidly processing

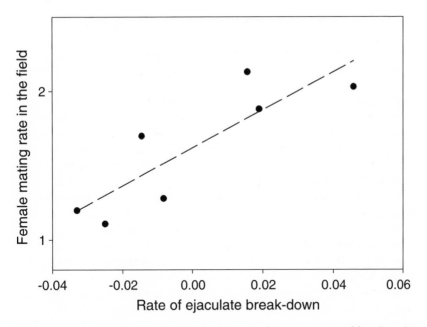

Figure 4.15. A rapid processing of male ejaculates may be a component of female resistance to male manipulation, by alleviating direct costs of ejaculatory substances and allowing females to remate sooner. Comparative data show that the rate of ejaculate breakdown in female butterflies might be positively related to the degree of polyandry across seven Pierid species. The former was determined in the laboratory, by weighing the deposited ejaculate at various times after mating, and the latter was estimated by counting spermatophores in wild-caught females. (Data from Lindfors 1998.)

ejaculates or by a reduced investment in antagonistic adaptations among males, female fitness is also elevated.

A few experimental studies are also consistent with a role for postmating sexual conflict. For example, four studies of gift-giving species (Wedell 1996, Karlsson 1998, Stjernholm and Karlsson 2000, Vahed 2003) have shown that females that receive larger volumes of ejaculate substances increase their lifetime egg production not only as a result of the larger amounts of "nutrients" transferred by males but also as a result of a higher female reproductive investment. This suggests a dose-dependent female response to male gonadotropic substances in the ejaculate, which sets the scene for male exploitation of the female reproductive system and the evolution of female resistance to male manipulations. Males could, in effect, attempt to manipulate females into allocating a higher than optimal share of resources to immediate gamete production. These conflicts over female reproductive rate could, of course, generate evolution of elaborated ejaculates by coevolutionary scenarios similar to those of the conflicts over female mating rate discussed above.

In conclusion, the term "nuptial gifts" for the protein-rich and voluminous ejaculates of many insects may be misleading, since the origin and maintenance of greatly enlarged and elaborated ejaculates may partly be explained by sexual conflict, manipulation, and resistance. For this reason, Arnqvist and Nilsson (2000) suggested that such ejaculates should perhaps be seen as rather sinister "Medea gifts" rather than amiable "wedding cakes" (see Michiels 1998, for a similar argument in hermaphrodites). However, since the relative importance of postmating sexual conflict will depend on female remating rate, one would expect that selection generated by the divergent interests of the sexes would diminish if females were completely monandrous. Under monandry, it is therefore much easier to understand how males might benefit from elevated offspring quantity or quality, since the level of paternity assurance is extremely high (cf. Wickler 1994). This in turn suggests that both selection for paternity investment and selection generated by antagonistic co-evolution may contribute to the maintenance of large and "protein-rich" ejaculates in males (see Wiklund et al. 2001). The relative importance of these, then, should depend on the level of female polyandry, so that the former dominates under monandry while the latter increases in importance with the degree of polyandry. This might also be reflected in the types of substances that make up these "gifts."

4.6 Are Male Postmating Adaptations Costly to Females?

Much of the discussion above hinges upon the assumption that various male adaptations incur costs to females. Unfortunately, direct evidence for costs to females from male adaptations that further male postmating interests is very scarce and often ambiguous or indirect. This may reflect the fact that female counteradaptations would tend to "hide" such deleterious effects, but also the fact that it is often easier to study measurable traits in males than to quantify fitness effects in females of various responses to such traits. Several concrete examples of costs to females have been discussed earlier in this chapter, and others will follow, but there are also a couple of more general points which deserve attention here. First, female postmating reproductive behavior in animals (remating, egg production, etc.) is physiologically regulated by an intricate endogenous regulatory system, which must be presumed to be the result of natural selection and thus to be adapted to respond appropriately to various extrinsically derived stimuli (including those provided by males). We have stressed in this chapter that such complex regulatory systems present males with a suite of exploitable preexisting biases in females. Male exploitation of any of these pathways is likely, at least initially, to move females off their natural selection optima. The fact that male ejaculate substances are often the same as (or mimics of) the signals which females produce endogenously to regulate their reproduction (see Eberhard 1996 and above) suggests that exploitation by males may be common. We feel that this is a viable general argument, and thus predict that resistance to male-derived stimuli should be a common component in the evolution of female reproductive receptors.

Second, the nature of female adaptations, both structural and physiological, often suggests that these have been selected at least in part to resist male stimuli. This indicates that stimulation is sometimes costly to females (i.e., represents manipulation). A potential example of male exploitation and female resistance to manipulation not discussed earlier in this book concerns the physiological short-term regulation of female egg laying in crickets and locusts. Prostaglandins are known to stimulate female reproduction in many insects, and it seems that male insects have frequently exploited this regulatory pathway. In many orthopterans, males produce an enzyme (prostaglandin synthease) in their reproductive glands that is transferred with their ejaculate to females, and converts arachidonic acid (a fatty acid) in the female to prostaglandin (Loher et al. 1981, Tobe and Loher 1983, Lange 1984, Murtaugh and Denlinger 1985). Since prostaglandin has a dose-dependent effect on female reproduction (Destephano and Brady 1977, Stanley-Samuelson and Peloquin 1986), the benefits to male crickets from thus increasing the short-term reproductive response of their mates are obvious. However, females of at least some species appear to resist this stimulation. For example, females of the cricket *Teleogryllus commodus* have evolved a compact cuticular lining of their genital chamber which apparently reduces the uptake or diffusion of substances from the ejaculate into the blood stream (Sugawara 1987), although radio-labeling studies have shown that this obstacle does not totally prevent uptake (Stanley-Samuelson et al. 1987). Males of this species, however, have apparently elevated the sophistication of manipulation since they not only transfer prostaglandin synthease to females, but have also added considerable amounts of the actual fatty acid substrate on which this enzyme acts to their ejaculate blend (Stanley-Samuelson and Loher 1983, Stanley-Samuelson et al. 1987). Because of this, high rates of prostaglandin biosynthesis occur already in the female genital tract (Loher et al. 1981, Stanley-Samuelson et al. 1987) and thus derived prostaglandins also appear in the female hemolymph (Ai et al. 1986, Stanley-Samuelson et al. 1987). Interestingly enough, female *Teleogryllus commodus* are able to quite efficiently reduce prostaglandin titers by eliminating prostaglandin from their circulatory system by specific excretory pathways (Stanley-Samuelson and Loher 1985), possibly in part representing a "second line of defence." It is worth noting, finally, that these insects are included among those where the ejaculate is considered to constitute a "nuptial gift" to females (see Vahed 1998).

Another type of observation indicating costs to females comes from comparative studies showing that female physiological responses to mating often evolve rapidly, as revealed by species-specific or interpopulation differences in the reproductive responses of mating to males of various forms. These differences constitute evidence for rapid evolution of female response to male ejacu-

Figure 4.16. Female Zeus bugs are equipped with a pair of dorsal glands, producing a secretion that, upon drying, creates two oblong, whitish patches (see arrows). Males, which are much smaller than females, lack these glands and ride females almost permanently. Experiments using radio-labeled females have shown that males feed on these secretions. The evolution of sex-role-reversed nuptial feeding in Zeus bugs may involve an unusual form of sexual conflict. (Copyright CSIRO 2001. Reprinted from Andersen and Weir 2001, with permission from CSIRO Publishing, Melbourne, Australia; photo by N. M. Andersen.)

late substances (Andrés and Arnqvist 2001), and females have in several cases been shown to respond more weakly to males with which they are coevolved than to other males. This is equivalent to female resistance to males and such divergence has been attributed to sexually antagonistic coevolution by several authors (Wilson et al. 1997, Price 1997, Clark and Begun 1998, Parker and Partridge 1998, Clark et al. 1999, Andrés and Arnqvist 2001, Knowles and Markow 2001, Hosken et al. 2002, Nilsson et al. 2002, 2003). However, this interpretation should be treated very cautiously as there may be other explanations for the pattern (see section 6.5.2).

A unique mating system, with unique costs to females, is found in the remarkable Zeus bugs (Heteroptera, *Phoreticovelia* sp.). In these little semi-aquatic insects, females, but not males, are equipped with a pair of dorsal glands that produce a waxlike secretion (Polhemus and Polhemus 2000, Andersen and Weir 2001). Males are much smaller than females and ride on top of females for extended periods: mean mating duration in the laboratory is over one week (see figure 4.16). By using females that had been fed radio-labeled food, Arnqvist et al. (2003) were able to show that males feed on the glandular secretions produced by the females' dorsal glands. The glands are located in the region where the male mouthparts reside during the male's extended association with the female. This unique case of sex-role-reversed nuptial feeding is made even more remarkable by the fact that females start producing this secretion already as juveniles (in their fourth and fifth larval instars) and that adult males start riding females in their juvenile stages. When females molt into adulthood, males simply crawl over from the cast skin of the larva to the newly emerged female and continue riding. The most puzzling aspect of this mating system is, of course, why females spend energy continually feeding males on their backs. Because females can store viable sperm for several weeks, and because Zeus bugs live in fairly dense populations with male-biased adult sex ratios, it seems unlikely that females are paying for an insurance against a shortage of viable sperm. Females simply do not need to mate very frequently to maintain a normal fecundity and fertility, and finding a mate when they do need sperm does not seem to be a problem (Arnqvist et al. 2003). A peculiar observation instead suggests that sexual conflict may be involved. Females, but not males, carry a variable number of scars on their back (0–50), and these are again located in the region where the male mouthparts are held during riding. Preliminary experiments have also shown that females carry more scars when kept with males compared to when kept alone (Arnqvist et al., unpublished). One astonishing conclusion of these observations is that males may be cannibalizing their mates. This could then have selected for "damage control" in females, provided that feeding males with glandular secretion is inexpensive compared to being pierced by males. According to this scenario, a male that is guarding a female has two options when faced with hunger: to leave the female in search for food or to literally take a bite out of his mate. Because of strongly male-biased adult sex ratios (i.e., intense competition among males for mating opportunities) and because the costs to females of being cannibalized may be minor, both of which result from the extreme sexual size dimorphism (figure 4.16), the second option may be more beneficial to males than the first. Seen this way, sex-role-reversed nuptial feeding may represent a female counteradaptation to a sexually antagonistic adaptation in males (i.e., sex-role-reversed sexual cannibalism). Future experimental work on Zeus bugs will, hopefully, reveal whether this hypothesis is supported.

4.7 It Takes Two to Tango: Sexually Antagonistic Coevolution in Fruit Flies

Fruit flies of the genus *Drosophila* may not be the most attractive animals in the world, but they have served as a model system in numerous biological disciplines. The study of sexual conflict after mating is no exception. Many of the questions and points discussed earlier in this chapter are perhaps best illustrated by studies of fruit flies, in particular *D. melanogaster*. Mating in fruit flies induces a series of changes in female physiology and behavior, including decreased attractiveness to other males, decreased receptivity to further matings, elevation of egg laying, storage and utilization of sperm, and decreased life span (see Chen 1996, Wolfner 1997, 2002, and Chapman 2001 for reviews). Most of these changes should obviously benefit males, but it is less clear whether they benefit females. Thanks to a very active research program employing the emerging tools of cell and molecular biology (Swanson 2003), we now have a relatively thorough understanding of the proximate causes of these effects. Most changes in female behavior and physiology are caused by a variety of seminal substances transferred to females during mating. These substances are peptides or proteins, produced in the accessory reproductive glands of males and termed Acps. To date, over 20 different genes coding for Acps have been characterized, and it is estimated that *D. melanogaster* males produce between 80 and 100 different Acps! The majority are relatively simple peptides that pass through the vaginal wall, bind to various receptors, and act as neuropeptides (Lung and Wolfner 1999, Lung et al. 2002).

To establish the effects and function of all male Acps is, needless to say, an enormous task. Nevertheless, the function of a few Acps in females is relatively well understood (see table 4.3). For example, the sex peptide Acp70A has several functions in females (Chen 1996, Wolfner 1997, Chapman 2001, Liu and Kubli 2003), one of which is to elevate egg production by stimulating the endogenous production of juvenile hormone (JH) in the female corpora allata (figure 4.2). The seminal fluid molecule Acp26Aa also acts to elevate egg laying, but this is a myotropic peptide which binds to and stimulates muscular activity in the female ovaries (Heifetz et al. 2000). Thus, while Acp70A stimulates the long-term production of eggs by the ovary, Acp26Aa exerts its effect by a more discrete increase in ovulation and egg deposition (Wolfner 1997). The largest known Acp is a glycoprotein, Acp36DE, which facilitates the uptake and storage of sperm. Acp36DE is unusual in that it localizes to the oviduct wall, close to the opening of the female sperm storage organs, where it apparently acts by "corralling" sperm into this strategic position (Wolfner 1997, Chapman et al. 2000).

Drosophila males, therefore, affect female reproduction in many ways that are beneficial for the males, but are these seemingly manipulative substances in any way costly to females? Two separate lines of evidence show that at least some of these substances are indeed detrimental to females. First, a few experiments have revealed such costs. Fowler and Partridge (1989) originally showed that mating itself reduced female life span in *D. melanogaster*, under conditions in which both the costs of egg production and exposure to males were controlled. Using transgenic males which produce and transfer very little seminal fluid but normal amounts of sperm, Chapman et al. (1995) were then

Table 4.3.
Effects of male accessory gland products on female behavior and physiology in *D. melanogaster*.

Effect in females	Male benefit	Acp responsible
Decreased attractiveness to other males	Reduced risk of sperm competition	Unknown
Decreased receptivity to further matings	Reduced risk of sperm competition	Acp70A
Increased egg production and egg laying	Increased number of offspring fathered	Acp70A, Acp26Aa
Increased sperm storage	Increased sperm competition success	Acp36DE
Associated with sperm displacement	Increased sperm competition success	Acp26Aa, Acp29AB, Acp36DE, Acp53Ea
Life span decrease	Inhibition of female "immuno-response" to ejaculate	Acp62F
Decreased fertility	Increased rate of ovulation in females	Acp26Aa
Formation of a mating plug	Reduced risk of sperm competition	PEB-me

Sources: http://flybase.bio.indiana.edu/ and references in text.

able to convincingly attribute the reduction in female life span to the transfer to the females of male accessory gland products. It seems that this "toxic" effect to females arises at least in part as a pleiotropic side effect of Acps that affect male defensive ability in sperm competition, since Civetta and Clark (2000) showed that female postmating survival is negatively genetically correlated with male defensive success in sperm competition. More recently, the illuminating experiments of Lung et al. (2002) singled out Acp62F as a candidate gene contributing to this deleterious effect. In fruit flies as in most other animals, male sperm and accessory substances are short lived in females, as they are attacked and digested by proteolytic enzymes in the female reproductive tract. The gene Acp62F codes for a protein which is one of many protease inhibitors (Wolfner 1997, 2002, Chapman 2001) transferred to females and may well protect male seminal proteins or sperm from attack by female enzymes. Various protease inhibitors are also found in the seminal fluid of many other organisms, including humans (see Hocini et al 1995, Kurpisz and Fernandez 1995). Lung et al. (2002) found that the Acp62F protein reduced the life span of adult flies when ectopically expressed, and suggested that the toxic effects of this protein may be due to deleterious interference with essential somatic proteolytic events in females because a fraction of the male-derived

Acp62F proteins passes through the vaginal wall into the females' hemolymph. It thus seems that females may be paying an incidental mortality cost as a result of male substances which protect male ejaculates against female enzymatic attack. A quite different cost was discovered by Chapman et al. (2001), who were able to document that the transfer of Acp26Aa carries subtle fertility costs to females by using mutant males. Evidently, Acp26Aa is overly efficient and causes the release of eggs from the ovaries by recently mated virgin females prior to sperm being stored. As a result of this premature ovulation, these eggs are inefficiently fertilized, an effect also noted by Prout and Clark (2000). All of these costs to females seem to represent incidental antagonistic side effects of substances whose primary effects are beneficial for males. There is currently no support for the suggestion that *Drosophila* males transfer substances which do nothing but cause harm to females (cf. section 4.2.1.3 above).

A second line of evidence for costs to females comes from two artificial selection experiments in *D. melanogaster*. Given that males transfer seminal fluid substances that act in males' interest and that these may not always coincide with those of females, we expect sexually antagonistic coevolution between seminal traits in males and female responses to these. As noted earlier, these coevolutionary chases can be very difficult to detect in comparative studies. In an ingenious artificial selection experiment, utilizing novel genetic techniques in *Drosophila*, Rice (1996b) was able to reveal a coevolutionary chase by preventing females from counteradapting while allowing males to adapt to a static female phenotype. When females were thus prevented from evolving, adapting males evolved an "upper hand" relative to control males in only 30 generations, by achieving more matings with previously mated females, through a reduced remating of the mates of adapting males and/or by an increased efficiency in sperm defence. Interestingly enough, females also suffered a greater reduction in life span when mating with adapting males, suggesting that the male beneficial adaptations had come at a cost to females. Rice (1996b) suggested that the seminal fluid of adapting males had become more toxic to the static females.

Postmating sexual conflict in polygamous mating systems basically stems from the fact that lifetime fitness of the two mates may differ dramatically, and what pays the male may not pay the female (Rice 2000). However, this is not true for cases in which a male and a female are each others' only lifetime partners (i.e., strict genetic monogamy). In such cases, we would actually expect sexually mutualistic coevolution as the fitness of the mates are equal and correlated across pairs. In another elegant artificial selection experiment, Holland and Rice (1999) exploited this fact by enforcing lifelong monogamy (no sexual conflict) in the otherwise promiscuous *D. melanogaster*. After only 35 generations, they found that test females mated to monogamous males lived longer, and had a greater net fitness, compared to those mated to polygamous control males. Males from the monogamous lines also evolved reduced persistence in courtship. Further, females from the monogamous lines showed higher mortality rates compared to polygamous females when housed with polygamous control males. In essence, the experiment of Holland and Rice (1999) is the reverse of that of (Rice 1996b) in that it demonstrates that when selection for sexually antagonistic adaptations is experimentally removed, both male persistence and female resistance are rapidly reduced. Given that sexually antagonistic adaptations should represent an "adaptational load" on both sexes and thus lower absolute fitness (Rice 1992), it is interesting to note that the net reproductive rate was markedly higher in monogamous lines compared to polygamous lines. We note, however, that results with regards to the population level fitness effects of imposed monogamy are mixed (Pitnick et al. 2001a, Martin and Hosken 2003b, Holland 2002). These two artificial selection experiments are marvellous illustrations of the potency of sexually antagonistic coevolution, and also provide evidence for the critical role of direct female costs in this process. Follow-up experiments by Pitnick et al. (2001a,b), using the lines of Holland and Rice (1999), supported a role for seminal substances in males and resistance to these in females. They confirmed that monogamy-line males evolved ejaculates that were less effective in inducing female refractiveness to mating, and that females mated to monogamous males had a higher offspring production compared to those mated to polygamous males. Pitnick et al. (2001a) also found that monogamy males evolved a smaller body size, suggesting that male body size may at least in part be mediating the observed antagonistic effects (see also Pitnick and Garcia-Gonzalez 2002, Friberg and Arnqvist 2003).

Given the unique background understanding that we have about the genetics and adaptive significance of postmating conflict in fruit flies, looking more closely at the evolution of the genes involved becomes an interesting task. Many such studies have now been performed, and they reveal three patterns: the genes involved are often highly polymorphic, several are exposed to selection, and they show rapid evolutionary change. It is clear from electrophoretic studies of gonadal proteins and from sequencing the genes themselves that genes encoding for Acps are generally more polymorphic within species than are nonreproductive proteins and genes (e.g., Begun et al. 2000, Tsaur et al. 2001). For example, ten different wild-type alleles of Acp26Aa have been described (Aguade et al. 1992) and no less than 39 of Acp29AB (Aguade 1999). At the protein level, it is also clear that reproductive proteins evolve faster than do other proteins (e.g., Civetta and Singh 1995). Similarly, several molecular genetic studies of closely related *Drosophila* species have also shown that divergence in Acp genes is generally more rapid than in other genes (e.g., Begun et al. 2000, Swanson et al 2001b; but see Fan et al. 1999, 2000), and this divergence has often been accelerated by selection (e.g., Tsaur and Wu 1997, Tsaur et al. 1998, Begun et al. 2000). The fact that Acp genes are often affected by selection is also validated by the fact that Acp genotype correlates with sperm competition success in fruit flies (Clark et al 1995).

Given these genetic patterns, what can be said about the processes involved in the evolution of Acp genes? Well, at a first glance, process and pattern seem to be somewhat contradictory. The fact that Acp genes are polymorphic within species seems to be incompatible with observations of strong selection on these loci, since strong selection should tend to exhaust such genetic variation. The most likely process by which such polymorphism could be maintained in the face of selection is balancing selection (Prout and Clark 1996). In particular, if there is negative frequency-dependent selection on Acp alleles, rare alleles would generally be favored and polymorphism could be maintained. What scenario could generate such negative frequency dependency? The first clue was given by Clark and Begun (1998) who demonstrated that there is extensive genetic variation among female fruit flies in their propensity to manifest sperm displacement. Hence, there is clearly genetic polymorphism among genes in females that determine how particular male Acps alleles are received and processed. The second clue was presented shortly thereafter by Clark et al. (1999), who showed that male and female genotypes interact with one another in determining the outcome of sperm competition, so that a given male genotype who was "best" with one female genotype did "worst" with another. These data, although not conclusive, are consistent with negative frequency dependence and can also explain the generally very low heritabilities of sperm competition success observed in fruit flies (e.g., Gilchrist and Partridge 1997). Further, Clark et al. (1999) found that males did poorly overall with their own females, suggesting that females are relatively resistant to males of their own populations. A similar pattern has been reported for several other insects as well (see section 6.5.2). These observations are consistent with a frequency-dependent "arms race," driven by postmating sexual conflict (Chapman and Partridge 1996a, Clark et al. 1999, Chapman 2001, Wolfner 2002). One can envision a scenario where certain male Acp alleles are more or less efficient at eliciting female responses depending on the specific corresponding receptor alleles in females. If reproductive responses are indeed costly to female fruit flies, female genotypes that are resistant to common male genotypes will be favored by selection. This, in turn, will generate selection for rare Acp alleles among males, so that Acp allele frequencies essentially "track" the evolution of female resistance. This could in theory lead to cyclic and perpetual limit cycles of allele frequencies within populations. While such a scenario of polygenic frequency-dependent selection can explain both the polymorphism observed (cf. Hamilton 1993) and the interactions between males and females, it fails to account for the rapid long-term evolution observed in the genes involved (Chapman 2001). In order to do so, we must assume that novel mutations occasionally enter into the system of interacting male and female alleles (Chapman and Partridge 1996a, Rice 1998a), thus changing the rules of the game. Such novel alleles would theoretically spread rapidly in a population, and affect selection on several interactive genes. The picture that emerges is complex and dynamic: male Acp alleles and female receptor/response alleles may be involved in a polygenic, cyclic, and frequency-dependent "arms race," which is periodically displaced and perturbed by the invasion of novel alleles and/or genes. Despite "an apparently unchanging sexual interaction, male and female [fruit flies] may be running frantically to stand still" in evolutionary time (Chapman and Partridge 1996a).

5

Parental Care and Sexual Conflict

In the two preceding chapters, we have argued that males and females often have conflicting interests over the initiation of mating, the interactions during mating, and the termination of mating. When either parent cares for eggs or offspring, with increased offspring fitness as a consequence, we refer to this behavior as parental care (see Clutton-Brock 1991). Parental care comes in many different forms, ranging from short-term protection from predators to the extended periods of nurturing and caring for altricial young observed in several vertebrate groups. At first glance, one might think that the relationship between male and female parents is free of conflict, given that both share a common genetic interest in their offspring and often appear to cooperate in raising their young. Yet this period can be a fleeting and uneasy alliance. Selfish behavior of one parent is often favored rather than punished, and parental care is really the scene of competing selfish behaviors by parents. It is true that the genetic interests of both parents in their offspring may be equal, but the genetic interests in each other may be near zero. It therefore comes as no surprise that the evolution of parental care is riddled with sexual conflict and a resulting evolutionary tug of war between males and females.

The study of parental care is one of the cornerstones of behavioral ecology, and there are many excellent general discussions of the subject as well as more focused reviews (e.g., Gubernick and Klopfer 1981, Zeh and Smith 1985, Adiyodi and Adiyodi 1990, Clutton-Brock 1991, Davies 1992, Baur 1994, Rosenblatt and Snowdon 1996, Sargent 1996, Westneat and Sargent 1996). Moreover, a recognition of conflicts between parents over care has played a central role in the development of this field (e.g., Trivers 1972, Dawkins 1976, Maynard-Smith 1977, Davies 1992, Westneat and Sargent 1996, Lessells 1999). It is, therefore, not our aim to provide a synthesis or detailed treatment of this very large field. Here we will introduce sexual conflict over care, and discuss a few illuminating cases of these conflicts.

5.1 The Basic Conflict

Sexual conflict over parental care rests on the fact that care is generally costly in terms of decreased future survival and reproduction of the caregiver (see Clutton-Brock 1991, Roff 1992, Rosenblatt and Snowdon 1996, Sargent 1996)

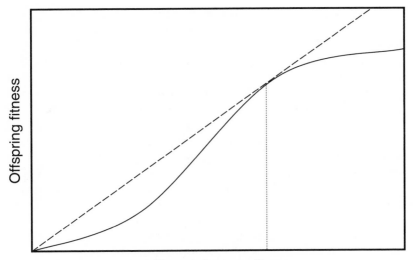

Figure 5.1. Relationship between the fitness of individual offspring and total parental expenditure. The optimal parental expenditure is denoted by the dotted line, above which parents should produce more young rather than invest more heavily in each. This model does not incorporate costs of parental care in terms of decreased future parental survival and/or reproduction, but nevertheless illustrates the general idea that there is some optimal level of parental care. (After Smith and Fretwell 1974.)

(figure 5.1). Individuals thus face trade-offs, where expenditure on parental care in a particular reproductive event must be traded against its cost to future reproduction. The best option for an individual male or female would always be to have someone else do their job in parental care. Even though alloparental care occurs in many groups (i.e., assistance in parental care by others than the parents, such as helping, and cooperative or communal breeding; see, e.g., Jennions and MacDonald 1994, Cockburn 1998, Wisenden 1999), the chief candidate for this role is usually the other partner. Sexual conflict over parental care occurs simply because each partner should generally prefer the other to "work harder," and the evolution of increased parental care by one sex will open up possibilities for reduced parental care in the other.

Until the early 1970s, ethologists focused on the common interests of the sexes, viewing males and females as "forming associations in which they cooperate in an altruistic manner" (e.g., Eibl-Eibesfeldt 1970). This view emphasizes their shared genetic interests in offspring, but ignores the obvious fact that the mates typically share no genetic interests in one another (Dawkins 1976). In the same vein, David Lack (1968) suggested that natural selection will always favor increased male parental care and cooperation whenever two parents can raise more young than can the female alone. Trivers (1972) recognized that parent's interests differ, and laid the foundations of our current theory of parental care.

Drawing on Bateman's (1948) pioneering work on the sexual difference in limits on net reproductive rate, Trivers reasoned that "even when ostensibly cooperating in a joint task male and female interests are rarely identical." Trivers' contribution was important in several ways (see Clutton-Brock 1991). For example, he suggested that adaptations that benefit one sex at the expense of the other (e.g., mate desertion) favor counteradaptation in the other sex to avoid being exploited. This scenario captures the essence of sexually antagonistic coevolution. Following Trivers, our understanding of mating systems and parental care has slowly and gradually shifted toward a framework based more on the conflicting interests of the sexes (see Clutton-Brock 1991). These ideas were strongly advocated by Dawkins (1976), and by Maynard-Smith (1977) who pioneered theoretical modeling of sexual conflicts over parental investment and parental care. In particular, Maynard-Smith (1977) stressed that the optimal parental care behavior by one parent will depend on the behavior of the other, and so applied game theory to tackle the problem. Following Maynard-Smith and important contributions by Houston and Davies (1985) and Parker (1985), most subsequent modeling of the evolution of parental care has been based on such an ESS approach (figure 5.2) (but see Wachtmeister and Enquist 2000, Wade and Shuster 2002).

In the following sections, we first discuss sexual conflict over who the caregiver should be in the first place. We then turn to conflict over the amount of care each sex should provide, first in species with variable mating systems and then in monogamous species.

5.2 Mate Desertion

Before Trivers (1972), cases of uniparental care were simply thought of as those where offspring did not benefit from having an additional parent providing care (e.g., Lack 1968). However, by recognizing that the caregiver pays all the costs for benefits that are shared between parents, Trivers reasoned that the caring sex may suffer from the desertion of their mates.

5.2.1 CONFLICT OVER CARE AND DESERTION IN UNIPARENTAL SPECIES

In most species with parental care, care is uniparental; only one parent tends the offspring (Clutton-Brock 1991). In general, we expect to see uniparental care in species where care by two parents is less than twice as beneficial to offspring as care by one (Maynard-Smith 1977, 1978). The distribution of parental roles varies considerably among taxa, with female-only care being most common among arthropods, reptiles, and mammals while male care predominates among fish and anurans. The question of which sex should

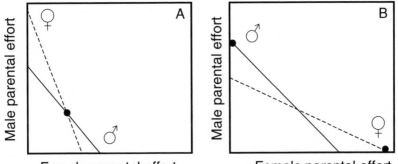

Figure 5.2. In general, when one sex evolves to increase its investment in parental care, the other is expected to decrease its current investment in favor of future reproduction and/or survival. This figure shows optimal female (dashed) and male (solid) parental expenditure plotted against the level of expenditure provided by the other sex. Game-theoretic analyses of such biparental games show that the conditions for evolutionary stable biparental care are fairly restrictive; the optimality functions must cross and the slopes of the functions at the point of intersection must be smaller than −1. This is the case in (A), where a single evolutionarily stable strategy for both parents exists (filled circle). If the lines do not cross at all, or if they do cross but at least one of the slopes is larger than −1 at the point of intersection (see, for example, B), then uniparental care by either sex will be the evolutionarily stable strategy. (After Chase 1980, Houston and Davies 1985.)

care and which should desert has consequently received a lot of attention in the literature.

Trivers (1972) suggested that the likelihood of mate desertion at any point in time should be related to past investment in offspring. This was criticized by Dawkins and Carlisle (1976) and by Boucher (1977), both of whom pointed out that future returns rather than past investment should determine this decision (the infamous "Concorde fallacy"). Both Trivers (1972) and Dawkins and Carlisle (1976), however, suggested that mate desertion may place the deserted sex in a "cruel bind": the deserting sex will place its partner in a difficult situation, since to desert after one's partner has already deserted is to condemn the offspring to death. Dawkins and Carlisle (1976) suggested that the sex first given an opportunity to desert should tend to be the deserting sex, and used this logic to explain the pattern of male and female uniparental care among taxa. In animals with internal fertilization, females are constrained to be the last in possession of the zygote and hence end up in the caring role. In animals with external fertilization, where females first deposit eggs which are then fertilized by males, females are in a situation to place males in a cruel bind, and uniparental male care of offspring is much more widespread. They thus argued that sexual conflict over parental care would determine the relative roles of the sexes, and sexually antagonistic adaptations in both sexes that would tend to give one sex an earlier opportunity to desert would be favored.

As elegant as it may seem, this idea has very little if any empirical support. Most importantly, there is no apparent association between the timing of gamete release and subsequent role in uniparental care of the sexes across various taxa of externally fertilizing species (Gross and Shine 1981, Gross and Sargent 1985, Beck 1998).

The apparent association between the fertilization system and sex roles in parental care may instead be related to the relative costs and benefits of parental care for the sexes, which may generally differ between externally and internally fertilizing species (see Clutton-Brock 1991 for a discussion). Sexual conflict therefore does not generally seem to be responsible for the evolution and maintenance of unilateral mate desertion in uniparental species.

In species with uniparental care where the relative parental roles are fixed, the timing of uniparental desertion may vary widely and there may nevertheless be sexual conflict over the timing of mate desertion (Székely et al. 1996). We would expect stronger conflict in the early phases of the parental care period, when both the cost of being deserted and the benefits of deserting are generally higher (Lazarus 1990, Székely and Williams 1994).

5.2.2 NEVER TRUST A PENDULINE TIT!

The relative parental roles of the sexes can vary within some uniparental species so that either sex may end up caring for the offspring (amphi- or ambisexual parental care; Maynard Smith 1977, Clutton-Brock 1991, Székely et al. 1996). However, even in these species, variation in various ecological factors affecting the relative trade-offs of the sexes, rather than the relative timing of opportunities for desertion, often seems to determine which sex ends up in the caring role (e.g., Blumer 1986, Bourne 1998). Nevertheless, sexual conflict over the caring role may certainly affect which sex cares, and cruel bind scenarios may contribute to the relative role assignment. One particularly interesting and well-studied case is the penduline tit (*Remiz pendulinus*) (figure 5.3). Breeding commences as in most other birds. The male begins to build an elaborate, roofed, and pendulous nest, and may or may not be joined by a female at any stage of construction. Incubation and offspring care are strictly uniparental, but either sex may care for the eggs and young (see Persson and Öhrström 1989, Franz 1991, Haupt and Todte 1992, Todte 1994, Valera et al. 1997). The female cares for the offspring in approximately 50–70% of all initiated breedings, while the male adopts the caring role in some 5–20% of the cases. The truly amazing aspect of the breeding system of the penduline tit is that these numbers do not add up to 100%: in 30–35% of all initiated pairings, both parents desert the nest with the eggs simply left to die! The deserted broods contain on average 3–4 apparently viable eggs, so this phenomenon does indeed represent a huge "waste" to both parents.

Figure 5.3. The roofed and pendulous nest of the penduline tit is primarily built by the male. Incubation and offspring care are uniparental, with the female caring in some 50–70% of the cases and the male in 5–20%. However, in 30–35% of all initiated breedings, both parents desert the nest with the eggs which are simply left to die. (Photo by J. Elmelid.)

What mechanisms could possibly maintain such an astonishing mating system? A closer look at the domestic affairs in penduline tits gives some hints. First, both sexes can potentially gain by deserting, if their partner continues to care for their offspring, because both sexes can remate and initiate other successful breedings. For example, Persson and Öhrström (1989) found that females that laid only one clutch, which they incubated and cared for, produced on average 4.6 fledglings, whereas those that deserted at least one clutch prior to caring for one produced a more variable but higher average number of fledglings (5.8). Similarly, some 25–35% of all males succeeded in securing a second female after deserting an initial mate, and thus increased their reproductive success by being polygynous (Persson and Öhrström 1989, Franz 1991, Valera et al. 1997). Second, among nests deserted by both parents, Persson and Öhrström (1989) found that males were often observed for at least one day after the last sighting of the female, but that the reverse situation never occurred. This observation suggests that females first desert nests that are subsequently deserted by both parents. Third, clutches incubated by females are considerably larger (average ≈ 5 eggs) than those incubated by males (average ≈ 3.5 eggs), and the latter do not differ from those deserted by both parents (Persson and Öhrström 1989, Franz 1991, Haupt and Todte 1992, Todte 1994). This pattern also holds true within polyandrous females, so that early clutches deserted by females are smaller than those cared for later by the same females (Persson and Öhrström 1989). Fourth, breedings that involve female care are initiated earlier in the season than are those involving male care, and the frequency of desertion of both parents decreases over the season (Persson and Öhrström 1989, Franz 1991, Haupt and Todte 1992).

These data collectively suggest that the bizarre mating system of penduline tits is indeed the result of a sexual conflict over parental care, where both sexes prefer the other to care for the clutch and so are simultaneously trying to achieve polygamy (Persson and Öhrström 1989, Franz 1991, Valera et al. 1997). Females seem to be the first to desert the offspring in clutches which are subsequently deserted by both parents. In some 30–50% of pairs, these females apparently leave the male with an "incomplete" clutch to care for, hoping that the abandoned male will actually care for the offspring. Data show that the probability that these females will be able to initiate a second (or even a third) breeding elsewhere is very high, due to a male-biased operational sex ratio. However, males that are caught in Dawkins and Carlisle's (1976) cruel bind most often desert the clutch, and it is primarily later in the season, when the remating probabilities for males are very low due to a limited availability of unpaired females, that deserted males tend to accept the caring role. Males desert females first in some 50–70% of pairs, and it seems as if females who find themselves placed in a cruel bind only very rarely desert their offspring.

Odd as it may seem, game-theoretical analyses have demonstrated that parental care systems such as that of the penduline tit can be evolutionarily stable, given a feedback between the frequency of deserters of both sexes in the population and the changing fitness payoffs of deserting to both sexes during the season (Webb et al. 1999, Székely et al. 2000; see Székely et al. 1996 for a review of the theory on mate desertion). Lazarus (1990) modeled situations where the desertion strategy of one parent depends on whether or not the other parent has already deserted, and showed that allowing for this can lead to preemptive desertion, where each parent is "racing against the clock" to leave the other in a cruel bind.

Given that mate desertion in penduline tits is so obviously costly for the deserted male or female that is "left with the tab," we expect both sexes to exhibit sexually antagonistic adaptations to promote their interests (Trivers 1972). Once again, it seems that Dawkins and Carlisle's (1976) focus on the timing of desertion is playing a role. For males, deserting a female too early means either that the female will fail to initiate egg laying altogether or will not complete the clutch, or that fertilization opportunities, especially for the last eggs in a given clutch, are lost to other males in the population (Schleicher et al. 1997). On the other hand, deserting too late will mean that valuable breeding opportunities with other females are lost and that the risk of being left in a cruel bind by the female increases. Under these circumstances it is critical for males to collect information on the temporal progress of reproduction (i.e., clutch status), in order to time their desertion decisions optimally, and we would, therefore, expect males to pay a close interest in the clutch status over the laying period. In accord, males stay close to and frequently enter the nest during the egg-laying period (Schleicher et al. 1993, Valera et al. 1997).

For females, deserting too early might reduce the probability of males staying to care for offspring while deserting too late will increase the risk of being left in a cruel bind by the male. Females lay the eggs and so are ultimately in control over information regarding clutch progress over time, and would clearly benefit from concealing information about egg laying in order to manipulate the male into underestimating the progress of the clutch. Females have apparently evolved two behavioral strategies in order to achieve this goal. First, females become highly aggressive to males that approach the nest during egg laying. Females that are present when their mate tries to enter the nest during this period aggressively attack the male, and are highly successful in preventing males from entering (Valera et al. 1997). Aggression between the mates during this phase can be very intense indeed, and observations in aviaries have shown that it can result in serious injury and even male death (Loehrl 1990). Second, most females conceal eggs by carefully burying one or more eggs in the bottom of the nest. Should the male enter the nest during the early phase of laying, he may be deceived and fail to gather reliable information about the clutch progress.

Egg burial behavior occurs in several birds, but a series of illuminating observations and experiments by Valera et al. (1997) have convincingly demonstrated that egg burial and mate aggression in penduline tits both likely represent sexually antagonistic female adaptations, aimed at increasing the probability of successfully leaving the male in a cruel bind. First, the longer the females kept the eggs covered, the longer the male stayed at the nest. Second, when eggs were experimentally uncovered, female aggression toward males near the nest increased dramatically but males nevertheless tended to desert females on the very same day. Third, females that carefully covered their eggs were more likely to succeed in becoming polyandrous than were those not covering their eggs carefully. Fourth, females ceased covering their eggs when males were experimentally removed. Thus, male and female penduline tits indeed seem to be involved in a conflict over control of information pertinent to the timing of mate desertion (see also Lazarus 1990), and both sexes have apparently evolved antagonistic behavioral traits which favor their interests.

5.3 "Partial" Mate Desertion and Sexual Conflict over the Mating System in Biparental Species

The examples discussed above refer to mating systems in which the deserting parent completely ceases care for the offspring at the time of desertion. However, among species with biparental care, a very common situation is that one of the parents in a pair seeks and takes opportunities to simultaneously engage in reproduction with an additional mate. The first mate, then, will find him- or herself having to share the parental care of his or her mate with a competitor of the same sex. It has long been recognized, among monogamous and biparental species, that both sexes can potentially gain from reproducing with more than one individual of the opposite sex (e.g., Orians 1969). This forms the basis for sexual conflicts over fidelity. All else equal, a female would benefit from having two or more males assisting her in caring for the offspring (polyandry). In contrast, a male would benefit from having two or more females caring for his offspring (polygyny) (figure 5.4). In theory, however, the sex with less parental investment has more to gain from such "adulterous" behavior (Bateman 1948). It therefore should come as no surprise that polygyny is more common than polyandry among biparental species with a variable mating system (Clutton-Brock 1991, Ligon 1999).

Why, then, should a female ever accept reproducing with an already "occupied" male? During the pre-Trivers (1972) era, the harmonious view of reproduction dominated, and scholars assumed that polygynously mated females must be compensated for the cost of reduced parental care with some other direct benefit (Verner 1964, Verner and Willson 1966, Orians 1969). Weather-

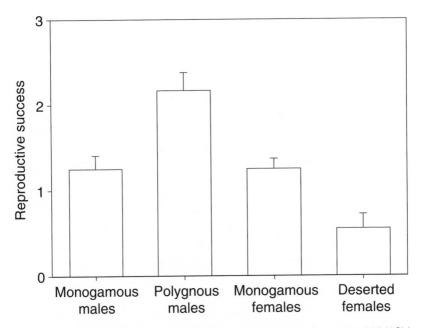

Figure 5.4. In the biparental and predominantly monogamous fish the convict cichlid (*Cichlasoma nigrofasciatum*), males aid females in protecting the eggs and young from predators. However, caring males frequently seek additional breeding opportunities and occasionally desert their females to spawn with a second mate, in which case the offspring of the first female receive little paternal care. Experimental data on the reproductive success of males and females show that there is sexual conflict over the mating system in this species. While polygynous males gain from attracting additional females, their partially "deserted" mates suffer reduced reproductive success. (Data from Keenleyside 1985, Keenleyside and Mackereth 1992.)

head and Robertson (1979) subsequently expanded this idea to include indirect benefits. The "polygyny threshold" model suggests that females adaptively choose the best possible option when settling with a mate. If the "quality" of a male is high enough, in terms of either associated direct resources (e.g., territory quality) or indirect genetic benefits (i.e., "sexy sons"; see Kirkpatrick 1985), the female may do as well or even better by becoming a secondary female to an already mated male as by initiating a monogamous breeding with an unoccupied male. This seemingly simple model has proven surprisingly hard to support with data (see Davies 1989, Alatalo and Ratti 1995, and Bensch 1997 for discussions), despite a massive amount of attention, primarily on birds (Orians' original contribution [1969] is cited in more than 700 journal contributions during the period from 1986 to 2001). These data from avian mating systems are largely correlational, and although the patterns are mixed (Pribil and Searcy 2001) most studies indicate that secondary females do not produce nearly as many fledglings/recruits as do simultaneously settling monogamous females (see Slagsvold and Lifjeld 1994 and Ligon 1999 for reviews). The differences in immediate reproductive success between these groups of females are frequently as great as 20–50%. These studies collectively suggest that there is commonly a substantial direct cost of polygyny to females, which may actually often be underestimated since some potentially detrimental future effects of polygyny (elevated maternal care effort and decreased offspring condition as a result of reduced total amount of care) to female fitness are not incorporated into this cost. Several studies also show that this results largely from shared/reduced parental care, and that females are not compensated for this cost by direct benefits.

Many other explanations for the occurrence of apparently costly polygyny in females have been offered, but most represent extensions or modifications of the original "polygyny threshold" model (Alatalo and Ratti 1995, Bensch 1997). A fundamentally different hypothesis was proposed by Alatalo et al. (1981), who suggested that female pied flycatchers (*Ficedula hypoleuca*) may actually be manipulated into breeding polygynously. Inspired by the work of Trivers (1972) and Dawkins (1976), they assumed sexual conflict in the mating system, and reasoned that males may have evolved a set of sexually antagonistic and deceptive behaviors that makes is difficult for females to correctly assess a male's breeding status (for example, establishing a second territory at some distance from the first: polyterritoriality). Females, in turn, were thought to have evolved counteradaptations, such as vigilance during the pair bonding and/or preoviposition period (Dale and Slagsvold 1994), to reduce the probability of deception. Thus, Alatalo et al. (1981) argued that sexual conflict over parental care generates sexually antagonistic coevolution, involving a variety of behavioral traits in the two sexes. Polygynous pairings might tend to represent cases of successful male manipulation of females, and monogamous breedings successful resistance by females (see also below).

This "female deception" hypothesis was originally met with some skepticism, perhaps because students of avian mating systems had difficulty accepting that individuals of one sex could be manipulated by the other into making suboptimal and costly decisions. The "female deception" hypothesis has more recently received some support in the bird literature (e.g., Catchpole et al. 1985, Searcy et al. 1991, Ratti and Alatalo 1993, Korpimäki 1994, Sandell and Smith 1996). For example, studies have documented an increased probability of attracting a second mate with an increasing distance between the first and the second territories. The results from other studies are less supportive (e.g., Dale and Slagsvold 1994, Slagsvold and Drevon 1999). In sex-role-reversed birds such as phalaropes, it has been suggested that females instead might conceal their reproductive status from males (i.e., "male deception"; see Whitfield 1989).

Given that costs of polygyny are well documented in birds, a central question is whether indirect genetic benefits can somehow compensate for these direct costs (Alatalo and Ratti 1995). While indirect benefits certainly have been documented in other contexts (see Jennions and Petrie 2000 for a review), the magnitude of such benefits required to offset the often large direct costs of polygyny to females would seem prohibitively high (Alatalo and Lundberg 1986; see also section 2.3). However, we are unaware of any relevant and long-term "economic" studies where simultaneous measures of direct costs and indirect benefits of polyandry have been performed. An exception is the collared flycatcher (*Ficedula albicollis*), where the direct cost of polygyny is substantial (some 25–35%). Here, indirect genetic benefits do not compensate for this cost. On the contrary, the detrimental effects of reduced care during the nestling phase seem to overwhelm whatever sexiness and genetic vigor may have been present in offspring of polygynous females: daughters of secondary females are more likely to become secondary females themselves, and sons of secondary females are less likely to succeed in becoming polygynous, compared to offspring of either primary or monogamous females (L. Gustafsson, personal communication).

Finally, we note that discussions of the "female deception" hypothesis for the occurrence of polygyny in the face of the cost of polygyny has been limited to polyterritorial species, where males locate territores far apart. However, it is possible that females more generally can be manipulated into making suboptimal reproductive decisions (e.g., Enquist and Arak 1993), such as breeding as the secondary female of a polygynous male even if the female potentially has access to "information" about the male's breeding status. Concealing their breeding status by being polyterritorial may merely be one of many sexually antagonistic adaptations in males that further their interests in sexual conflicts over paternal care. Various forms of male signals could, for example, be used to manipulate prospective mates into making suboptimal pairing decisions. Comparative data documenting correlated evolution between various male signals and social mating systems in birds are in general agreement with this possibility, but such patterns have been interpreted in a different framework (e.g., Owens and Hartley 1998, Dunn et al. 2001). In any case, sexually antagonistic coevolution may play a more central role in the origin and maintenance of polygyny in biparental species than had earlier been believed. On the other hand, recent studies showing that females are sometimes compensated for the costs of polygyny (e.g., Pribil and Searcy 2001) illustrate the fact that no single model is likely to explain all cases of polygyny even among biparental birds (see also Ligon 1999).

Irrespective of the strategy used by a paired male to attract a second female, and irrespective of whether or not it is in the best interest of "secondary" females to become polygynous females, it is clear that sexual conflict between primary females and monogamous males should be virtually ubiquitous in species with biparental care. It is obviously in the primary female's interest to monopolize the reproductive effort invested by her mate, and consequently we expect the evolution of sexually antagonistic adaptations aimed at keeping her mate from attracting additional females. These antagonistic adaptations appear to be common among animals with variable mating systems. Primary females, for example, often interfere with male mate attraction by soliciting superfluous copulations (Petrie 1992, Eens and Pinxten 1995, 1996) or, more commonly, by directly attacking or otherwise obstructing displaying males (e.g., Davies 1992, Trumbo and Eggert 1994, Kempenaers 1995). In the burying beetle *Nicrophorus defodiens*, for example, males attract females using a pheromonal signal and provide parental care and a food resource (carrion) for the developing offspring. Following copulation with a female, males often resume pheromone emission, but primary females interfere by various more or less aggressive behaviors directed toward mates attempting to attract additional females (Trumbo and Eggert 1994) (figure 5.5). Females mount and/or bite males, and may also displace them from their signaling perches. Experimental tests using tethered primary females, rendered unable to obstruct male signaling, have demonstrated that physical interference by females, to some extent, imposes monogamy on males (Eggert and Sakaluk 1995). Perhaps like the pied flycatchers, male beetles often choose less accessible sites at some distance away from the carcass for polygynous signaling (Trumbo and Eggert 1994).

It is also conceivable that females might be able to keep their mates from attracting additional females by elevating their own current reproductive effort, thus increasing the optimal paternal investment. For example, Komdeur et al. (2002) found that male European starlings (*Sturnus vulgaris*) mated to females with enlarged clutches spent less time attracting secondary females and consequently were less successful in doing so. The extent to which increased reproductive effort by females contains elements of sexual antagonism and manipulation, however, will depend on the relative fitness payoffs to both sexes by such elevation. It should also be noted here that Barta et al. (2002) suggested that females may be able to manipulate males into providing care by strategically reducing their own energy reserves available for care, rather than elevating their reproductive effort. By keeping their reserves below those required for successful uniparental care, females may render mate desertion a less viable option for males.

A more direct strategy that primary females commonly do use to keep their mate from attracting additional females is aggression toward those females, particularly during the pair formation period. These behaviors seem especially common in birds with facultative polygyny. For example, male starlings can attract one to four mates and clearly gain by being polygynous. In contrast, already mated females suffer a cost if their mates attract additional females because of reduced parental care. Experimental tests have convincingly shown that primary females do use aggression toward other females to

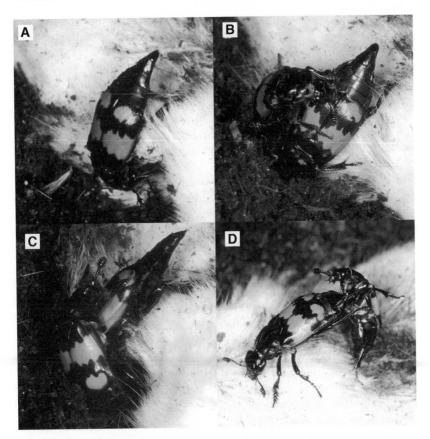

Figure 5.5. Male burying beetles, *Nicrophorus defodiens*, emit pheromone signals in a characteristic "headstand" position (A) to attract additional females. Their mates, however, interfere with this behavior (B and C). In (D), the male (left) is kicking back at the interfering female as she attempts to bite him (note the flared mandibles of the female). (Photos by A. K. Eggert.)

prevent these secondary females from establishing a pair bond with their mates (Sandell and Smith 1996, 1997). Further, experimental data also suggest that variation in aggression from the primary female, rather than in male courtship behavior, determines whether a female will settle with an already mated male (Sandell 1998). These and other experimental studies on both birds (e.g., Veiga 1990, 1992, Slagsvold 1993, Kempenaers 1995, Liker and Székely 1997, Slagsvold et al. 1999) and fish (Walter and Trillmich 1994, Schradin and Lamprecht 2000, Kokita and Nakazono 2001, Kokita 2002) collectively demonstrate that aggression toward other females can be an efficient way for the already paired female to impose monogamy on her mate. However, among harem-holding cichlids, males of some species have evolved a counteradaptation to this sexually antagonistic female strategy: by physically intervening in aggression by harem females directed toward new potential "recruits" to the harem, males are able to keep the harem together and even increase its size against the interests of resident females (Schradin and Lamprecht 2000, Walter and Trillmich 1994).

In summary, among species with biparental care, sexual conflict over the mating system is very well documented in many different groups of animals, including birds, fish, and insects. Empirical data often show that the social setting of reproduction is a major determinant of the reproductive success of both sexes in such species. Commonly, males gain by simultaneously reproducing with many females (polygyny) while females gain by monopolizing the parental care invested by their mate. There are also several examples of adaptations in males which aid in attracting more females and of counteradaptations in females which make it more difficult for males to achieve polygyny. This evidence shows that males are sometimes able to achieve polygyny against the interests of the females involved, while females are also often able to effectively restrict male polygyny. In most cases, however, we know relatively little about traits in both sexes which might be involved in such sexually antagonistic coevolution, and most empirical results relate to behavioral strategies. The potential for sexual conflicts over parental care to more generally fuel coevolution between male signals and female resistance to these has not been explored empirically to our knowledge. Needless to say, the realized mating system in any given situation will be influenced by a range of different ultimate and proximate factors. Several authors have stressed that variation within and between populations/species in biparental mating systems should be seen as the combined result of the relative efficacy of antagonistic adaptations in males and females (relating to several sexual conflicts), other sexual selection processes, and ecological factors (e.g., Davies 1989, 1992, Westneat and Sargent 1996, Alonzo and Warner 2000b). While more traditional views focus on variance in the external ecological setting as the force driving mating system dynamics (Verner and Willson 1966, Orians 1969, Emlen and Oring 1977, Arnold and Duvall 1994), recent data suggest that sexually antagonistic coevolution of male and female traits may internally generate evolution of mating system components, even in the absence of any change in the ecological setting (figure 5.6).

5.4 Sexual Conflict over the Relative Amount of Care in Biparental Monogamous Species

In species where there is strict, lifelong genetic monogamy, there is no room for sexually antagonistic adaptations, since the fitness of the partners is perfectly correlated across breeding pairs (Rice 2000). However, when there is

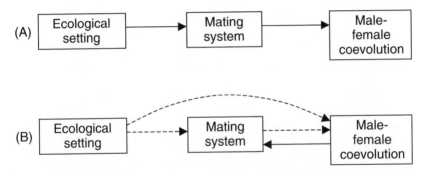

Figure 5.6. Traditional views of mating system evolution (A) suggest that the external ecological setting (distribution and amount of resources, climate, interspecific resource competition, etc.) dictates the prevailing mating system. The mating system then sets the "rules" for any male-female coevolution that may occur. In contrast, alternative views (B) hold that male-female coevolution, for example in the form of sexually antagonistic coevolution, may internally generate evolution of mating system components even in the complete absence of any changes in the ecological setting. The ecological setting is thought to have a weaker effect on mating system dynamics, and may also affect such dynamics indirectly by influencing costs of male and female coevolutionary adaptations. Viewed in this way, male-female coevolution is more a driver of mating system evolution than an indirect consequence of changes in the ecological setting.

even a slight deviation from this condition, each sex should prefer the other to work harder in caring for the offspring. Therefore, even in socially monogamous species, where both provide care to only the shared offspring in a given reproductive season, we may expect conflict. If one mate can somehow manipulate the other into making a larger parental care effort, this opens up an opportunity for reduced care and consequently increased future reproduction for the manipulator. Sexual conflict among socially monogamous species over the amount of parental care invested has received much less attention, but there are reasons to believe that it has affected the evolution of parental behavior in many socially monogamous species. If this is true, we expect to see various adaptations in both sexes which aim at shifting the parental work load toward the other sex.

Parental care behavior of both sexes is regulated by various endogenous hormones (see Rosenblatt and Snowdon 1996 for reviews). In vertebrates, the hormones with the greatest influence on parental care are lactogenic hormones, but estrogens, progesterone, testosterone, and several other hormones are also involved. Several experimental studies, on a variety of species, have shown that injection with prolactin induces and increases incubation behavior, and affects several other parental behaviors in birds (see Buntin 1996). For example, prolactin-treated female willow ptarmigans (*Lagopus lagopus*) exhibit markedly elevated levels of a series of different brood defence behaviors (Pedersen 1989). Interestingly, this prolactin-mediated increase in parental risk taking also translated into greater reproductive success: prolactin-treated females produced more than twice the mean number of chicks produced by control females. Similarly, paternal care in the form of offspring feeding behavior is markedly reduced in passerine birds when males are given exogenous testosterone, but increased when they are given antiandrogens (see Ketterson and Nolan 1994, Buntin 1996). The reduction in paternal care is apparently a result of increased time spent on other activities, such as mate guarding and territorial defence. Again, several experiments have shown that androgen-induced reduction of paternal care is associated with a decrease in fledgling survival. Clearly, altered levels of hormones can have dramatic effects on parental care and, ultimately, fitness.

The production of hormones that regulate parental care behavior is triggered by a series of external stimuli. These involve abiotic cues, such as photoperiod and temperature, but also a range of social stimuli. For example, several experimental studies of birds have shown that visual, auditory, olfactory, and tactile stimuli, emitted by the offspring, elevate prolactin titers in their parents (see Buntin 1996). An obvious additional possibility is that signals emitted by their mates affect the titers of parental care hormones such as prolactin. While this is very well established in mammals (see Wang and Insel 1996, Ziegler 2000), empirical research on birds addressing this is more limited (Buntin, personal communication). Several experimental studies have shown that plasma levels of various steroid hormones are directly affected by signals emitted by their partners (e.g., Bluhm et al. 1984, 2000, Shields et al. 1989, Wingfield et al. 1989, Wingfield 1994, Massa et al. 1996), and the results of a few studies indicate that prolactin may not be different. Studies of pigeons (*Columba livia*, Patel 1993) and ring doves (*Strepropelia risoria*, Friedman and Lehrman 1968) both showed that prolactin secretion was markedly reduced in individuals that were separated from their mates, unless they were allowed to view and hear their incubating mates from across a glass partition. To the extent that parental care hormones are affected by signals emitted by their partners, there is an opportunity for mutual "sensory exploitation" between the partners in monogamous species with biparental care. An adaptation in one sex that successfully manipulates individuals of the other sex into elevating their parental care behavior beyond what is optimal for the receiver will be favored by selection, provided that elevated signaling is not too costly. We then expect individuals of the other sex to evolve resistance to such stimuli, leading to sexually antagonistic coevolution between signals and response to these.

This basic scenario has been used to explain the evolution of courtship behavior in socially monogamous birds by Wachtmeister and Enquist (2000) and Wachtmeister (2001). In these species, both sexes often engage in courtship rituals that may extend far beyond pair formation into the entire breeding period. Explanations for such extended display behavior cannot rest entirely on mate choice. Classical justifications are instead based on the ethological concepts of either strengthening the "pair bond" or synchronizing reproductive behavior of the mates (e.g., Eibl-Eibesfeldt 1970), but it is unclear if these notions can be framed in a rigorous evolutionary framework (Birkhead et al. 1987, Wachtmeister 2001). These display behaviors have also been suggested to be honest signals of parental "quality," which would allow both sexes to adjust their parental effort adaptively according to the direct or indirect benefits provided by their mates (see section 3.1.3 above). Wachtmeister and Enquist (2000) and Wachtmeister (2001) instead argued that, rather than conveying information beneficial to the reciever, courtship behavior may have evolved to manipulate the partner into making a greater investment in parental care. Using neural network simulations, Wachtmeister and Enquist (2000) demonstrated that courtship signals in males can evolve to exploit response biases in the "neural machinery" of females and to successfully manipulate females into breeding earlier than is optimal for females, even in the presence of evolving female resistance. Interestingly enough, endocrinological studies have shown that poor within-pair stimulation leads to depressed hormone levels, and seems to be a key factor in whether or not individual birds delay breeding (e.g., Schoech et al. 1996). The simulations of Wachtmeister and Enquist (2000) suggest that various courtship rituals and other signals may evolve as one sex exploits perceptive biases in the other in an effort to affect the outcome of sexual conflict over parental care, and that adaptations in one sex may thus depress the long-term reproductive output of the other. In a subsequent review of display behavior, Wachtmeister (2001) concluded that sexually antagonistic coevolution best explains the temporal pattern, context, and form of display observed among socially monogamous birds.

One curious effect of antagonistic "negotiations" between breeding parents over parental care is that it may actually lead to reduced amounts of total parental care per offspring in biparental species (Parker 1985), if parents show high responsiveness to any deficit by the partner (McNamara et al. 2003) or if there are parental costs of "negotiations" which translate into reduced care. Offspring may, therefore, suffer reduced fitness resulting from sexual conflict between parents. This possibility was recently examined by Royle et al. (2002) in a study of zebra finches where the parental workload of single females was compared to that of pairs, when the number of offspring per parent was constant. They found that offspring received greater per capita parental investment from single females than from both parents working together. Likewise, offspring of single parents fared better in subsequent reproductive competition. Thus, in zebra finches, offspring do indeed seem to suffer from sexual conflict between their parents.

Slagsvold and Lifjeld (1989) suggested that part of the sexual conflict over parental care may be mediated by the age and size distribution of the offspring in monogamous species. How even aged the offspring in a clutch or litter are may affect the relative parental contribution of the mates. Slagsvold et al.

(1994, 1995) tested this hypothesis in blue tits (*Parus caeruleus*). By experimentally manipulating the degree of hatching asynchrony among eggs within nests, they showed that males contributed more to parental care in even-aged broods. Most importantly, the postbreeding survival of breeding females increased when offspring were even aged, while the reverse was true for males. In blue tits, females alone incubate the eggs and create relatively even-aged broods by delaying incubation until most eggs in the clutch have been laid. To some extent, this behavior can thus be seen as a sexually antagonistic adaptation aimed at manipulating males into providing more paternal care (Slagsvold et al. 1994, 1995).

In summary, we suggest that various male and female "signals" in biparental species may function at least in part to elevate the parental care invested by their mates, contrary to their best interests. Sexual conflict over parental care, and subsequent sexually antagonistic coevolution of signals and perception of these, may have contributed significantly to the diversity of displays observed among socially monogamous species. Obviously, the evolutionary profitability of such manipulative strategies will depend not only on the costs of the "signals" but also on how easily resistance to exploitation of parental effort evolves in the exploited "victim." For those who might doubt that parental care can be manipulated, consider that interspecific brood parasites, such as the cuckoo (*Cuculus canorus*) and the brown-headed cowbird (*Molothrus ater*), have obviously succeeded in a more dramatic and highly effective exploitation of their hosts' investment in parental care. Likewise, conflicts of interest between parents and their offspring over care are common to many species. Here, it seems that offspring have frequently succeeded in manipulating their parents into providing parental care above the parental optimum (see Trivers 1974, Mock and Parker 1997).

5.5 The Dunnock: Family Life in Cambridge University Botanic Garden

In the early 1980s, Nick Davies set out to study the mating system of an inconspicuous passerine bird: the dunnock (*Prunella modularis*). Inspired by the work of Trivers (1972), Davies was determined to understand the maintenance of the curious mating system of this little bird. The extraordinary research on the dunnock reported by Davies and his coworkers during the last 20 years (reference is to Davies 1992, unless otherwise stated) has perhaps taught us more about sexual conflict over parental care and its consequences than has research on any other system, and the family life of dunnocks illustrates many of the topics discussed in this chapter.

The mating system of the dunnock is unusually variable: even within a single small population, one may find monogamous pairs, a single male breeding with two different females (polygyny), a single female nesting with the aid of two or even three males (polyandry), and commonly even more complex associations involving two or three males sharing several females (polygynandry). On Davies's study site, in the Cambridge University Botanic Garden, about equal portions of the females were involved in monogamous, polyandrous, and polygynandrous breedings, while polygyny was rare. Thus, most males find themselves sharing, with other males, the reproductive output of one or two females. When this occurs, one male (the alpha male) becomes socially dominant to the other (the beta male). While alpha and beta males both attempt to keep intruding males away, this becomes an uneasy alliance as soon as females become receptive to copulations. The copulation frequency in dunnocks is extraordinarily high, averaging 1–2 copulations per hour during the ten-day mating period. Males spend a significant proportion of their time guarding the female during this period: alpha males spend about 80% of their time within 5 meters of the female, while the corresponding figure for monogamous males is some 50%. Competition for copulations between the alpha and beta males is obvious and can be intense. Copulations are actively "solicited" by the female, who initiates copulations by a precopulatory display, involving exposing her cloaca to the male who then pecks at it for about a minute. During this cloacal pecking, females frequently eject a droplet of previously deposited ejaculate, perhaps providing space for new. Interestingly, polyandrous and polygynandrous females often seem to attempt to "escape" the guarding alpha male and to actively solicit copulations with beta males. They do this by fleeing from the guarding alpha male, responding to song from the beta male, actively soliciting copulations from beta males when encountered, and producing a male-attracting song when alpha males are absent (Davies et al. 1996, Langmore and Davies 1997). However, females show no interest in copulating with neighboring males, having only been observed soliciting copulations from resident males. As we will argue below, these data suggest that females use copulations to induce beta males to provide paternal care.

Dunnocks exhibit biparental care. However, the extent to which males contribute depends to a large degree on their social role and on their perceived level of paternity. The feeding rate of alpha males is comparable to that of monogamous males, but beta males tend to feed the young only if they have had access to the female during the copulation period. A beta male's share of paternal care has been shown to increase with his share of the copulations, and fingerprinting analyses have shown that this also correlates well with his share of the genetic paternity. In polygynandrous groupings, the males involved are very discriminating with regard to which brood they feed, and choose to feed the offspring of the female with which they have had a greater share of the copulations. Thus, polyandrous and polygynadrous dunnock males follow a clear-cut principle when investing parental care: the more a given male has copulated with a female, the more he contributes to feeding the young.

The reproductive output of dunnocks follows a very simple rule: the more collective effort that is spent on parental care, the more offspring are fledged, and the higher is the weight of those fledglings. The productivity of polygynous breedings, where two females share the paternal care effort of a single male, is about half of that of polyandrous breedings where both the alpha and the beta male contribute to feeding the offspring (see table 5.1). The data on individual reproductive success in dunnocks beautifully illustrate one of the main themes of this chapter: the reproductive success of individuals varies with social setting, and the mating system that yields the highest fitness for individuals of one sex does so because it successfully exploits the parental care of the other. As would be expected from simple logic, data on dunnocks show that the reproductive success of polygynous males is about twice that of polyandrous males, while the exact opposite is true for females (table 5.1).

Given such intense sexual conflict over the mating system, we expect both sexes to evolve traits that further their interest in this conflict. In dunnocks, many behavioral traits can be interpreted as sexually antagonistic adaptations. First, females are very aggressive toward other females, both prior to and after initiating breeding. From the female's point of view, this can be seen as an attempt to monopolize the paternal care of at least one male. Rivalry and aggression between polygynandous females is frequently so intense that it results in one female deserting her clutch, which is clearly costly to the males involved. As in harem-holding cichlids (see above), however, dunnock males physically intervene in aggressive encounters between females, in what can be seen as a male counteradaptation to neutralize the deleterious effects of female-female aggression. Second, since male dunnocks provide care in relation to copulation frequency, we expect females to exploit this by luring males into elevated copulation rates. The copulation rates in dunnocks are extraordinarily high, as a result of frequent female solicitation. It is also clear that polyandrous females "encourage" copulations by the beta males. Since copulation with the beta male increases the probability of receiving additional parental care from the beta male, this behavior is clearly beneficial for females. For the alpha male, in contrast, this behavior is costly because of shared paternity, and because of the requirement of intense mate guarding and aggression toward the beta male during the copulation period.

Davies and coworkers have also performed an in-depth comparative study of the mating system of the congeneric alpine accentor (*Prunella collaris*), which is almost exclusively polygynandrous and breeds in groups of up to four males and four females (Davies et al. 1995, 1996, Hartley et al. 1995, Langmore et al. 1996). The mating system of the alpine accentor is fairly similar to that of polygynandrous dunnocks, and sexual conflict is also apparent, in that females gain through increased male help by giving shared paternity to several males, while alpha males do best by monopolizing all copulations.

Table 5.1.

Tremendous variation of postbreeding survival, reproductive output, and reproductive success per individual (calculated as seasonal averages of the number of young fledged weighted by paternity for males) with the mating system in dunnocks.

Mating system	Frequency among females	Postbreeding survival of females	Postbreeding survival of males		Number of young fledged per nest	Female reproductive success	Male reproductive success
Polygyny (♂ ♀♀)	7%	—	—		1.43	3.8	7.6
Monogamy (♂ ♀)	32%	38%	67%		1.95	5.0	5.0
Polygynandry (♂♂ ♀♀)	38%	48%	57%		1.29	3.6	alpha = 5.0 beta = 2.2
Polyandry (♂♂ ♀)	23%	52%	45%	Only alpha ♂ feed offspring	1.21	4.4	alpha = 4.4 beta = 0
				Both ♂♂ feed offspring	2.95	6.7	alpha = 3.7 beta = 3.0

Source: After Davies 1992.

Many behavioral adaptations of males and females are also shared between the two species.

To summarize, it is clear that the optimal mating system differs between the sexes in dunnocks. The detailed observations made by Davies and his coworkers during the last 20 years have painted a remarkably complex and variable picture, where sexual conflict is at the heart of male-female interactions. Davies (1992) suggested that the variable mating system of the dunnock reflects the different outcomes of conflicts between the sexes, sometimes resulting in males achieving their favored mating system (polygyny) and sometimes females (polyandry with several males feeding the young). Monogamy represents cases where neither sex has succeeded in attracting a second mate, and polygynandry may be seen as a "stalemate." The relative ability of any given male and female to achieve their preferred mating system is obviously dependent on the effectiveness of a series of behavioral adaptations and counteradaptations in both sexes, but also to some extent on ecological conditions. The work by Davies and his colleagues has also provided two additional insights. First, the mating system dynamics in dunnocks and alpine accentors can be understood solely in terms of direct fitness effects to males and females, and there is no evidence in either species that differences in genetic "quality" between members of either sex affect the reproductive decisions made. Second, the variable mating system seems to be maintained by sexual conflict, through adaptations in both sexes that promote their interests, rather than by ecological factors (cf. figure 5.6).

6

Other Implications of Sexual Conflict

In chapters 3–5, we have discussed reproductive conflicts between the sexes based on whether they occur prior to, during, or after mating. However, there is also a range of topics that fall somewhat outside this classification scheme but are nevertheless highly relevant to this book. In this chapter, we will discuss some of the more important of these topics, and will also look a bit more closely at the potential role of sexual conflict in the evolution of reproductive isolation.

6.1 The Evolution of Genomic Imprinting

In species in which there is strict and lifelong genetic monogamy, the total genetic contribution to the mutual offspring produced (and hence to fitness) is equal in males and females within each monogamous pair. A mutation that benefits one individual in a pair at the expense of his/her mate is therefore inconceivable. However, whenever there is not lifelong genetic monogamy, various sexual conflicts will open up due to relatedness asymmetries between males and females to offspring. While a female will always be equally related to all of her offspring, each male will not be equally related to these offspring, since offspring will have different fathers either within or across litters and reproductive episodes. In the previous chapter, we emphasized that when both parents contribute to the care of the offspring, both should generally prefer the other to invest above its optimum in parental care for these reasons.

When mothers alone provision the offspring in nonmonogamous systems, there may be sexual conflict over the flow of maternal resources to individual offspring. Let us assume that the optimal maternal investment in each individual offspring, for the female, is X. A lower investment will be punished because of poor offspring quality, while a higher investment will lower future reproductive success of the female. A male who fathers any one of her offspring will benefit from maternal investments that are greater than X (closer to the offspring's optimal level), simply because this particular male has no genetic interest in offspring he does not father. As a consequence, a mutation which causes a maternal investment that is higher than X to the offspring bearing it will be favored by its father, despite any net costs experienced by its mother and present/future maternal half-sibs. To put things crudely, genes acting to "crank up" maternal provisioning to individual offspring will be favored when residing in males, while selection will favor genes in females which resist this (offspring provisioning at a lower [X] and more evenly distributed rate).

This form of sexual conflict has been invoked to explain the evolution of one of the oddest genetic "quirks" known: genomic imprinting. Genomic imprinting is the unequal expression of genes in offspring depending on the sex of the parent from which they were inherited. The maternal copy of a gene may, for example, be effectively "silenced" while the paternal copy may be expressed normally (see Ohlsson et al. 1995, Reik and Surani 1997, Ohlsson 1999). This form of parent-of-origin-specific expression leads to a non-Mendelian inheritance and even monoallelic expression of genes, and the evolution of genomic imprinting is puzzling considering the benefits of diploidy (e.g., buffering against deleterious recessive alleles; see Otto and Goldstein 1992). Haig and Westoby (1989) originally suggested that the evolution of genomic imprinting can be understood in the light of sexual conflict over offspring provisioning in those species where offspring are nourished directly from maternal tissues, such as in most mammals and flowering plants. Haig and coworkers have subsequently elaborated on the original idea in a series of papers (e.g., Haig and Westoby 1991, Moore and Haig 1991, Haig 1997, 2000). This ingenious hypothesis is based on the fact that genes that affect the amount of resources transferred from a mother to a particular offspring will do so at the expense of other current and future offspring of that female (see above). A mutant allele which causes elevated rates of maternal provisioning only when it is paternally derived (i.e., it is imprinted) will be favored by selection. Therefore, genes that code for embryonic/fetal growth promoters (GP) should tend to be paternally expressed but maternally silenced. As a result of female resistance to superoptimal provisioning, however, we would expect to see genes coding for products which cause embryonic/fetal growth suppression (GS) to tend to be maternally expressed, but paternally silenced. Moore and Haig (1991) described this evolutionary process as a parental "tug of war," and the hypothesis has since been labeled the conflict hypothesis for the evolution of genomic imprinting. Set in a larger framework, however, this hypothesis is firmly rooted in Trivers' (1972, 1974) original treatment of sexual conflicts over parental investment (see chapter 4). But here, conflict between males and females over maternal care is played out in the interactions between mothers and their growing offspring, leading to different forms of antagonistic adaptations and counteradaptations (i.e., maternally and paternally derived genes within the offspring). Maternal silencing of GP's, as well as the evolution of GS's, can then be seen as the evolution of female resistance to male manipulation of embryonic growth, while paternal silencing of GS's can be seen as escalated manipulation by males (see Wilkins and Haig 2001).

What is the current evidence in favor of the conflict hypothesis for the evolution of genomic imprinting? Both theory and observation support the connection between sexual conflict over provisioning and the evolution of genomic imprinting. Mathematical models have confirmed that sexual conflict favors imprinting of GP and GS genes, in particular when the relatedness asymmetries between males and females to offspring are great (Haig 1992, Mochizuki et al. 1996, Hurst 1999, Kondoh and Higashi 2000, Wilkins and Haig 2001). As would be expected, imprinting of GP and GS genes has so far been described only in placental mammals and flowering plants, in which offspring provisioning from maternal tissues occurs. It does not seem to occur in taxa where the potential for this type of sexual conflict is absent, such as in nonplacental mammals and birds (O'Neill et al. 2000, Killian et al. 2001, Nolan et al. 2001, Yokomine et al. 2001). Moreover, genes that affect early offspring growth are strongly overrepresented among the genes that we currently know are imprinted in mammals (Hurst and McVean 1998, Tilghman 1999). Perhaps most importantly, the parental pattern of genomic imprinting in mammals is such that maternal or paternal expression at imprinted loci is typically associated with reduced or increased maternal provisioning of offspring, respectively (Moore 2001). The "direction" of the imprint seen is often what would be expected under the conflict hypothesis for the evolution of genomic imprinting. For example, the insulin-like-growth-factor/insulin pathway is an important regulatory pathway for fetal growth in mammals. This regulatory pathway contains several imprinted genes in several different species. The list includes a variety of paternally expressed GP genes and maternally expressed GS genes, but the most compelling example may be the interaction between the products of Igf2 and Igf2r known to be imprinted in mice and humans (see Bartolomei and Tilghman 1997, Tilghman 1999). Igf2 codes for an insulin-like fetal growth factor, and only the paternally derived copy is expressed in most tissues. Igf2r codes for an antagonistic receptor to the products of Igf2, which targets these for degradation and hence neutralizes the growth promoting effects of Igf2. The maternal copy of Igf2r is expressed while the paternal is at least partly silenced.

Studies of genomic imprinting in flowering plants also offer some support. In angiosperms, two female cells are fertilized by two identical "sperm" nuclei. The fertilized haploid egg forms the embryo, and a second typically diploid female cell is fertilized and forms the endosperm. The triploid endosperm, carrying two copies of each gene from the mother and one from the father, develops into a tissue that provisions the growing embryo much like the placenta in mammals. It seems that imprinted genes in angiosperms are expressed primarily in the endosperm, rather than in the embryo (Moore 2001). The parental pattern of genomic imprinting also seems to conform to predictions of the conflict hypothesis (Alleman and Doctor 2000). First, studies of interploidy crosses in plants that produce endosperm with different doses of maternal and paternal genomes have shown that paternal genomic excess in the endosperm is associated with unusually vigorous early growth of the endosperm, and maternal excess with the opposite (Haig and Westoby 1989, 1991). These data suggest that the paternally derived copies of genes are acting to elevate provisioning relative to female-derived genes. Second, detailed documentation of expression patterns also provides support. For example, the imprinted MEDEA gene of *Arabidopsis* suppresses endosperm development. As expected under the conflict hypothesis, maternal MEDEA alleles are expressed in the endosperm while the paternal allele is silenced (e.g., Kinoshita et al. 1999).

In light of the often "detrimental" population effects of evolutionary conflicts between males and females (see chapters 2 and 7), it is interesting to note that genomic imprinting persists in mammals despite apparently creating severe developmental problems. Imprinting appears to underlie several human genetic diseases, which result from imbalances in the expression of imprinted genes during fetal growth and development (Tilghman, 1999). Further, loss of maternal silencing of Igf2 has been demonstrated in more than a dozen different mammalian tumor types (Yun 1998), suggesting that some forms of cancer might represent negative pleiotropic effects of genes mediating male-female conflict. The possibility that some forms of cancer represent an evolutionary "cost" of alleles which are favored in sexually antagonistic coevolution over maternal provisioning of offspring is, indeed, intriguing.

The conflict hypothesis for the evolution of genomic imprinting is, however, not universally embraced. Actually, some 14 different hypotheses address the evolution of genomic imprinting (see Hurst 1997 for a review) and the status of the conflict hypothesis therefore remains controversial. In particular, it has been pointed out that not all observations of imprinting fit simple predictions of the conflict hypothesis (Hurst and McVean 1998). For example, few imprinted genes seem to evolve rapidly, as would be expected for interacting and antagonistically coevolving loci (McVean and Hurst 1997). However, it is difficult to assess the gravity of such observations as it is unclear exactly which regions of the genes should evolve rapidly and under which circumstances this should be observed (see Haig 1997, Tilghman 1999). It is certainly also possible that the evolution of imprinting has multiple explanations (Hurst and McVean 1998, Ohlsson 1999).

In the examples discussed above, the maternal and paternal alleles in any given individual have different phenotypic effects but both are later passed on to offspring. Genomic imprinting of a different kind occurs in several animals other than placental mammals. These cases are characterized by parent-of-origin-specific behavior of entire chromosomes, or even sets of chromosomes (also termed chromosome imprinting; Chandra and Brown 1975). For example, the first case of imprinting was actually described in a fungus gnat (Diptera, *Sciara coprophila*) where the entire paternal set of chromosomes is eliminated altogether from the germ line and partly from the soma (Crouse 1960). Similar forms of imprinting have been described to occur naturally in several other insects, such as in mealy bugs (Pseudococcidae) (Chandra and Brown 1975), the parasitic wasp *Nasonia vitripennis* (Nur et al. 1988), and the bark beetle *Hypothenemus hampei* (Brun et al. 1995). These cases might be best understood as dramatic examples of meiotic drive generated by more conventional selfish genes, whereby a selfish gene (or chromosome) increases its rate of transmission to the next generation by essentially deleting its homologue from the germ line (see Nur et al. 1988, Herick and Seger 1999).

Finally, it should be stressed that the sexual conflict over the flow of resources from the mother to an individual offspring discussed in this chapter is only one of several simultaneous evolutionary conflicts which more generally should affect the pattern of parental investment. Family affairs and kinship considerations are often complex indeed. For example, the limited degree of relatedness between individual siblings of a given female ($r = 0.5$ for full sibs and $r = 0.25$ for half-sibs) sets the stage for conflict between siblings over the division of parental resources (Mock and Parker 1997). Similarly, conflict between parents and their offspring is virtually ubiquitous in species that exhibit any form of parental care (Trivers 1974) and may even result in antagonistic coevolution between acquisition genes expressed in offspring and allocation genes expressed in parents (Mock and Parker 1997, Agrawal et al. 2001). The relative importance of these distinct but simultaneous conflicts between "family members" for the evolution of various aspects of parental care and parental investment is a very complex and poorly understood issue (e.g., Parker et al. 2002).

6.2 Sexual Conflict, Sex Ratios, and Sex Allocation

A large body of empirical research has demonstrated that the operational sex ratio affects sexual selection and sexual conflict in many different ways. The operational sex ratio affects the fitness payoffs to interacting males and females, through its effect on intrasexual competition and the frequency of interactions between males and females (see Andersson 1994, Rowe et al. 1994, Kvarnemo and Ahnesjö 2002). It is, however, much less clear whether genes in males and females may generally disagree on offspring sex ratios.

A suite of different genetic conflicts play roles in the evolution of sex determination and sex allocation. Evolutionary conflicts that influence offspring sex ratios may, for example, occur between cytoplasmic genes or parasites on the one hand and nuclear genes on the other, between parental sex ratio genes and sex-determining genes in offspring, or between sex chromosomes and autosomes. It is, however, not our purpose to review this large body of literature here (see Werren and Beukeboom 1998, Hurst and Werren 2001, Jaenike 2001, Majerus 2003 for reviews). We restrict ourselves to asking whether there are situations in which paternal and maternal sex ratio genes might be in conflict.

Fisher (1930) concluded that an equal investment in male and female offspring by parents is evolutionarily stable, simply because offspring of the rarer sex will have the higher average reproductive success. When one sex is more costly to the parent to produce than the other, however, natural selection will favor the parent overproducing the less costly sex. This logic successfully explains both the general tendency to produce a 1:1 sex ratio as well as deviations from this ratio. To see whether selection might act differently on sex ratio genes in males and females, consider a polyandrous mammal with no paternal care of offspring, where sons are more costly to produce for mothers than are daughters. Genes coding for an excess production of daughters will be favored when residing in these females, and a resulting female-biased offspring sex ratio is thus "adaptive" from a Fisherian point of view. In contrast, sex ratio genes coding for an even offspring sex ratio (or even a male-biased sex ratio, if sons have higher average reproductive fitness than daughters due to a deviation from a 1:1 sex ratio) will be favored when residing in males, since the (minimal) cost to males of producing sons and daughters is the same. The evolutionary interests of males and females in terms of offspring sex ratios can therefore differ, and antagonistic coevolution between male and female sex ratio genes is certainly possible (cf. Werren and Beukeboom 1998, Lessels 1999). Many mammals meet the assumptions outlined in the above scenario. Perhaps an evolutionary tug of war between males and females over offspring sex ratios might contribute to the fact that there is little evidence of female-biased sex ratios in mammalian species where mothers invest more heavily in sons (e.g., Clutton-Brock et al. 1981, Clutton-Brock and Iason 1986, Festa-Bianchet 1996, Packer et al. 2000, Cockburn et al. 2002). However, the dynamics of genetic conflicts over sex ratio have not been explored theoretically, to our knowledge. Several factors need to be considered, including the relative production costs of sons and daughters, the pattern of parental investment and care, relatedness asymmetries between fathers and mothers to offspring, and the degree of polyandry. In species where production costs do not differ for sons and daughters, where there is symmetric and biparental care of offspring and/or there is lifelong genetic monogamy, it is difficult to see how selection on parental sex ratio genes could differ depending on the sex in which they reside.

An important exception to this conclusion occurs in those species with haplodiploid sex determination, such as social hymenopterans (see Boomsma 1996, Sundström and Boomsma 2001). In this group, male offspring develop from unfertilized eggs and parental males will therefore be related to their daughters but not to any male offspring. This creates a dramatic relatedness asymmetry between the mates to the offspring, which generates sexual conflict

Figure 6.1. In some marine flatworms, sexually interacting individuals engage in what have been described as "penis fencing duels." This picture shows two interacting individuals of *Pseudobiceros bedfordi*, in a characteristic upright posture that is both defensive (protecting their back) and offensive (allowing them to strike with their paired penises). In this particular species, sperm are not injected into, but rather deposited upon, the skin of the partner. (Photo by N. K. Michiels.)

between the parents over the sex ratio of offspring: males will be selected to bias offspring production toward an increased production of daughters (Trivers and Hare 1976, Brockman and Grafen 1989).

6.3 Dueling Worms and Stabbing Snails: Sexual Conflict within Hermaphrodites

The marine flatworm *Pseudocerus bifurcus* is, like almost all flatworms (Platyhelminthes), a simultaneous hermaphrodite. When two sexually interacting (mating) individuals meet, they rear up facing one another in what can be described as a penis fencing duel (see Michiels and Newman 1998). They simultaneously evert their stylet-shaped penises and literally try to stab each other anywhere on the body, while simultaneously attempting to avoid being stabbed by their dueling partner. The mating posture adopted during duels appears highly efficient for both striking and parrying (figure 6.1), and these duels last about 20 minutes. When one animal is successful at penetrating the skin of the other, it holds on with its stylet embedded in its partner's epidermis for several minutes while its sperm are injected into the body. Most insemina-

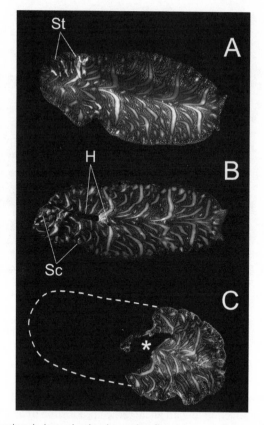

Figure 6.2. Hypodermic insemination in marine flatworms can be very injurious to the recipient. Picture shows tissue damage resulting from ejaculate receipt in *Pseudoboceros bedfordi*. (A) An individual shortly after having received multiple streaks of sperm (St) from its partner. (B) Within 30 minutes, components of the ejaculate dissolve the skin and underlying tissue of the receiver, sometimes resulting in holes in the body (H). These wounds will heal, but leave scars (Sc) and dents in the body. (C) In some cases, when ejaculates are received in the mid-front of the body (*), mating results in a complete loss of the posterior part of the body (hatched outline). Such individuals can regenerate and are found in natural populations. (Photo by N. K. Michiels.)

tions in *P. bifurcus* are unidirectional, where one individual ends up in the male role and the other in the female role. Needless to say, both striking and hypodermic insemination often cause considerable injury, occasionally leaving large holes in the body (figure 6.2). Injurious mating interactions such as this, including overt conflict, avoidance behaviors, and hypodermic insemination, are the rule among free-living flatworms of the order Polycladida (see Michiels 1998). The strong avoidance to being inseminated suggests that it is generally beneficial to inseminate the partner while avoiding being inseminated in these animals. This makes perfect evolutionary sense, since injecting sperm offers fertilization possibilities whereas sperm reception is associated with the costs of injury. The mating interactions of *P. bifurcus* show that overt reproductive conflicts extend beyond species with separate sexes, and that the study of hermaphrodites may tell us something about the nature of sexual conflicts.

As emphasized earlier in this book, sexual conflict is not about males or females per se but rather is an inevitable consequence when two unrelated individuals make a joint investment in reproduction. Since males of most species have a higher potential reproductive rate, males are often selected to exploit female investment in reproduction and females are selected to avoid being exploited. In simultaneous hermaphrodites (from now on, hermaphrodites), the same individual plays both reproductive roles at the same time and on average derives half of its fitness from the male and half from the female function. Sexual conflict among hermaphrodites stems from the fact that interacting individuals may disagree over what roles to play, and the way in which these roles are played. Seen in this way, sexual conflict in hermaphrodites highlights the fact that conflict and potential for mate exploitation is ubiquitous when two individuals engage in joint sexual reproduction.

6.3.1 PREMATING CONFLICT IN HERMAPHRODITES

In any given encounter, hermaphrodites may act as either males (sperm donors), females (sperm recipients), or both. Although some hermaphroditic animals are able to use their own sperm to fertilize their eggs (i.e., selfing), most cannot, and sperm reception in either case can be beneficial for at least three reasons. First, outbreeding may increase offspring fitness (Uyenoyama et al. 1993). Second, sperm digestion occurs among hermaphrodites (see Michiels 1998, Greeff and Michiels 1999a, Greeff and Parker 2000) and may have an important nutritional value (see below). Third, the ejaculate may contain other substances that are beneficial to the recipient, such as various "signals" which trigger reproductive events (Baur and Baur 1992). As illustrated by the mating habits of the marine flatworm *P. bifurcus*, however, sperm reception can also be associated with severe costs in hermaphrodites (see also van Duivenboden et al. 1985). Sperm donation, on the other hand, will potentially lead to fertilizations, but the profitability of sperm donation will depend on factors such as ejaculate production costs, the patterns of sperm use, the degree of multiple mating, and sperm digestion.

The relative costs and benefits of playing a sperm donor and/or a sperm recipient in mating will therefore depend on a number of different and interacting factors, many of which are likely to change over evolutionary time. Michiels (1998) has suggested that mating interactions between pairs of hermaphrodites may be characterized by anything from complete confluence to conflict (see figure 6.3). Conflicts between interacting individuals should be

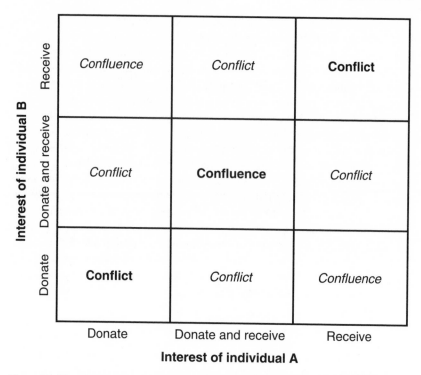

Figure 6.3. The relative interests of interacting hermaphroditic individuals can be classified according to a scheme describing the compatibility of interests. The relative interests can be either confluent or divergent. Italicized compartments represent cases where the interests of the interacting individuals are different. Since such cases are probably rare in hermaphrodites, it has been argued that mating conflicts (lower left or upper right compartment) are central in hermaphroditic mating systems. (After Michiels 1998.)

particularly strong when both benefit from only donating or only receiving sperm (Charnov 1979, Leonard 1990), a situation which should be common, since the interests of interacting hermaphroditic individuals are likely to be similar more often than they are dissimilar. Just like species with separate sexes, sexual conflict in hermaphrodites arises when two interacting individuals disagree over the outcome of the interaction. Intriguingly, such sexual conflicts over mating in hermaphrodites may result in either reciprocal cooperation or overt conflict.

One possible solution to these conflicts over reproductive roles is reciprocal exchange of gametes, such that individuals essentially trade gametes. In theory, however, conditions for this cooperative exchange are restrictive. Individuals could cheat by playing the preferred role, and then desert the partner. One possible solution is the evolution of successively reciprocal mating, in which mutual gamete exchange is achieved a little at a time. This was first described by Fischer (1980) in a serranid sea bass, *Hypoplectrus nigricans*, and he coined the term "egg trading" to describe the phenomenon. In this hermaphroditic fish, individuals form pairs which can be said to take turns at playing the male and female role, and the sperm donor role seems to be preferred (Leonard 1993). A pair can go through up to ten spawning events in a day, releasing only a few eggs each time. Since reciprocal mating interactions are repeated many times, Axelrod and Hamilton (1981) suggested, in their now classic paper, that serranid fishes may provide a good example of cooperation based on a "tit for tat" strategy. Egg trading has since been observed in several other serranids (see Leonard 1993 for a review) and has also been observed among hermaphroditic polychaetes (Sella et al. 1997, Sella and Lorenzi 2000). In general, since sperm might be cheaper to produce than eggs, Charnov (1979) expected the sperm donation role to be the preferred role in hermaphrodites (see also Greeff and Michiels 1999a; but see below). It is, however, certainly also possible that the sperm recipient role is preferred (Leonard 1990, Michiels 1998), and in these cases we would instead expect to see reciprocal "sperm trading." This seems to be the case in at least some interactions between individual sea slugs *Navanax inermis*, where partners repeatedly alternate in exchanging sperm during a copulation bout (Leonard and Lukowiak 1991, Michiels et al. 2003). In summary, gamete trading during repeated and alternating interactions is one way of "resolving" conflicts over sex roles in hermaphrodites. It is worth noting, however, that reciprocal insemination is also the rule among internally fertilizing hermaphrodites, despite the fact that successively reciprocal mating in the form of gamete trading is rare. Instead, courtship is often extended, and individuals seem to spend much effort assessing each other's intentions (see Baur 1998, Michiels 1998). It is currently not entirely clear what maintains reciprocity of mating interactions in this group of organisms (see Leonard 1990, 1993, Michiels and Streng 1998, Vreys and Michiels 1998, Michiels and Bakowski 2000 for discussions). One possibility to achieve reciprocal cooperation might be to assume that it is more important for both individuals to play complementary sex roles than it is to play a particular preferred role at any given interaction (see Crowley et al. 1998, Michiels and Kuhl 2003).

An alternative evolutionary outcome to alternating gamete trading, when one sex role is preferred over the other, could be that individuals frequently or even typically try to "cheat," such that cooperative reciprocity may not be the rule. In theory, it is easy to envision a scenario where coevolution between the individual's ability to play the preferred role (persistence) and the ability to avoid playing the nonpreferred role (resistance) leads to escalated and deceptive interactions (see section 6.3.3. below). If sperm donation is preferred over sperm receipt, the result might be coercive "hit-and-run" strategies using hypodermic insemination countered by various avoidance behaviors. Such mating interactions are indeed common among several groups of hermaphrodites, for example, leeches and certain groups of flatworms (see the description of *P. bifurcus* above) (Michiels 1998). This suggests that conflicts over sex roles, which are unique to hermaphrodites, are responsible for the evolution of such coercive mating interactions. If sperm receipt is preferred over sperm donation, the result might instead be elaborate and extended interactions where individuals are stimulating the partner into donating sperm while avoiding donation themselves. Escalated bouts of stimulation and assessment characterize mating interactions in several hermaphroditic groups, and may be the result of such conflicts (see Michiels 1998). Of course, there is in theory no difference between persistence adaptations based on coercion or stimulation on one hand, or between resistance adaptations in the form of avoidance or assessment on the other. It is simply the relative efficacy of various adaptations that will determine the evolutionary path.

The examples of conflict discussed above are all based on the assumption that one sex role is generally preferred over the other. While it is seemingly reasonable that such situations can persist, two problems cast some doubt on this possibility. First, hermaphroditic reproduction is favored in situations where one sex role is not generally preferred, but where the reproductive success of a hermaphrodite is higher than that of a pure male or female (Charnov 1982). It is therefore difficult to see how simultaneous hermaphroditism could be stable in a situation where one sex role is generally preferred. Second, theory predicts that several mechanisms (e.g., sperm digestion; see below) will lead to an increased investment in male function, even to the point where hermaphrodites should allocate resources roughly equally to the male and female functions (Greeff and Michiels 1999a, Pen and Weissing 1999). Again, this seems incompatible with the idea that one sex role is "cheaper" and thus generally preferred.

6.3.2 POSTMATING CONFLICT IN HERMAPHRODITES

True genetic monogamy is probably very rare among hermaphrodites, as individuals of many taxa engage in multiple mating with multiple partners (see Baur 1998 and Michiels 1998). Because sperm competition resulting from multiple mating can be intense, we expect analogous postmating sexual conflicts to those discussed in chapter 4 to occur in hermaphrodites. Manipulative adaptations in a given individual, including those that cause its partner to preferentially use its sperm for fertilization (rather than to digest it), elevate egg production, or induce a period of unreceptivity in its partners, might be favored even when these adaptations come at a net cost to the partner. Therefore, there is certainly potential for postmating conflicts to drive sexually antagonistic coevolution in hermaphrodites (see below). However, there is a lack of experimental data on the function of potentially antagonistic adaptation (see section 6.3.4 below). For example, the common earthworm *Lumbricus terrestris* uses special copulatory setae to pierce through the skin of the partner during mating and inject a secretion, originating from glands at the base of the setae. It appears that this substance affects sperm uptake and possibly induces a refractory period in the partner, but this has proven difficult to demonstrate conclusively (Koene et al. 2002, Koene, personal communication). It is also not certain whether this seemingly traumatic mode of transmitting these substances comes at any cost to the partner (Koene, personal communication).

In species with separate sexes, individuals never face trade-offs between male and female functions. Hermaphrodites do, and are expected to evolve to allocate resources so that the relative investment in male and female reproductive functions is in some sense optimal (see Charnov 1982, 1996). This fact creates one additional and potentially very influential type of sexual conflict, which is truly unique to hermaphrodites (see Michiels 1998): there should be selection in hermaphrodites for postmating adaptations that manipulate the sex allocation of their partners. Whatever the optimal sex allocation of a focal individual hermaphrodite may be at a given occasion, its partner will generally benefit from shifting this allocation toward producing eggs since it has no genetic interests in the future sperm of our focal individual. Thus, individuals that essentially feminize their partners should enjoy a selective advantage over those that do not. We are unaware of any clear examples of traits in hermaphrodites that serve this manipulative function, apart from the general observation that sperm or substances transferred at mating stimulate egg production (Khan et al. 1990, Baur and Baur 1992, Butt and Nuutinen 1998, Bishop et al. 2000) and affect growth/function of the genital tract (Bride and Gomot 1991a) in much the same way as in many separate-sex species (see section 4.1.1). In gastropod mollusks, for example, neurosecretory cells produce endogenous neuropeptides (gonadotropic "egg-laying hormones") which elevate egg production (e.g., Wiens and Brownell 1994, Li et al. 1999). The secretion of such hormones, however, is apparently triggered in a dose-dependent fashion by signal substances produced by accessory glands and transferred at mating (Saleuddin et al. 1989, Khan et al. 1990, Bride and Gomot 1991a,b). The possibility that an individual may affect regulation of sex allocation in its partner should open up possibilities for exploitation and manipulation. One way to counter such manipulation would be for hermaphrodites to exhibit a male-biased investment at maturity, which is then shifted toward a female sex allocation during their reproductive life as a result of the receipt of feminizing agents from their partners (Michiels 1998).

The spectacular mating habits of *Ariolimax* spp. banana slugs (see Mead 1943, Harper 1988) may provide another example of individuals feminizing their partners. Mating in these slugs is prolonged, and typically involves reciprocal copulation and sperm exchange. The female reproductive tract is equipped with musculature that is apparently able to grip the partner's penis internally. Consequently, individuals frequently "get stuck" during mating and fail to separate despite repeated vigorous attempts to do so. At this point, one or both partners may resort to apophallation, where the trapped penis is simply bitten off (or rather gnawed off) and sometimes even consumed by the "recipient" (i.e., the individual in the female role). An individual that has lost its penis is unable to mate in the male role again. Michiels (1998) and Reise and Hutchinson (2002) have suggested that apophallation may therefore benefit the "recipient," provided of course that mating was reciprocal, by forcing a diversion of resources from sperm production to egg production in the amputee. The specialized musculature in the female reproductive tract which allows "penile pinching" in *Ariolimax* spp. is, therefore, a candidate for a sexually antagonistic adaptation aimed at feminizing the partner, though there is apparently no experimental evidence in support of this possibility.

Another type of postmating conflict occurs as a result of sperm digestion (i.e., spermaticide), which seems to be particularly common in hermaphrodites. Some taxa even possess special structures to this end (see Michiels 1998, Greeff and Michiels 1999a, Greeff and Parker 2000). Obviously, provided that an individual has access to enough viable sperm to fertilize its eggs, digesting excess sperm might be beneficial, especially if digested sperm has a significant nutritional value. At the same time, sperm digestion is most likely very costly to the donor, since digested sperm will obviously not fertilize eggs. Greeff and Michiels (1999a) showed that this can in theory lead to an antagonistic arms race between sperm digestion (female function) and sperm production (male function) in hermaphrodites, which is further fueled by sperm competition. The resulting arms races are predicted to generate either large ejaculates or ejaculates which are well protected against spermaticide. Incidentally, spermaticide is common also in species with separate sexes, and Greeff and Parker (2000) showed that it can likewise lead to intersexual arms races in these species.

6.3.3 SEXUAL SELECTION AND ANTAGONISTIC COEVOLUTION IN HERMAPHRODITES

A central question in this context is whether, and in what way, sexual selection and sexually antagonistic coevolution may be different in hermaphrodites than in species with separate sexes. Several authors have argued that there are good reasons to expect premating sexual selection, or selection due to variance in number of mates acquired, to be very low in hermaphrodites. First, Charnov (1979) pointed out that the conditions under which hermaphroditism is profitable (relative to separate sexes) include very restricted mating opportunities (e.g., low population density and parasitic or sessile life styles), and this limits the intensity of sexual selection. Second, while Morgan (1994) actually saw few conceptual differences between hermaphrodites and separate-sex species in this regard, he pointed out that the evolution of sexual dimorphism should be constrained in hermaphrodites since sex-specific trait expression is essentially impossible. This should greatly limit the elaboration of traits which confer a reproductive advantage for one sexual function because they will have a cost to the other. Third, Greeff and Michiels (1999a) argued that several factors can lead to an equal, or near equal, optimal sex allocation pattern in hermaphrodites (i.e., 1:1). Whenever this is the case, fitness through the male function may no longer be primarily limited by mate acquisition and both fitness components can instead be limited by the amount of resources allocated to gamete production. This is an important point, since it implies that Bateman's (1948) principle, suggesting that female fitness is limited by resources and male fitness by number of mates, may not be valid for some hermaphrodites (but see Pongratz and Michiels 2003). If the male and female functions indeed have similar potential reproductive rates in hermaphrodites, there will be little opportunity for premating sexual selection. However, there is very little reliable empirical data on absolute allocation to male and female functions in hermaphroditic animals. Some studies suggest that investment in male and female functions is indeed similar (e.g., De Visser et al. 1994), but others indicate that male gametic investment is much less than 50% (e.g., Locher and Baur 2000, < 5%). This would imply that one sex function can be generally preferred (see also Leonard 1993), and the argument of Greeff and Michiels (1999a) would perhaps not be widely applicable. Fourth, Greeff and Michiels (1999b) argued that a mutation yielding a reproductive advantage for one sexual function must be twice as advantageous in a hermaphrodite to yield the same selective advantage as in a separate-sex species. This is because hermaphroditic individuals on average derive only half their fitness from each of the two sexual functions while a separate-sex individual derives it from one. Consequently, variance in reproductive success between the two sex roles is constrained to be equal between the sexes, since everyone is male and female simultaneously. Greeff and Michiels (1999b) therefore argued that the opportunity for sexual selection in hermaphrodites should be at most half that of separate-sex taxa. However, this effect will be counterbalanced by the fact that any given allele (with the exception of those located on sex chromosomes) spends half of its time in males and half in females in separate-sex taxa, whereas it is always exposed for selection on male function in hermaphrodites. Therefore, this argument may not hold (Greeff and Michiels, personal communication). Fifth, if all individuals in a hermaphroditic population invest equally in mate acquisition, the optimal investment in mate acquisition is lower than in separate-sex species where only half of the individuals typically do this "job" (Greeff and Michiels 1999b). This further reduces the opportunity for sexual selection via differential mate acquisition.

These arguments all suggest that classic premating sexual selection via the male function is relatively weak in hermaphrodites. But we believe that it would be premature to use these arguments to claim that the opportunity for sexual selection, defined more broadly as individual variance in mating and fertilization success, is generally low in hermaphrodites. Theory in this field seems incomplete, and empirical data are very scarce (Morgan 1994, Greeff and Michiels 1999b, Pongratz and Michiels 2003). Most importantly, the opportunity for postmating sexual selection may actually be relatively high in hermaphrodites, not only because sperm competition should be as important as in any other species (see Baur 1998, Michiels 1998, Greeff and Michiels 1999a, Landolfa et al. 2001) but also because there are added components of variance in postmating success in hermaphrodites. First, sexual selection due to differential allocation (see section 3.1) could be larger in hermaphrodites, because the variance in number of female gametes produced per reproductive bout should be higher. This is because variance in egg production rate across individuals due to sex allocation trade-offs will be added to that due to general life history trade-offs (e.g., current versus future reproduction). The former component of variance is absent in taxa with separate sexes. This added variance may even outweigh the effect discussed by Greeff and Michiels (1999b), and the number of eggs laid by an individual's partner after mating may thus be a major fitness component in hermaphrodites (Michiels 1998). Second, sperm digestion (see above) opens up possibilities for differential success both in the ability to protect one's ejaculates from being digested rather than used for fertilization as well as in the ability to digest received ejaculates. This may constitute an important and unique component of variance in fitness in several groups of hermaphrodites. Third, variance in the relative amount of resources allocated to male and female functions will certainly contribute to individual variance in postmating reproductive success among hermaphrodites, as the number of sperm transferred at a given mating and the number of eggs laid following a mating are both important determinants of reproductive success. Seen this way, individual variance in sex allocation is a component of sexual selection that is absent in species with separate sexes. Taken together, individual variance in reproductive success might well be as high in hermaphrodites as it is in taxa with separate sexes, although we know of no empirical data directly addressing this issue.

Irrespective of the opportunity for sexual selection, it is nevertheless interesting to consider the fate of sexually antagonistic alleles in hermaphrodites. First, let us consider an intralocus conflict: an (autosomal) mutation confers a reproductive advantage for one sexual function but is disadvantageous when expressed in the other. In species with separate sexes, Rice (1984) argued that a mutation such as this will become established simply when selection favoring the allele, when expressed in one sex, is stronger than selection against the allele when expressed in the other. Since the average fitness of the male and female functions must be the same in an outbreeding hermaphrodite, the very same rule applies to hermaphrodites. A sexually antagonistic allele will spread whenever the positive selection in one sexual function outweighs the negative selection in the other (Morgan 1994). One difference here is that, while species with separate sexes can subsequently "resolve" intralocus conflicts by second-arily evolving sex-limited expression of these alleles (Rice 1984), hermaphro-dites cannot. Put in quantitative genetic terms, the genetic correlation between male and female expression of a trait is constrained to be one in hermaphro-dites. Because the two sex roles often require different adaptations to be played well, hermaphrodites may suffer maladaptation from intralocus conflict, be-cause none of the sexual functions are free to evolve toward their fitness peaks (see Morgan 1994). To use an analogy with motor vehicles, hermaphrodites may be doomed to be an evolutionary compromise between a Ferrari sports car and a John Deere tractor.

Next, consider an interlocus conflict: a mutation that affects only male func-tion causes a dramatically elevated egg production rate in its partners, counter to the long-term interests of the latter. In species with separate sexes, such a mutation will spread if the benefit to males from an increased reproductive output of its mates outweighs the cost to males from reduced viability of its mates (Parker 1979). When direct costs to females from mating with males carrying these mutations outweigh any indirect benefits to females that might occur, they are expected to evolve resistance at other loci setting off sexually antagonistic coevolution (chapter 2). In hermaphrodites, things should again be fairly similar. The same basic logic applies, namely, that we expect such a mutation to spread when the reproductive advantage is greater than the repro-ductive disadvantage due to costs experienced by partners (i.e., when net selec-tion on the male function is positive). The conditions under which individuals should resist the effects of this mutation might, however, be a bit different. Assuming that the negative direct cost affects only or primarily the female function, a hermaphroditic individual mated to a bearer of this mutation will suffer a smaller direct fitness loss than a separate-sex female since only half of total individual fitness is derived from the female function. At the same time, any indirect "sexy sons" benefits to this individual will be the same as for a separate-sex female: only half of its offspring and grand-offspring (those produced through the female germ line) can express the "sexy sons" gene. This is directly comparable to a female of a species with separate sexes, where half of its offspring and grand-offspring (sons and grandsons) can express the "sexy sons" gene. Thus, it is possible that manipulation might be more apparent in hermaphrodites, since a manipulated individual could get the same indirect benefits at a lower direct cost. On the other hand, this would imply that escalat-ing antagonistic coevolution is less likely in hermaphrodites since the condi-tions under which we expect resistance to evolve may be more restrictive (cf. Parker 1979). It is also important to note that direct costs to individuals re-sulting from "male persistence adaptations" in their partners may also affect both male and female functions in the former. If this is the case, there will be selection for resistance through both the male and the female functions.

Michiels (1998) suggested that obviously coercive mating strategies, involving hypodermic insemination and associated with obvious harm to partners, are apparently more common in hermaphrodites compared to species with separate sexes. The considerations above suggest that this is perhaps more likely to be the result of a unique form of conflict in hermaphrodites (that over sex roles played in mating interactions) rather than any inherent differences in the potential for sexually antagonistic coevolution between these two groups of organisms (see also Michiels and Newman 1998). On the other hand, Michiels and Koene (2004) pointed out that hermaphrodites should be more willing to engage in costly mating interactions which depress fitness through the female function, compared to females in species with separate sexes, since such mating costs can be counterbalanced in all individuals by a potential gain in fitness through the male function. For this reason, higher mating costs are required to keep a hermaphrodite from remating, and they suggested that sexual interactions may, consequently, have escalated to become particularly harmful in hermaphrodites. An additional difference between hermaphrodites and many animals is the absence of sex chromosomes in the former. It is, however, not clear if and how a lack of chromosomal sex determination could lead to more or less sexually antagonistic coevolution (see section 6.6 below).

6.3.4 THE LOVE DART IN SNAILS—A SHOT AT PATERNITY?

In many hermaphroditic gastropod snails, individuals produce a sharp and calcified "stiletto" in the female part of the reproductive system (Koene and Schulenburg 2003). These sizable structures come in a variety of different shapes and have become known as "love darts" (figure 6.4), despite the fact that they are not actually shot or thrown. Rather, the dart is stabbed and/or forcefully pushed into the partner during courtship, where it is then left in most species. One dart only is transferred in a mating, and it takes some time for snails to regenerate a new dart (Tompa 1982). This spectacular mating habit has generated a lot of discussion, and many hypotheses concerning the function and evolution of love darts have been formulated (see Landolfa 2002 for a review). It is, however, only during the last few years that more detailed experimental studies have shed light on this bizarre mating habit.

The function of the love dart has been best studied in snails of the genus *Helix*. Here, mating starts with an extended and elaborate courtship phase, after which the animals evert their penises and eventually copulate simultaneously and reciprocally. In *H. aspersa,* the courtship phase lasts about 0.5 hour and the copulation for about 7 hours (see Adamo and Chase 1988). During copulation, the snails exchange spermatophores that are deposited in the genital tract of the partner. From the site of spermatophore deposition, sperm can either travel up the spermoviduct to the site of sperm storage and fertilization, or be caught in a canal leading to an organ (bursa copulatrix) with the sole

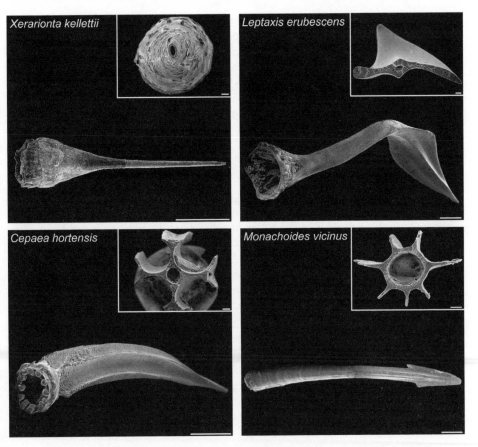

Figure 6.4. Many land snails produce sharp "stilettos" which are forcefully pushed into the partner during courtship. The elaboration of these "love darts" varies across species and has been found to coevolve with changes in the sperm-receiving organ which promote sperm digestion but hamper sperm storage. (From Koene and Schulenburg 2004; photos by J. Koene.)

function of digesting sperm. The vast majority of sperm are, in fact, digested: approximately 0.1% of the sperm transferred escape digestion in *H. pomatia* (Lind 1973) and even less in *H. aspersa* (Rogers and Chase 2001).

Darts are exchanged toward the end of the courtship phase. One animal pushes against its mate, and the dart is forced into the body of the partner. Usually, the second animal places its dart shortly after the first animal has done so. Darts are not required for successful mating in *Helix* snails, as sperm can be transferred by virgins (who have no darts), recently mated individuals (who have not yet regenerated new darts), and in cases where darts are not successfully transferred. Virtually all snails that are carrying a dart will attempt to transfer it at mating, but many end up missing their target. Adamo and Chase (1988) reported that only 46% of the darts successfully penetrated the body wall of the partner in *H. aspersa*; the remainder either missed the partner completely or did not penetrate the body wall. Similarly, Giusti and Lepri (1980) found that 43% of the darts did not penetrate the body wall in *H. aperta*.

So what is the function of this mysterious dart? All available experimental data suggest that the dart serves as a vehicle for signaling substances that are transferred between individuals at mating. The love dart is covered with a mucus secretion carrying biologically active polypeptides that are produced by special glands in the reproductive tract (Chung 1986, Adamo and Chase 1990). Three recent studies of *H. aspersa* collectively indicate that the love dart constitutes a sperm competition adaptation. First, Koene and Chase (1998a) applied dart mucus to the reproductive system in vitro. They showed that the mucus caused muscle contractions that should facilitate the uptake of the spermatophore and that it tended to render the canal leading to the sperm digestion organ less accessible. Obviously, this should increase the probability that a dart donor's sperm are used to fertilize the eggs of the recipient rather than being digested by the recipient. Second, Rogers and Chase (2001) were able to corroborate these findings. They showed that snails that were hit by a dart indeed stored more sperm from a given mating than did snails that were not hit. Third, Landolfa et al. (2001) mated virgins sequentially to two different partners, while monitoring dart placing success and subsequently assigning paternity of offspring by allozyme markers. They showed that snails that successfully forced a dart into their partner had significantly higher postmating paternity than individuals that failed to do so. In summary, mucus carried on the love dart in *H. aspersa* contains substances that act to increase the use of the shooter's sperm for fertilization of eggs.

Many gastropod snails are known to mate multiply and can store viable sperm for very long periods of time (Baur 1998). Sperm competition thus clearly occurs in many snails. The love dart is in theory no different from some of the accessory gland substances transferred with the ejaculate in most groups of animals (see chapter 4). Both represent adaptations that function to increase postmating fertilization success of the donor and that can potentially compromise the fitness interests of the recipient. In the case of the love dart, several suggestions regarding its evolution have related to "male" manipulation and conflicts over sex roles (Chung 1987, Leonard 1992, Adamo and Chase 1996, Koene and Chase 1998b). This is no doubt due to the seemingly damaging mode of transfer of substances via love darts, but also because they appear to steer sperm use toward fertilization and away from digestion. Because the former effect is in the interest of the donor and the latter is possibly in the interest of the recipient (Greeff and Michiels 1999a, Greeff and Parker 2000), the love dart may well be an instrument of manipulation resulting from sexually antagonistic coevolution. However, several authors have pointed out that individuals which allow themselves to be "manipulated" by love darts may gain indirect benefits in terms of elevated offspring fitness (Charnov 1979, Pomiankowski and Reguera 2001, Landolfa 2002) and that conflict may not be involved in the evolution and maintenance of love darts.

Given multiple mating and sperm digestion, it is easy to see how a dart mutant could invade a dartless snail population: the mutant would have higher postmating fertilization success. Using the love dart as a vehicle for manipulative substances may be more efficient than other modes of transfer, such as in the ejaculate, for several reasons (evading ejaculate digestion or adaptations in the reproductive tract of the recipient aimed at neutralizing the effects of such substances). Assuming direct costs to recipients, resistance to the dart would be favored, leading to coevolution between dart use (or the physiological effects of the dart) and resistance to darts. Two forms of direct costs to recipients of love darts appear to be significant. First, it is difficult to imagine that it is not traumatic to have a dart forced into your body, as they are frequently very large relative to the body of the recipients. For example, in *H. aspersa*, the love dart is about 1 cm long while the adult shell width is about 3.5 cm. Unfortunately, observations on these direct costs are only anecdotal and have not been properly quantified. Baur (1998) does note that love darts have been reported to sometimes wound and kill the partner, and Koene (unpublished) has recorded darts apparently penetrating neural tissue. Second, if sperm digestion is beneficial to the recipient, then interference with sperm digestion is a direct cost. Snails penetrated by their partner's darts have been found to store more than twice as much sperm as those not penetrated by darts (Rogers and Chase 2001), so this effect could be substantial. In this view, the love dart could actually be interpreted as a resistance adaptation aimed at reducing the deleterious effects of sperm digestion (cf. Greeff and Michiels 1999a, Greeff and Parker 2000).

Even though the function of the love dart is now well established, at least in *H. aspersa*, it does not currently seem possible to draw any firm conclusions about its evolution (Pomiankowski and Reguera 2001, Landolfa 2002). However, the fact that fewer than 50% of the darts successfully penetrate the partner indicates that snails do resist being stabbed, and the mere size of the potential direct costs to victims of darts (see above) suggests that this resistance is primarily under direct selection. Consequently, several authors (e.g., Chung 1987, Adamo and Chase 1996, Koene and Chase 1998b, Koene and Schulenburg 2003) have suggested that dart shooting ability and resistance to darts are both primarily results of sexually antagonistic coevolution, generated by conflict over the fate of donated sperm.

Koene and Schulenburg (2003) recently presented an illuminating comparative study of the coevolution of love darts and other reproductive structures in hermaphroditic land snails. First, they revealed a complex of coevolving traits, all relating to dart elaboration. This evolutionary "axis" of love dart elaboration included evolution of the number of darts, their structural complexity, the number of dart-producing organs, and the number and size of the glands that are associated with mucus production. Second, the evolution of love dart elaboration was associated with changes in the sperm-receiving organ which promote sperm digestion but hamper sperm storage. Koene and Schulenburg (2003) concluded that the pattern of coevolution is consistent with an arms race between dart persistence and dart resistance.

6.4 Sexual Conflict in Plants

This book is largely about animals. This fact may admittedly be due in part to our own (i.e., the authors') taxonomic bias, since we are zoologists, but it is also true that the literature on sexual selection has a much longer and richer history in animals than in plants (Willson 1994). After all, plants generally lack the elaborate secondary sexual traits that have spurred so much research interest in sexual selection in animals, and Darwin himself never applied sexual selection thinking to flowering plants (Queller 1987). The vast majority of angiosperms are also hermaphroditic, and most of what was said about simultaneous hermaphrodites above (section 6.3) is also directly applicable to these plants (Morgan 1994). Although there is good evidence that competition between pollen donors over fertilization (i.e., sexual selection on male function) has been central in the evolution of many traits which affect pollen delivery, such as flower and/or inflorescence characteristics (e.g., Willson and Burley 1983, Willson 1994), it is difficult to envision much conventional premating sexual selection in plants. Plants are typically sedentary and do not show any behavior in the traditional sense, and which male gametes end up on a particular stigma is often a chance event (Charnov 1984, Charlesworth et al. 1987). We will not, however, provide a general discussion of the potential for sexual selection in plants here, but merely discuss a few of the evolutionary conflicts over the reproductive interests of males and females that do occur in plants (see also Queller 1994, Shaanker and Ganeshaiah 1997).

When they end up on a "compatible" stigma, male pollen grains germinate and make their way to the female ovules via a pollen tube that grows through the cellular matrix of the style. Angiosperms exhibit a peculiar "double fertilization," whereby the pollen nucleus divides by mitosis. One nucleus fertilizes the haploid ovule and the other fuses with a diploid female cell to form the triploid endosperm. The endosperm is a tissue that provides nutrition to the developing zygote, and is in this respect analogous to the mammalian placenta. Because the maternal genome is equally related to all of its offspring, which is not necessarily the case for a particular paternal genome (Delph and Havens 1998), there will be parental relatedness asymmetries to the growing zygotes (see section 6.1). Charnov (1984) suggested that double fertilization itself may actually be a result of sexual conflict over zygote provisioning, in that it represents paternal manipulation of maternal provisioning in favor of its own offspring. The "doubling" of the maternal contribution to the endosperm may then have restored some maternal control of offspring provisioning (Härdling and Nilsson 2001). These interesting but contentious suggestions have little direct support, to our knowledge, apart perhaps from studies of interploidy crosses showing that paternal genomic excess in the endosperm is associated with unusually vigorous early growth of the endosperm and maternal excess with the opposite (Haig and Westoby 1989, 1991). In any case, studies of genomic imprinting in plants do suggest that sexual conflict over maternal provisioning of zygotes is indeed consequential in plants (see section 6.1 above).

In plants, perhaps the most interesting potential battleground for sexual antagonism is the reproductive parts of the flower. Here, pollen grains and growing pollen tubes (paternal tissue) interact with the stigma and the style (female tissue) prior to fertilization. The most obvious sexual conflict may arise as a side effect of pollen competition. Given that pollen of several donors frequently competes over the fertilization of ovules, we expect pollen tubes to be "aggressive" much like sperm in animals (see Delph and Havens 1998). Pollen tubes should thus be selected to rapidly grow toward and penetrate the ovules and to inhibit germination and pollen tube growth in competing pollen (see Walsh and Charlesworth 1992, Snow 1994, Stanton 1994, Willson 1994, Marshall et al. 1996). Both of these male pollen competition adaptations might create costs to females as side effects, in the form of suboptimal fertilization. Overly efficient pollen tubes might create problems with polyspermy while successful pollen inhibition may lead to insufficient fertilization of the often many ovules in each ovary. We would then expect to see counteradaptations in the stigma and style which slow down access to the ovules, spurring sexually antagonistic coevolution between pollen/pollen tube traits and stigma/style/ovule traits. This scenario is conceptually identical to the conflict between sperm and eggs in animals discussed at some length earlier (see discussion in section 4.2.2.2) (Howard 1999, Frank 2000).

There is not much direct evidence at present for this hypothesis, but some indirect evidence is at least suggestive. First, the rates of "infertility" in natural plant populations are often very high. The prevailing view holds that angiosperms produce a large "surplus" of ovules, and several benefits have been suggested to counterbalance the costs of producing this excess (see Stephenson 1981, Sutherland 1986). These include allowing the exploitation of rare seasons with unusually favorable environmental conditions, maximizing male fitness through increased pollen donation success, and allowing reallocation of resources from offspring of low genetic quality to those of high quality by selective abortion of viable but inferior embryos. Despite scattered evidence for and against these putative benefits in the literature (see Melser and Klinkhamer 2001), our understanding of the proximate causes of infertility/abortion in natural populations of plants is limited. Second, fertilization and seed set often increase as a result of increased pollen deposition in natural populations of angiosperms, a phenomenon known as pollen limitation (see Burd 1994, Larson and Barrett 2000, Wilcock and Neiland 2002). This suggests that antagonistic and inhibitory interactions between pollen, perhaps as a result of pollen competition (Marshall et al. 1996), could come at a cost to females as they might reduce fertilization rates. Third, polyspermy certainly does occur in angiosperms (e.g., Maheshwari 1950), but little is known about its frequency under natural conditions. Fourth, the length and shape of the style are often such that the pollen tube has to grow a "long and treacherous" way to reach the ovules. This has been seen as an adaptation which allows the plant to assess and choose among pollen tubes (e.g., Queller 1987), but it may also provide a barrier against polyspermy. Fifth, evidence indicating that interactions between the pollen/pollen tube and the stigma/style are mediated by intercellular communication systems much as in animals is rapidly accumulating (Shimizu and Okada 2000, Wheeler et al. 2001, Johnson and Preuss 2002, Nasrallah 2002). For example, recent studies have revealed the existence of several distinct pollen-specific genes that code for receptors that presumably receive cues from the style during pollen tube growth (Mayfield et al. 2001, Kim et al. 2002). As is the case for genes encoding gamete-surface proteins in animals (Swanson and Vacquier 2002), these genes are often highly polymorphic and at least some evolve very rapidly (Mayfield et al. 2001, Kim et al. 2002, Nasrallah 2002), as would be expected if sexually antagonistic coevolution is involved. In summary, though the direct evidence is very limited, several facts at least suggest that sexual conflict between pollen and the stigma/style may contribute to the evolution of reproductive characters in angiosperms, and may thus also contribute to the evolution of reproductive isolation and speciation in this group. Conspecific pollen from different populations has frequently been reported to differ in their relative fertilization success on styles from different populations, showing that genetic differentiation across populations in traits which mediate pollen-style interactions is common (see Delph and Havens 1998, Howard 1999).

It is worth noting that "female" postmating discrimination against male gametes based on relatedness, or other forms of genetic compatability, has been considered a form of natural selection in the plant literature (e.g., Charlesworth et al. 1987). Thus, research on postmating mechanisms to avoid inbreeding, such as the evolution of self-incompatibility, has largely been framed outside the domain of sexual selection. This is not the case in the animal literature, where both overt and cryptic female choice for unrelated males has been considered a form of female choice driven by particular genetic benefits to offspring (e.g., Bateson 1983, Bishop 1996, Zeh and Zeh 1997, Tregenza and Wedell 2000, 2002).

A topic that deserves special attention in plants is intralocus sexual conflict, which should be especially consequential in angiosperms because of the predominance of hermaphroditism. This form of conflict occurs when the optima of a given phenotypic trait, determined at the same locus in males and females, differ for the two sexes or sex roles. Intralocus sexual conflict has received considerable attention in species with separate sexes (Parker and Partridge 1998, Rice and Chippindale 2001), but its potential significance is much larger in hermaphrodites since males and females are by definition unable to evolve toward separate fitness peaks in these groups (section 6.3.3; Morgan 1994).

Two instances of intralocus sexual conflict that are particular to plants are the determination of flower number and morphology. Queller (1987) suggested that a large number of flowers per individual is primarily the result of selection on male function. Indeed, many hermaphroditic angiosperms produce more flowers than fruits and there is experimental evidence that large floral displays increase male success (Queller 1987, Delph and Havens 1998). One can imagine scenarios in which such "excess" of flowers makes no contribution to fruit set and hence does not directly increase female reproductive success. In cases where a large floral display increases the transfer of pollen to stigmas of different flowers on the same plant (i.e., geitonogamy), or where the production of flowers involves direct costs, fruit set may even be depressed by "excess" of flowers (Barrett 2003). If this is true, there will be antagonistic selection on flower number by male and female functions, respectively. It should be stressed, however, that females may also benefit from an "excess" of flowers because it may reduce pollen limitation or may allow selective maturation of fruits of superior quality (see Burd 1998). Further, Barrett (2002) drew attention to the possibility that different flower designs may be optimal for pollen removal and pollen deposition. If this is the case, selection through male and female functions may be antagonistic on any flower characteristic that affects reproductive success. Flower design, including the spatial arrangement of anthers and stigmas, has traditionally been interpreted in terms of adaptations to reduce the harmful effects of self-fertilization. The fact that selection on flower design mediated by male and female functions may be antagonistic raises the possibility that at least some floral adaptations are constrained by intralocus sexual conflict (Barrett 2002).

6.5 Sexual Conflict, Speciation, and Extinction

The processes by which new species are formed are at the heart of biology, and understanding such processes is of fundamental importance. Research into speciation has a very long and multifaceted history (see Howard and Berlocher 1998 and Magurran and May 1999 for recent reviews), and much research is now aimed at testing and understanding the diversifying processes which lead to cladogenesis. The splitting of a lineage into two or more separate lineages occurs as a result of divergence in traits that mediate reproductive isolation between members of different lineages. The fact that closely related species are typically more different in reproductive traits compared to other types of traits has been recognized for a long time (e.g., Darwin 1871, Eberhard 1985, Butlin and Ritchie 1994, Turner 1994, Civetta and Singh 1995, Arnqvist 1998, Wells and Henry 1998). This observation has motivated the idea that sexual selection can initiate and propel rapid and divergent evolution of reproductive characters in allopatric populations and hence be an engine of speciation (Ringo 1977, Lande 1981, West-Eberhard 1983, 1984, Rice 1998a, Price 1998, Questiau 1999, Boake 2000, Gray and Cade 2000, Seehausen 2000, Panhuis et al. 2001, Slabbekoorn and Smith 2002). A great number of theoretical models have also confirmed that intersexual selection should be particularly efficient at generating new species (see Turelli et al. 2001 and Kirkpatrick and Ravigné 2002 for recent reviews of theory). There are several related reasons for this. First, sexual selection by definition involves selection on reproductive traits in both sexes. Such traits are, needless to say, particularly likely to affect reproductive isolation. Second, there are elements of arbitrariness involved in many models of sexual selection, such that any of several male traits is equally likely to become elaborated or become elaborated in slightly different ways. Such male-female coevolution will tend to increase the probability of allopatric speciation, even in the absence of environmental differences between populations, because the coevolutionary trajectories can differ among diverging populations. Third, sexual selection is more powerful than natural selection in causing reproductive fission in a population because nonrandom gamete fusion helps bring together favorable combinations of alleles at different loci. Even if reproductive isolation and speciation are often by-products of adaptation by natural selection to new adaptive zones (Mayr 1963, Schluter 2000), there are many good reasons to believe that sexual selection often plays a central role when novel species are generated. An additional reason is that empirical studies have revealed a surprisingly simple genetic architecture of signals involved in species mate recognition (Ritchie and Phillips 1998, Henry et al. 2002, Price 2002). This observation is inconsistent with the view that reproductive isolation is an incidental side effect of polygenic adaptation to novel environmental conditions.

In the most general case, we can expect to see divergence in reproductive traits exposed to sexual selection across allopatric populations even if the environment does not vary (Lande 1981, West-Eberhard 1983, 1984). Such "internal" coevolution in the absence of externally imposed selection can occur because of the inherent arbitrary element of sexual selection, especially if multiple traits are involved, and because of different mutational input in diverging lineages. In contrast, it is also possible for sexual and natural selection to interact in generating evolutionary divergence in sexual signals and perception of, or response to, such signals. If male-female communication systems adapt to different external signaling environments in different populations (e.g., light regime, food availability, acoustic regime, etc.), male traits and female preferences may diverge as a result of an interaction between natural and sexual selection. This process, known as sensory drive (Endler 1992), seems to have been involved in the evolutionary divergence of several sexual signal/response systems (see Boughman 2002) and hence to speciation in some groups. However, there is at least one reason to believe that other processes are involved. This is the fact that sexual signals which are neither emitted nor perceived in an external environment evolve at least as rapidly and divergently as those which are. For example, male genitalic morphology can be seen as a signal to females which functions exclusively within the female reproductive tract. Yet animal genitalia probably represent the set of traits that evolve most rapidly and divergently by intersexual selection (Arnqvist 1998) and which most generally differ even between closely related species (Eberhard 1985). This is also true for seminal fluid substances that are transferred to females at mating (e.g., Chapman 2001), tactile copulatory courtship behavior (West-Eberhard 1984, Eberhard 1994), gamete proteins which function in gamete binding and fusion (e.g., Swanson and Vacquier 2002), and contact pheromones (Etges 2002). In all these cases, communication between the sexes apparently does not occur in, or is not affected by, an external environment, and it is thus difficult to see how sensory drive could affect the coevolution of these male signals and female receptors.

Recent molecular evidence has indicated that selection has been involved in the evolution of differences in reproductive traits between closely related species (e.g., Civetta and Singh 1998, Aguade 1999, Swanson et al. 2001a, Swanson and Vacquier 2002), as would be expected if sexual selection had caused changes in these traits, but comparative evidence is more equivocal. Several studies, performed on birds, mammals, and insects, have searched for an association between net speciation rate and various indices of sexual selection (e.g., mating system characteristics or sexual dimorphism). Many comparative studies have indeed documented a positive relation between sexual selection and species richness (Barraclough et al 1995, Møller and Cuervo 1998, Mitra et al. 1996, Owens et al. 1999, Arnqvist et al. 2000, Katzourakis et al. 2001, Masta and Maddison 2002, Misof 2002), but others have failed to find such a relationship (e.g., Gage et al. 2002, Morrow et al. 2003b). One reason for the lack of consistency across these studies may be that some forms of sexual selection also could promote extinction, by a sexual selection "load" (Dawkins and Krebs 1979, Tanaka 1996, Houle and Kondrashov 2002, Kokko and Brooks 2003, Morrow and Pitcher 2003, Morrow et al. 2003b). All models of sexual selection predict that males should become more or less maladapted as they accumulate traits that are costly in terms of natural selection (Andersson 1994). This will, however, affect population fitness only if males become so poorly adapted that females suffer from, for example, a shortage of mating opportunities or from reduced paternal care. This is because it is typically female productivity that limits population growth. Many models of intersexual selection do involve the evolution of costly female preference/resistance, and these costs can sometimes be expected to depress population fitness (e.g., Gavrilets et al. 2001). However, several models show that intersexual selection processes can also elevate population fitness by increasing the rate of fixation of beneficial alleles or by decreasing the frequency of deleterious mutations (Proulx 1999, Whitlock 2000, Agrawal 2001, Siller 2001, Lorch et al. 2003). There is some empirical support for the suggestion that sexual selection can increase the risk of extinction. First, comparative studies have shown that male survival rates, relative to those in females, are lower in sexually dimorphic species compared to sexually monomorphic species (Promislow et al. 1992, 1994, Moore and Wilson 2002). Second, Morrow and Pitcher (2003) found that the risk of extinction among extant bird species (threat status) was positively related to an index of postmating sexual selection. Third, several artificial selection experiments in insects have studied the effects on population fitness of relaxing or removing sexual selection, with mixed results. Some have indeed found that the opportunity for sexual selection depresses absolute fitness (Rice 1992, Holland and Rice 1999) but others have found no or even the opposite effect (Promislow et al. 1998, Holland 2002, Martin and Hosken 2003b). Fourth, introductions of birds to islands have shown that extinction rates are higher for sexually dimorphic compared to sexually monomorphic species (McLain et al. 1995, Sorci et al. 1998, McLain et al. 1999). Thus, sexual selection may well be a double-edged sword: while increasing the rate of evolution of reproductive isolation, it may sometimes also increase the probability of extinction in diverging lineages by lowering their absolute fitness.

It should also be stressed here that intralocus sexual conflicts also can depress population fitness considerably, since each sex will constrain adaptive evolution in the other such that neither sex may be able to reach its adaptive peak for any particular trait (Lande 1981, Chippindale et al. 2001, Rice and Chippindale 2001). The evolution of sex-limited expression of genes under sexually antagonistic selection (i.e., sexual dimorphism) may also, in itself, involve considerable costs (see Rice 1984). Again, the selection load that this imposes upon populations (Rice 1992) should increase the probability of extinction. On the other hand, the evolution of sexual dimorphism could rid populations of fitness depressing intralocus conflicts, thus allowing an increase in population fitness. It is unclear if and how these two effects would counterbalance in affecting extinction risk.

6.5.1 SEXUAL CONFLICT AS AN ENGINE
OF EVOLUTIONARY DIVERGENCE

It has been argued throughout this book that sexually antagonistic coevolution can be seen as one form of intersexual selection (i.e, female choice) (chapter 2). It describes the coevolution of male sexual traits (persistence) and female traits (resistance) that bias reproductive success toward certain male phenotypes. Although the idea that sexual conflict and intersexual "arms races" may be important in speciation was published more than two decades ago (Dawkins and Krebs 1979, Parker 1979), it was not until more recently that this possibility has been fully appreciated (Rice 1996b, 1998a, Price 1997, Parker and Partridge 1998, Howard 1999, Gavrilets 2000, Martin and Hosken 2003a). However, all forms of intersexual selection can promote the evolution of reproductive isolation (see Eady 2001, Panhuis et al. 2001, Turelli et al. 2001, Kirkpatrick and Ravigné 2002 for reviews). Yet we suggest that sexually antagonistic coevolution may be a more potent "engine" of coevolutionary divergence and speciation than are other forms of intersexual selection, for several related reasons.

In general, we expect evolution of female resistance/preference to be relatively rapid because selection on females is exclusively direct, rather than relying on indirect benefits to balance direct costs (cf. Kirkpatrick and Barton 1997, Gavrilets et al. 2001, Kirkpatrick and Ravigné 2002). More specifically, all models of preference evolution can account for the spread of novel male traits if one assumes that the preference exists before invasion of the male trait. It is easy to see that any male trait that elicits a favorable response in females should rapidly become established in a population. For example, one would expect the evolution of male traits that exploit "hidden" female preferences (West-Eberhard 1984, Ryan 1990, Arak and Enquist 1993). Yet it is not obvious why female preference would evolve in response to these novel male traits.

Most models predict an evolutionary response in female preference only if its expression is correlated from the outset with indirect or direct benefits provided to females (cf. Fisher 1930, Wu 1985, Berglund et al. 1996). This limits the number of traits that will elicit an evolutionary response in females. In contrast, increased female resistance to novel male traits, by sexually antagonistic coevolution, should evolve under a broader range of conditions: novel manipulative traits in males which exploit "hidden" female preferences are often likely to displace females from the optima to which they have adapted in the absence of the novel male trait, thereby favoring an altered response to such signals in females. Such female evasion of male traits may in theory drive both allopatric (Gavrilets 2000, Gavrilets et al. 2001, Martin and Hosken 2003a) and sympatric speciation (Gavrilets and Waxman 2002).

Models also differ with regard to the conditions under which we expect novelty in female perception to evolve given that male traits and female response to these have reached some form of equilibrium. Under conventional models, we expect novel, or modification of existing, female perceptive abilities only if they match male traits or signals that are associated with indirect or direct benefits provided by males to females (cf. Fisher 1930). The number of evolutionary routes open to females is thus relatively restricted, and female perceptive abilities which respond to nonexisting male signals can, for example, never be favored by this type of selection. In contrast, under sexually antagonistic coevolution, any change in female perceptive ability which decreases, lessens, or shuns the response to male traits/signals can be favored by selection. This means that the number of profitable evolutionary routes that female perception might take is very large, if not virtually infinite. To put it simply, the fact that females are evading, rather than tracking, male traits in the coevolutionary chases which characterize sexually antagonistic coevolution makes the evolution of novelty more likely. Therefore, coevolutionary alterations in male and female traits should be more rapid.

Male "signals" are often numerous, and typically include several different types of stimuli (e.g., visual, auditory, olfactory, and/or physiological). For example, consider fruit fly males, which not only court females vigorously, using a complex of many different visual, chemical, acoustic, and tactile signals (see Rybak et al. 2002), but also transfer an ejaculate containing some 80 different proteins and peptides many of which function as "signals" to females (see chapter 4). Under conventional models of male-female coevolution, it is very difficult indeed to understand evolution of such multiple male signals and multiple female preferences for these, given that preferences are costly (e.g., Schluter and Price 1993, Andersson 1994, Rowe 1999, Pryke et al. 2001, Andersson et al. 2002). The reason is simply that once a single reliable indicator of male direct or indirect mate quality and female preference for this trait are established in a population, costly additional signals and preferences do not easily invade even if these are more accurate indicators of male quality (Iwasa and Pomiankowski 1994). We expect this to happen only if novel preferences carry very low costs (Pomiankowski and Iwasa 1993, Iwasa and Pomiankowski 1994), when additional male traits are reliable indicators of independent and additional "aspects" of male quality (Johnstone 1995) and/or when the various male signals are intended for different receivers (Pryke et al. 2001, Andersson et al. 2002). In contrast, sexually antagonistic coevolution actually predicts the evolution of multiple male signals under a wide range of conditions. Assuming that females exhibit a large number of hidden preferences (Arak and Enquist 1993), it is easy to see how males can evolve signals or other traits which exploit/manipulate multiple female perception biases and how females then evolve partial resistance to each of these multiple male traits (Arak and Enquist 1995, Gavrilets et al. 2001). Hence, in a more general sense, intersexual selection driven by sexual conflict predicts that male-female coevolution should often involve multiple traits in both sexes. Male-female coevolution should thus be multidimensional and there should be a greater number of coevolutionary trajectories along which coevolution could proceed. Needless to say, this alone would tend to promote population differentiation and allopatric speciation.

At this point, it is interesting to compare these arguments with similar arguments made by students of other forms of coevolutionary interactions. The claim made here, that sexually antagonistic coevolution should tend to be more rapid and divergent than other modes of sexual coevolution, to some extent reflects the literature on coevolution between interacting species: coevolutionary scenarios where species are exploiting/evading each other do differ from those where species have convergent interests. Models of interspecific coevolution between antagonistically interacting species (e.g., parasite/pathogen-host or predator-prey systems) are generally characterized by predicting complex and endless coevolutionary chases between the victim and the exploiter (e.g., Gavrilets 1997), which have become known as "red queen" processes (van Valen 1973) or coevolutionary "arms races" (Dawkins and Krebs 1979). A particularly interesting analogy is found in the literature on the evolution of mimicry (see Turner 1987, Gavrilets and Hastings 1998, Holmgren and Enquist 1999, Mallet 1999). Models of purely mutualistic mimicry, such as classic Müllerian mimicry, where both coevolving species are unpalatable and benefit from coevolving resemblance, predict (i) relatively low rates of coevolutionary convergence toward (ii) a stable equilibrium state that is (iii) spatially monomorphic. In contrast, coevolutionary models of exploitative mimicry, such as in Batesian mimicry, where a palatable mimic coevolves with an unpalatable model species, predict (i) higher rates of divergent coevolution resulting in (ii) complex dynamics and (iii) geographic polymorphism in the traits that mediate interactions between coevolving species. Spatially and temporally dynamic coevolutionary interactions thus seem to be a general emergent property of antagonistically coevolving systems (Nuismer et al. 2000, Gavrilets et al. 2001).

It should also be noted that the potency of sexually antagonistic coevolution in generating divergence and speciation is not in principle limited to interactions between alleles expressed in the two parents (i.e., prezygotic interactions). For example, Kondoh and Higashi (2000) showed that sexual conflict over maternal provisioning of offspring can result in a dynamic arms race between imprinted growth promoters (paternally expressed in offspring) and growth suppressors (maternally expressed in offspring). They concluded that a mismatch in such growth promoters/suppressors is likely to result in the evolution of postzygotic reproductive isolation via hybrid inviability.

Sexual conflict may also play a quite different role when allopatric populations which have partly diverged, for whatever reason, come into secondary contact. In such cases, there may be sexual conflict over mating due to the differential fitness effects to males and females of hybridization between incipient species (see Parker and Partridge 1998). If "hybrid" offspring suffer depressed fitness, and mating costs to males are small, it may pay males but not females to mate in encounters between individuals from different populations. Parker and Partridge (1998) stressed that the "balance" between sexually antagonistic adaptations in the two sexes (cf. Arnqvist and Rowe 2002a) at the time of secondary contact may greatly affect the probability of reproductive isolation in such cases. If male persistence adaptations are relatively efficient, males will be able to fertilize the eggs of heteropopulation females, leading to genetic intermixing between populations. In contrast, if female resistance adaptations are efficient, speciation is more likely as males may be unable to fertilize the eggs of heteropopulation females. This mechanism has been invoked to explain the absence of speciation in the guppy, *Poecilia reticulata*, where male harassment of females and coercive matings are believed to have prevented the evolution of reproductive isolation (see Magurran 1998).

6.5.2 POPULATION CROSSES—INFERRING PROCESS FROM PATTERN

It has been suggested that results from crosses of allopatric populations on traits such as mating rate, offspring production rate, sperm displacement, and survivorship could be used to illuminate processes of male-female coevolution (Parker and Partridge 1998, Clark et al. 1999, Andrés and Arnqvist 2001). Recent theory casts doubt on some of the hypotheses derived from this reasoning (Rowe et al. 2003). Nevertheless, the ideas have motivated a number of studies and data (Brown and Eady 2001, Knowles and Markow 2001, Hosken et al. 2002, Nilsson et al. 2002, 2003) and these do suggest interesting patterns in male-female coevolution.

Specifically, Andrés and Arnqvist (2001) argued that patterns of statistical interactions between male and female origin in their effects on female reproductive response could be used to identify a role of sexual selection, and to distinguish among alternative modes of sexual selection. Assuming that more than one signal receiver pathway was involved in determining a trait (e.g., remating rate), they reasoned that mismatches may emerge in divergent populations. For example females from population A may be more sensitive to signals of males from population B than from population A. Likewise females from population B may be more sensitive to signals of males from population A than from population B. This would be indicated by an interaction between male and female origins. In contrast, if divergence involved only variation in a single male signal, variation in both efficiency/strength of the signal and the sensitivity of female perception might be observed, but males from different populations would rank similarly among females from different populations, and we would see no interaction. We also note that since interactions demonstrate that females from different populations vary in their response to male signals, they have been used as one line of evidence for female choice or cryptic female choice (see chapter 4).

The pattern of the interaction might also offer insights into the processes under which these male signals and female receptors coevolve. Apart from random processes, such as founder effects and/or genetic drift, there are at least two adaptive evolutionary scenarios that can lead to divergence. The divergence can have evolved either by sexually antagonistic coevolution (Rice 1998a, Parker and Partridge 1998) or by male-female coevolution driven by selection on females to secure indirect or direct benefits. Clark et al. (1999) and Andrés and Arnqvist (2001) both suggested that the pattern of male-female genotypic interactions should differ under these processes. In short, random processes should result in a variable pattern where females' relative reproductive response to males from their own population/strain should not differ overall from that to males from other populations/strains. In contrast, as originally pointed out by Parker and Partridge (1998), since females should evolve resistance to males with which they are coevolved if divergence is driven by sexually antagonistic coevolution, due to fitness costs of antagonistic male adaptations, they may respond more weakly than average to males with which they are coevolved (figure 6.5). Under the alternative adaptive hypothesis, divergence through indirect benefits, females should evolve preference for male signals, thereby responding more strongly than average to males with which they are coevolved. However, the absolute magnitude of divergence between populations or genotypes used in the experiment will play a crucial role: as qualitative coevolutionary modifications accumulate over time, sexually antagonistic coevolution will also result in a situation where females respond more strongly than average to males with which they are coevolved (see figure 6.6). Moreoever, recent theory suggests that the reasoning used by Andrés and Arnqvist (2001) is limited to those cases where the shape of the female response function is fixed (Rowe et al. 2003). If the sensitivity of preference/resistance functions (the slope) is free to evolve, then their predictions do not hold. Coevolution of single signal receiver systems can lead to interactions in population crosses, and the pattern of these interactions cannot be used to distinguish among alternative modes of male-female coevolution.

The data from population crosses are mixed. Interactions between the sexes are quite common, though absent in some cases (e.g., Clark et al. 1999, Brown and Eady 2001, Knowles and Markow 2001, Hosken et al. 2002, Nilsson et al. 2002, 2003). Yet these studies do suggest rapid evolution of male signals and female receptors. The fact that interactions are common suggests that populations are evolving along divergent rather than parallel paths. Finally, the signals are mediating reproductive traits (e.g., remating rate, oviposition rate) that are likely to have strong direct selection operating on them. Collectively,

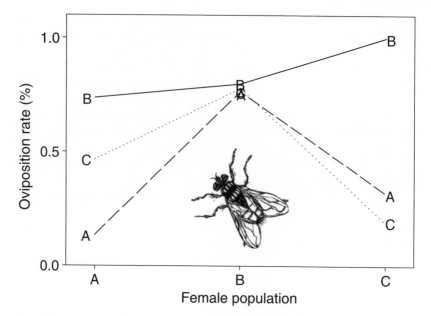

Figure 6.5. Females sometimes tend to exhibit relatively weak reproductive response to males with which they are coevolved, such that interactions are observed when reciprocal population crosses are performed. This figure shows the rate of egg laying two days after mating among house fly females from three different populations mated to their "own" as well as two "foreign" males. Letters inside the panel represent male populations. Note that the stimulatory effect that mating has on female egg laying is weakest in within-population crosses in two out of three cases (after Andrés and Arnqvist 2001). Such a pattern is consistent with sexually antagonistic coevolution, where females more effectively resist males with which they are coevolved, but there are also alternative explanations.

these data are congruent with sexually antagonistic coevolution (section 6.5.1), although they do not rule out other modes of male-female coevolution.

6.6 Sexual Conflict and Sex Chromosomes

The evolution and genetics of sex chromosomes is a very large and active field of research in evolutionary biology, most of which is outside the scope of this book. However, we will briefly discuss the fact that intralocus sexual conflict is a key element in the favored hypothesis for the evolution of sex chromosomes and, conversely, that the genetic architecture of sex determination has important implications for sexual selection and for the evolutionary dynamics of sexual conflict.

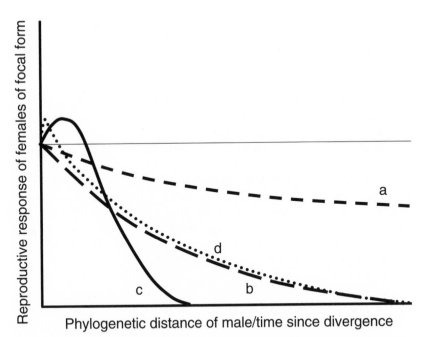

Figure 6.6. For females of a given focal population, we expect any form of reproductive response to a "signal" transmitted by allopatric males to decrease on average with decreased relatedness, until females no longer respond and populations are effectively reproductively isolated. Here, the reproductive response could represent, for example, probability of mating, sperm precedence, or reproductive effort. It has been suggested that the shape of this relationship should differ slightly under different processes of allopatric divergence, although this proposition has not been verified. When divergence results from random processes, as an indirect result of natural selection, or from intrasexual selection, relatively slow rates of gradual divergence may be expected (a). Traditional models of female choice predict a more rapid divergence (b). Several authors have suggested (i) that female choice generated by sexually antagonistic coevolution should generate the most rapid divergence and (ii) that one possible outcome is that females may be more responsive on average to signals of closely related allopatric forms (c) (see text). A pattern similar to this might also be expected if females are selected to avoid inbreeding by choosing males that are optimally unrelated (d). (Bateson 1983.)

According to the prevailing theory of sex chromosome evolution, sex chromosomes evolved from an autosomal pair carrying a genetic sex-determining gene (see, e.g., Rice 1996a). In this ancestral species, sexually antagonistic alleles expressed in both sexes would become established only if the benefit to one sex more than outweighed the cost to the other. This is not true, however, if these alleles occur at loci tightly linked to the sex-determining locus. The reason is that linkage would allow sexually antagonistic alleles to co-occur with the sex-determining gene coding for the gender in which they are beneficial. Such linkage would cause sex-biased gene transmission, permitting the accumulation of even sexually antagonistic alleles that are highly deleterious to one sex while only moderately beneficial to the other. Once sexually antagonistic alleles start accumulating on the primitive sex chromosomes, there will be selection for reduced rate recombination between these. This is simply because crossing over would sometimes cause a mismatch between the gender-determining allele and the linked sexually antagonistic alleles. With suppressed recombination, and subsequent degeneration of the nonrecombining chromosome, the result would be functional sex chromosomes. Under this hypothesis, therefore, it is the fact that certain alleles have antagonistic fitness effects in the two sexes that has led to the evolution of sex chromosomes.

There are also reasons to believe that the genetic architecture of sex determination, and thus the genetic "rules of the game," has a major impact on the evolutionary dynamics of sexual conflict. In particular, we would often expect the heterogametic sex to have a coevolutionary "edge" over the homogametic sex with regard to intralocus sexual conflicts. First, theory predicts that the X chromosome in an X/Y system, or the Z in a Z/W system, may be enriched with alleles that favor the heterogametic sex (see Rice 1984). The reasons for this are as follows. A rare sexually antagonistic allele of an autosomal locus that is expressed in both sexes will increase in frequency in a population only if the advantage to one sex is larger than the disadvantage to the other (i.e., positive net selection). In contrast, a sexually antagonistic allele that is recessive and favors the heterogametic sex will spread when rare if located on X (or Z), even if the disadvantage to the homogametic sex is much larger than the benefit to the heterogametic sex. The logic behind this reasoning is fairly straightforward. The deleterious effects of an X/Z-linked sexually antagonistic allele will rarely be expressed when the allele is uncommon in a population, since phenotypic expression is masked by dominance in the homogametic sex, while benefits to the heterogametic sex will be enjoyed by all carriers of this allele due to hemizygous expression of X/Z-linked genes. As such, X/Z-linked recessive sexually antagonistic alleles increase in frequency in a population; however, counterselection in the homogametic sex will halt the spread of the allele due to an increased frequency of homozygotes.

A very similar argument applies for sexually antagonistic alleles that are dominant but instead favor the homogametic sex (Rice 1984). Here, a rare mutant would be expressed equally in the two sexes, but because the X spends twice the time in females as in males, the effect of selection is greater in females. This form of frequency-dependent selection should promote polymorphism at X/Z-linked sexually antagonistic loci (Gibson et al. 2002). In accord, studies of species in which males are heterogametic indicate that genes relating to male reproductive function and/or under sexual selection have accumulated on the X (Reinhold 1998, Hurst and Randerson 1999, Saifi and Chandra 1999, Wang et al. 2001). One study has indicated that a large fraction of the genome-wide sexually antagonistic genetic variation is located on the X (Gibson et al. 2002). Conversely, at least a few studies of species in which females are heterogametic have demonstrated that genes affecting female reproductive traits, such as mating preferences (Iyengar et al. 2002), pheromone production (Sperling 1994), and egg production (Fairfull 1990), are located on the Z. A possible case of a Z-linked allele which is beneficial to females but deleterious to males was recently reported by Lee et al. (2002) in moorhens, *Gallinula chloropus*.

Second, one could argue that the presence of a unique chromosome in the heterogametic sex (i.e., Y/W) would allow the accumulation of novel mutations encoding traits favorable to the heterogametic sex on these chromosomes, since these mutations would instantly be sex limited in their expression and, therefore, freed from counterselection in the homogametic sex (e.g., Reinhold 1999, Roldan and Gomiendo 1999, Chippindale and Rice 2001, Reeve and Pfennig 2003). This could then further strengthen the coevolutionary "edge" of the heterogametic sex. However, because suppressed recombination has led to the subsequent degeneration of the Y/W (e.g., Rice 1996a), it is not clear how important this argument is: much of the Y in mammals and the W in birds consists of presumably nonfunctional heterochromatin. Interestingly enough, this genetic wasteland does contain some functional genes. For example, several genes related to male sex determination and gonad function are located on the Y in mammals (e.g., Willard 2003) while genes for female sex determination and gonad function seem to be located on the W in birds (Ellegren 2001, Reed and Sinclair 2002).

Many authors have noted a correlation between sexual dimorphism and male/female heterogameity. For example, birds and butterflies are female heterogametic and often have colorful males, in contrast to mammals and most insects in which males are heterogametic. In the most ambitious quantitative comparative study to date, Reeve and Pfennig (2003) showed that this pattern holds true across a range of taxa. To some extent, the direction of sexual dimorphism therefore seems to stem from differences in the sex chromosome system. Reeve and Pfennig (2003) suggested that the rapidity with which a genetic correlation between a male trait and female preference for this trait evolves may play a key role. Cases in which both the trait and the preference are Z linked are most potent in this regard, since sons can then inherit both the trait and the preference from their father. This fact could then contribute to ZZ/ZW systems being most conducive to sexual dimorphism. We note, however, that there are several possible explanations for this interesting pattern (see above).

7

Concepts and Levels
of Sexual Conflict

Like most quickly growing fields, the study of sexual conflict is rich in debate over concepts, assumptions, and interpretation, and much of the debate revolves around terminology. One striking feature of this field is the diversity of scholars that have become interested in questions relating to sexual conflict. Contributions have been made by evolutionary geneticists, game theoreticians, behavioral ecologists (students of sexual selection and mating systems), developmental biologists, evolutionary psychologists, molecular biologists, reproductive physiologists, and evolutionary biologists interested in speciation. Although the broad relevance of the field makes it a very interesting one, it has also led to problems because academics from different domains tend to think and write in distinct ways. This is reflected, for example, in the lack of a consistent terminology in the field and in the fact that a series of distinct definitions of "sexual conflict" have been proposed (table 7.1). Early work on sexual conflict was dominated by game theory thinking, where sexual conflict was referred to as a "battle of the sexes," in which males and females were employing various "armaments" in an "arms race" or a "war of attrition" (e.g., Dawkins 1976, Dawkins and Krebs 1979, Parker 1979, 1983a). This terminology has gradually been replaced by one using more explicitly genetic or evolutionary terms. In this chapter we discuss a few problematic concepts and views in the literature on sexual conflict (see also chapter 2). We try to find commonalities between different terms/views and attempt to frame a few remaining issues.

7.1 Levels of Analysis

If we assume that the central goal of our research is to understand how, and to what extent, sexual conflict accounts for both the origin and maintenance of the diversity of sexual traits and male-female interactions we see today, then we must have an understanding of coevolutionary history. We can imagine some ancient and sexually reproducing isogamous lineage where competition occurs between like individuals. Even here, the genetic interests of two core-producing individuals will not be identical (Parker et al. 1972), illustrating the point that conflict between mates is ubiquitous (Lessels 1999). Following the

Table 7.1.

Examples of definitions of sexual conflict

Definition of sexual conflict	References
A conflict between the evolutionary interests of individuals of the two sexes	Parker 1979
Arises because each parent's fitness is generally maximized if it invests less and the other parent invests more than would maximize the other parent's fitness	Lessels 1999
When the genetic interests of males and females diverge	Chapman et al. 2003a
When an allelic substitution at one locus expressed in one sex compromises the genetic interests of individuals of the other sex and thus selects for a new allele at an interacting locus expressed in the latter sex	Rice and Holland 1997
A sex x genotype interaction for fitness	Rice and Chippindale 2001
When an increase in the average fitness in one sex translates into a reduction in the average fitness in the other	Pizzari and Snook 2003
Different evolutionary interests of the two sexes	Parker and Partridge 1998
Sex difference in the covariance between promiscuity and offspring numbers	Shuster and Wade 2003
When the slopes of any relation between either (i) a single trait expressed in both sexes or (ii) an outcome of male-female interactions on one hand, and fitness on the other, differ in sign in the two sexes	Gavrilets et al. 2001
When traits favored by reproductive competition within one sex are costly for individuals of the other sex, when expressed either phenotypically or as a result of male-female interactions	Parker 1984, Rice 1998a

definition of Parker (1979), however, sexual conflict requires not only sexual reproduction but also the existence of sexes. With the evolution of anisogamy, there are suddenly two classes of individuals. As Trivers (1972) emphasized, differences in reproductive investment between these classes lead to distinct sex roles and evolutionary interests of individuals belonging to each of the two (Parker et al. 1972, Bulmer and Parker 2002), and those interests are often in conflict (Parker 1979, 1984). By analogy with the literature on parent-offspring conflict (see Godfray 1995, Mock and Parker 1997), this basic state of sexual conflict has been referred to as the battleground level (Lessels 1999). Sexual conflict at this level drives sexually antagonistic coevolution, and this shapes the sexes. Divergence of the sexes may be characterized by the evolution of interacting sexually antagonistic adaptations, or armaments, in both sexes. Past conflicts and antagonistic traits will, however, become more and more obscured as adaptations and counteradaptations are piled upon one another. If we stop this process at any time, we may assess the economics of particular sexual interactions resulting from the current balance of sexually antagonistic adaptations. These outcomes will be occurring in extant populations that can be seen as the observable level where we can address the economics of reproduction experimentally.

The economics of a particular intersexual interaction (e.g., mating rate, reproductive rate, parental care) in an extant population does not inform us about the existence or nonexistence of sexual conflict at the battleground level. Conflict exists because there are two sexes, and therefore will be present in all anisogamous species, and has neither an evolutionary starting point nor an end. Similarly, the current outcome of an interaction between the sexes, or the adaptations currently employed by sexes in this interaction, are not informative of where the battle has come from or where it is going: adaptations change continually. Studies of extant populations are empirically tractable, but theory suggests that economic studies at this level may often indicate little or no residual sexual conflict despite past or even ongoing sexually antagonistic coevolution. Defining sexual conflict as something occurring exclusively within extant populations (see table 7.1), essentially equating sexual conflict with the presence of sexually antagonistic selection (e.g., Shuster and Wade 2003) or the existence of alleles with sexually antagonistic effects (e.g., Rice and Holand 1997), would therefore underestimate the role of sexual conflict in male-female coevolution. The general problem of sexual conflict being "hidden" from our eyes has been discussed by several authors (Chapman and Partridge 1996a, Rice 1998a, 2000, Härdling et al. 2001, Arnqvist and Rowe 2002a, Arnqvist 2004) (see figure 2.5). A related point concerns coevolutionary dynamics: although episodes of sexually antagonistic coevolution may be largely responsible for the origin and elaboration of a suit of interacting traits in both sexes, there are reasons to believe that other coevolutionary processes (e.g., indirect selection on female preference/resistance; see chapter 2) may be important during extended periods of maintenance of these traits. The role that sexual conflict may have had would then be very difficult to detect by empirical studies in extant populations.

In any coevolving lineage, there may be multiple issues over which the interests of males and females diverge (chapters 3–5), and these conflicts may be interrelated. The evolution of a sexually antagonistic adaptation in one sex, employed in any one intersexual interaction, will likely change selection among members of the other sex for that particular interaction as well for other interrelated conflicts. For example, imagine a case where males acquire a novel seminal gonadotropin that more efficiently elevates the current female reproductive rate. Such an adaptation would be likely to then also affect the dynamics of sexual conflicts over the mating rate. In other words, changes in the dynamics of one particular conflict may be rooted in changes in another (Rowe and Arnqvist 2002).

The challenges of inference from present to past, and from one conflict to another, are not restricted to sexual conflict but are a property of all evolutionary questions. Nevertheless, they may be exacerbated in sexually antagonistic coevolution because of the speed of coevolution and its unpredictable direction. As we have shown in the previous chapters, there are several ways in which clever experiments, comparative studies, and studies of experimental evolution can unveil sexual conflict. However, we do need to be fully aware of the limitations of projecting backward and forward.

7.2 Resolution of Sexual Conflict

The term "resolution," or "resolved," is often used in the context of sexual conflict. In human society and politics, the term resolution describes a variety of processes aimed at nonviolent conflict prevention/solution. These processes involve negotiation, mediation, and arbitration. Used in this context, conflict resolution has a dual aim: to make conflict fade away and to do so in a peaceful manner (Galtung et al. 2002). The use of the term conflict resolution in proximate studies of animal behavior is similar (e.g., Aureli and de Waal 2000). Yet these processes and implications are very different from the meaning of evolutionary resolution (cf. Kennedy 1992). Resolution of evolutionary conflicts has no aim, but refers to the process by which players adapt to forces that originate in the conflict itself (Parker 1979, 1983a, 1983b, Yamamura and Higashi 1992, Godfray 1995, Härdling et al. 2001). In sexual conflict, it refers to the process of adaptation and counteradaptation that occurs during sexually antagonistic coevolution. In theory, the dynamics of sexual conflict resolution will depend on factors such as the relative strength of sexually antagonistic selection and the relative efficacy of sexually antagonistic adaptations in the two sexes (e.g., Parker 1979, 1983a, Clutton-Brock and Parker 1995, Härdling et al. 2001, Shuster and Wade 2003).

It is important to note that resolution does not in any real sense make sexual conflict disappear or even fade. There is no solution to sexual conflict that somehow causes the evolutionary interests of individuals of the two sexes to coincide. Further, sexually antagonistic coevolution can be a complex and never-ending dynamic process which makes prediction beyond the short term, even in carefully circumscribed interactions, difficult and imprecise (Parker 1979, 1983a,b, 1984, Hammerstein and Parker 1987, Arak and Enquist 1993, 1995, Clutton-Brock and Parker 1995, Härdling 1999, Gavrilets 2000, Gavrilets et al. 2001, Härdling et al. 2001, Enquist et al. 2002).

It is also important to recognize that sexually antagonistic coevolution can involve two distinct forms of conflict resolution. One is characterized by quantitative and reciprocal changes of persistence and resistance abilities, and involves the accumulation and loss of novel traits in both sexes. This is captured in the description of a classic arms race between the sexes, involving continual escalation and deescalation of adaptations that have largely additive effects on the outcome of interactions (Parker 1983a,b, Härdling 1999, Gavrilets et al. 2001). The coevolution of grasping and resistance adaptations in water striders discussed in chapter 3 can serve as an empirical example of this form (Arnqvist and Rowe 2002a). Another form occurs when specific persistence and resistance alleles or traits interact in their effect on the outcome of intersexual interactions. Imagine, for example, that a resistance allele R_1 is more resistant against the persistence allele P_1 than it is against P_2, while the opposite holds true for another allele R_2 in the population. This will result in frequency-dependent selection that promotes the maintenance of polymorphism in resistance/ persistence. The coexistence of multiple female morphs in damselflies, which use distinct strategies to evade male harassment (see chapter 3), is perhaps the best worked out example of this form (Andrés et al. 2002). Although these two forms of conflict resolution differ in coevolutionary dynamics, they are not mutually exclusive.

7.3 Winners and Losers of Sexual Conflict?

By analogy with many conflicts of interest in human societies (wars, sports events, etc.), interlocus sexual conflict is often thought of as something that one sex can "win" and the other "lose." Throughout the literature on sexual conflict, adaptations in one sex are often referred to as being harmful to the other sex, and what is beneficial for one sex under sexual conflict is costly for the other. Likewise, on the grander scale, sexual conflict is often analogized as the "battle of the sexes." This term was popularized by Dawkins (1976) as a description of evolutionary conflicts between males and females, but was introduced much earlier in the classic book on game theory by Luce and Raiffa (1957). They used it to describe a situation where a man and a woman disagree over the choice for an evening's entertainment. This lexicon of sexual conflict has tempted some to envision male and female fitness as evolving independently, such that an increase in the average fitness of one sex is translated into a reduction in the average fitness of the other (e.g., Pizzari and Snook 2003; table 7.1). Parker originally employed the concept of winning and losing sexual conflict in his discussion of game-theoretic models (Parker 1979, 1983a,b). Here, each sex had a discrete strategy set and the possible payoffs were variants of win, lose, or draw. In narrowly defined games such as these, it can sometimes be useful to think about conflict between individual males and females as something that can be won or lost. Nevertheless, this logic cannot be translated to discussions of the evolving fitness of the two sexes.

The average fitness of males and females is not independent because each offspring has a mother and father, and, in fact, must be equal to one another when the primary sex ratio is the typical 1:1 (see Fisher 1930, Queller 1997, Getty 1999, Rice and Holland 1999, Rowe and Arnqvist 2002, Shuster and Wade 2003, Arnqvist 2004). Therefore, one sex cannot "win" a conflict in the sense that its fitness has increased at the expense of the fitness of the "losing" sex: whatever happens to the average fitness of one sex will also happen to the average fitness of the other. Likewise, there cannot be a "battle" between males and females in the sense that these two classes of individuals are striving toward victory. In this sense, sexually antagonistic coevolution is quite different from antagonistic coevolutionary interactions between species (e.g., parasite-host and predator-prey systems), because in these cases the population fitness of one species can increase at the expense of the other. Wade and Shuster (2002) have criticized earlier ESS models of parent-offspring conflict for a very similar reason (see also Queller 1997).

If the sexes cannot win or lose, who can? Specific resistance and persistence alleles can. In an evolutionary sense, the dynamics of alleles most precisely describes sexually antagonistic coevolution. In interlocus conflict, the coevolving units are alleles at interacting loci that code for male persistence and female resistance traits. For example, the invasion of a given female resistance allele has the potential for two effects; one in other alleles at the same locus expressed in females, and another at different loci expressed in males. First, if favored because of its resistance properties, it by definition will depress the relative fitness of other resistance alleles at the same locus. Second, it can alter the relative fitness among a set of alleles coding for persistence at an interacting locus (or loci) expressed in males. In this way, one can think of an antagonistic allele expressed in one sex as depressing the fitness of an allele expressed in the other sex. However, this depression in allelic fitness is relative to other alleles at that same locus, and must be accompanied by the increase in fitness of one of its (within sex) competitor alleles. Genetic conflicts such as these can indeed be "won," in the sense that the spread of one allele in a population can lead to the exclusion of another. All the while, the average fitness of one sex remains the same as that of the other (Rice and Holland 1997, 1999). Sexual conflict can depress the fitness of a population (section 6.5), but never one sex independently of the other.

The literature on sexual conflict is riddled with phrasings such as "it is in the interest of males but not females" and "females benefit from resisting males," and this book is certainly no exception. The first phrase abbreviates the fact that the sign of selection on a specific trait or outcome is different in the two sexes. The phrase "females benefit from resisting males" means that, all else equal, females that resist males more efficiently are favored over other females.

7.4 Sexual Conflict over the Control of Interactions

It is sometimes said that there is sexual conflict over control of the outcome of reproductive interactions between the sexes (e.g., Eberhard 1996, 1998). Strictly speaking, however, selection can act only on the outcome of specific interactions (e.g., mating, egg production, parental care, etc.) and not on the control of such outcomes. Theory suggests that we should instead see control of the outcome of sexual interactions as a reflection of sexual conflict resolution (e.g., Parker 1979, Härdling et al. 2001), or the balance between sexually antagonistic adaptations that are employed by the two sexes, rather than something over which there can be sexual conflict. Control, then, is the dynamic result of sexually antagonistic coevolution and not a generator of such coevolution.

A related issue concerns the concept of evolutionary "power" in conflict resolution. In theory, we can choose either of two standpoints when predicting the outcome of sexual conflict resolution. On the one hand, we could predict that the outcome should simply reflect the relative strength of antagonistic selection in the two sexes. This has been referred to as the "relative value of winning" in game theory models (see below; Clutton-Brock and Parker 1995). If, for example, selection to achieve matings in males is more intense compared to selection to resist matings in females, we would expect population level mating rates to become elevated. This view, the spirit of which is advocated by Shuster and Wade (2003), implicitly assumes that neither of the sexes has an inherent "edge" in conflict resolution. This may not generally be true. In his game theory models, Parker has suggested that the evolutionary "power" of the sexes may frequently differ and that this will affect conflict resolution (see Clutton-Brock and Parker 1995). There can be several related reasons for such a discrepancy. First, the absolute costs of producing/bearing persistence adaptations may differ from those of resistance adaptations. A costly morphological trait in males may, for example, be countered by a slight and more economical behavioral modulation in females. Second, the efficacy of persistence and resistance adaptations may differ. It may, for example, sometimes be easier to efficiently avoid harassing males for females than to pursue reluctant females for males. Third, the origin of novel adaptations may be more likely in one sex than the other, because of differences in the number of genes involved in determining persistence and resistance or because of other idiosyncratic details of the biology of persistence/resistance. This second standpoint puts focus on the various constraints that might be important to the resolution of sexual conflict (see also Beekman et al. 2003).

7.5 The Intensity of Sexual Conflict

A central concept in the literature is the strength, or intensity, of sexual conflict. It is intuitive that this should somehow capture the "power" of the engine that drives sexually antagonistic coevolution. It would also be immensely useful for comparative studies if we could assess the strength of sexual conflict in different populations or species. Despite the appeal of the concept, it has rarely been defined and it is not clear whether we can ever hope to measure it in an encompassing manner. In contrast, we can directly assess both the opportunity and strength/intensity of selection in natural populations (Lande and Arnold 1983, Endler 1986). These methods can be employed to measure the strength and sign of selection on sexually antagonistic traits, and on the interactions in which they are employed.

The only empirically tractable definition of the strength of sexual conflict suggested so far employs a phenotypic selection gradient approach (e.g., Shuster and Wade 2003). This would involve estimating male and female fitness functions for a given outcome of male-female interactions (e.g., mating rate, number of eggs laid at time t, sperm precedence, etc.). If the sign of these functions differs, then there is sexual conflict over the outcome of that interaction. The intensity of the conflict would then be some measure of the difference in selection gradients. This metric is analogous to the "relative value of winning" a conflict to individuals of the two sexes, a concept used by Parker in several game theory models of sexual conflict. (Parker 1979, 1984, Clutton-Brock and Parker 1995).

It makes sense for a measure of the intensity of sexual conflict to relate in some way to the broader overall conflict between the sexes. Metrics based on the phenotypic selection gradient approach would give a measure of residual sexual conflict in extant populations, but would not relate to the role conflict has had in shaping adaptations in the sexes. Consider, for example, a case where fitness functions were found to be of the same sign in the two sexes. One could only fairly conclude that there was no *current* sexual conflict over this single interaction. Male and female fitness functions may be coinciding because of accumulated antagonistic adaptations in both sexes. With this caveat in mind, it may prove to be an empirically useful measure of the current strength of sexually antagonistic selection.

One measure of the strength or intensity of sexual conflict could be the cost to population fitness of sexually antagonistic adaptations. As we have discussed in previous chapters (2 and 6), there is much less opportunity for sexual conflict in monogamous species compared to polygamous species. Evolutionary experiments that contrast monogamous and polygamous lineages offer a method to assay this measure of sexual conflict ("reproductive load"; see Rice 2000). Several experiments have contrasted lines, derived from a common ancestor, that have evolved under monogamy with those evolved under polyandry. Although some of these experiments have seen fitness depressed in polyandrous lines (Rice 1992, Holland and Rice 1999), others have not (Promislow et al. 1998, Holland 2002, Martin and Hosken 2003b).

7.6 Sexual Conflict over Mate Choice

Mating is associated with a series of costs and benefits to both sexes (chapter 3). If the costs and benefits of mating are asymmetric to males and females and are largely unrelated to the phenotype/genotype of their mates, there will be sexual conflict over, for example, the mating rate per se or over the ecological setting of mating (e.g., Holland and Rice 1998, Gavrilets et al. 2001) (see section 2.7.3). If costs and benefits of mating are additionally determined by the phenotype/genotype of the other sex, we expect the evolution of mate choice in either or both sexes (see chapter 2). Basic theory tells us that differences in variance in mate "quality" and in the potential reproductive rate between the sexes will together generate an asymmetry in the optimal degree of mate selectivity between males and females, such that it will pay one sex more than the other to exhibit preference (Bateman 1948, Trivers 1972). Parker (1979, 1983b) pointed out that, to the extent that mate preference or mate selectivity is determined by the same set of genes in the two sexes, this will result in intralocus sexual conflict over mate choice. This might be the case when the direct fitness payoff of mating to one sex depends on the other sex (e.g., Parker 1983b, Price et al. 1993, Tregenza et al. 2000), but may also occur when payoffs depend on the particular genes passed on to offspring. In theory, indirect costs and benefits may differ between the sexes with regard to, for example, incestuous mating (Dawkins 1976, Parker 1979, 1983b, Hammerstein and Parker 1987) or mating between incipient or closely related species (Parker 1979, 1983b, Hammerstein and Parker 1987, Parker and Partridge 1998). We note, however, that this is the topic of traditional models of sexual selection, and the evolutionary outcome of these cares has been dealt with by models of optimal mate choice (see chapter 2).

Distinct from this reasoning is the suggestion that there may be sexual conflict over female choice (see Gowaty 1997b, Moore et al. 2001, 2003). The motivating idea here is that it is in the best interest of females to freely exercise mate choice, in pursuit of direct and/or indirect benefits, while it is not in the best interest of males to be chosen among. Males are said to be under selection to "override" or "circumvent" female choice, or cryptic female choice, and this should be costly to females and counter to their genetic interests. We see conceptual ambiguities with this line of reasoning (see also section 4.2.2.4). Female choice is not the outcome of an interaction nor the property of an individual, but is any process which leads to nonrandom reproductive success among males, no matter what forces of selection are acting on female preference traits. Sexual conflict is one of the processes that may lead to the evolution of preferences (chapter 2). Therefore, quantitative or qualitative aspects of the preference can be affected by sexual conflict, but it is unclear how there could be sexual conflict over female choice itself or why males should suffer costs from being chosen among.

Further, an adaptation in males that gives bearers a mating advantage over same-sex individuals lacking this trait will be sexually antagonistic if it results in elevated direct costs of mating for females (chapter 2). The "conflict over female choice" logic instead posits that such a trait is sexually antagonistic if it reduces the direct or indirect benefits of choice to females. It is presently unclear whether anything important can be gained by essentially regarding the reverse of a regular mate choice scenario as sexual conflict, and the boundary, if there is one, between traditional models of sexual selection and this view has not yet been carefully explored theoretically (Chapman et al. 2003a, Cameron et al. 2003). In many ways, "conflict over female choice" seems indistinguishable from conventional sexual selection theory (see chapter 2) and to a large extent merely reiterates the problems involved with the maintenance of honesty in male signals of quality (e.g., Zahavi 1975, 1977, Dawkins and Krebs 1978, Hamilton and Zuk 1982, Grafen 1990a, Maynard Smith 1991, Johnstone and Grafen 1992, Johnstone 1995, Kokko 1998, Proulx et al. 2002). Although the spread of dishonest signaling of quality among males (i.e., "cheating") may indeed be associated with a decrease in female fitness, this has been dealt with in conventional models of female choice. Here, more theoretical work is warranted (see chapter 2).

8 Concluding Remarks

The main aim of this book has been to build the case that we may need to revise the traditional view of coevolution between genes expressed in the two sexes and the consequences of this process for the differences we now see between the sexes. Reproduction in sexual species is rife with conflicts of interest between males and females. Most of this book has been a journey through the natural history of a selection of adaptations in both sexes that we believe have been shaped by such conflicts (chapters 3–6). Even though the combined weight of all these examples convincingly demonstrates a role for sexual conflict, most of the inferences we have made along the way, implicit or explicit, are based chiefly on the current function of traits and the overt nature of intersexual interactions. In some sense, this is unsatisfactory. Descriptions of this kind can certainly be indicative of sexual conflict (chapter 1), but this does not exclude the possibility that alternative processes may also have played a substantial role (chapters 2 and 7). At this point, we might ask what is needed in order to better understand the role of sexual conflict in generating the diversity of adaptations and species observed in nature.

It is clear that creative empirical research programs are needed to better test many of the propositions made in this book. Alternative forms of male-female coevolution differ primarily in the economics of intersexual interactions, and projection from present economics to the past can be problematic; therefore, disentangling alternatives is not an easy task. Some of the best-worked-out case studies, including damselflies, dunnocks, fruit flies, guppies, penduline tits, snails, and water striders, have taught us four things. First, it is certainly possible to empirically reveal a role for sexual conflict, past and present, in shaping intersexual coevolution. Second, the most successful research programs have been those employing a diverse set of empirical methods in their study of male-female interactions. This suggests to us that important insights can be gained by integrating the combined results of as many empirical strategies as possible (see the list in section 2.10). Any single experiment is unlikely, by itself, to reveal the role of sexual conflict in any one system. Third, we clearly need to pay much more attention to selection on females. At the heart of sexually antagonistic coevolution lies the supposition that male alleles, traits, or strategies incur costs in females (or males when the sex roles are reversed). Again, the most successful research programs discussed in our book have been able to document direct costs to female fitness, or at least components thereof. Fourth, research efforts in sexually antagonistic coevolution have been substantially biased toward relatively few species, particularly species of insects. However, we see no obvious reason why this coevolutionary process should play a less prominent role in other groups. Research programs employing a diversity of alternative study organisms would be particularly valuable for this reason.

There is also a need for more theoretical work, aimed at integrating sexual conflict into the current framework of male-female coevolution. Following are example questions that would benefit from formal modeling: To what extent are the various models of female preference/resistance evolution distinct from one another? Is the reasoning underlying the maintenance of honest signaling of male quality different from the reasoning that male traits denying females indirect benefits from mate choice represent a form of sexual conflict over choice? Is sexually antagonistic coevolution different from other forms of male-female coevolution in generating reproductive isolation and, if so, why? How does sexually antagonistic coevolution, relative to alternative processes, affect population fitness and the risk of extinction? Is sexually antagonistic coevolution more or less likely than other processes to lead to the accumulation of multiple sexual traits/signals in males? Finally, the models we use to understand sexually antagonistic coevolution rely to a large extent on the assumption of equilibrium (e.g., game theory and quantitative genetic models). Yet there are reasons to believe that many of the sexually antagonistic adaptations we see are evolving in a nonequilibrium manner. Numerical simulations, for example, using neural networks, may be able to identify the conditions under which the predictions of nonequilibrium models differ from those of their equilibrium analogues.

The field is undergoing a rapid growth and transformation, because more and more researchers are now entertaining the idea that sexual conflict may play a role in their study systems. Hopefully, this book will help to hasten and guide this transformation by stimulating novel and critical work on reproductive conflicts between the sexes and their evolutionary implications.

References

Abernathy, R. L., P.E.A. Teal, J. A. Meredith, and R. J. Nachman. 1996. Induction of pheromone production in a moth by topical application of a pseudopeptide mimic of a pheromonotropic neuropeptide. *Proceedings of the National Academy of Sciences of the United States of America* 93: 12621–12625.

Adamo, S. A., and R. Chase. 1988. Courtship and copulation in the terrestrial snail *Helix aspersa*. *Canadian Journal of Zoology* 66: 1446–1453.

Adamo, S. A., and R. Chase. 1990. The "love dart" of the snail *Helix aspersa* injects a pheromone that decreases courtship duration. *Journal of Experimental Zoology* 255: 80–87.

Adamo, S. A., and R. Chase. 1996. Dart shooting in Helicid snails: An 'honest' signal or an instrument of manipulation? *Journal of Theoretical Biology* 180: 77–80.

Adiyodi, R. G., and K. G. Adiyodi (eds.). 1990. *Reproductive Biology of Invertebrates. Vol. IV P.B. Fertilization, Development and Parental Care*. Wiley, Chichester.

Agrawal, A. F. 2001. Sexual selection and the maintenance of sexual reproduction. *Nature* 411: 692–695.

Agrawal, A. F., E. D. Brodie, and J. Brown. 2001. Parent-offspring coadaptation and the dual genetic control of maternal care. *Science* 292: 1710–1712.

Aguade, M. 1999. Positive selection drives the evolution of the Acp29AB accessory gland protein in *Drosophila*. *Genetics* 152: 543–551.

Aguade, M., N. Miyashita, and C. H. Langley. 1992. Polymorphism and divergence in the Mst26A male accessory gland gene region in *Drosophila*. *Genetics* 132: 755–770.

Ahnesjö, I., C. Kvarnemo, and S. Merilaita. 2001. Using potential reproductive rates to predict mating competition among individuals qualified to mate. *Behavioral Ecology* 12: 397–401.

Ahnesjö, I., A. Vincent, R. Alatalo, T. Halliday, and W. J. Sutherland. 1993. The role of females in influencing mating patterns. *Behavioral Ecology* 4: 187–189.

Ai, N., S. Komatsu, I. Kubo, and W. Loher. 1986. Manipulation of prostaglandin-mediated oviposition after mating in *Teleogryllus commodus*. *International Journal of Invertebrate Reproduction and Development* 10: 33–42.

Aiken, R. B. 1992. The mating behavior of a boreal water beetle, *Dytiscus alaskanus* (Coleoptera: Dytiscidae). *Ethology Ecology & Evolution* 4: 245–254.

Aiken, R. B., and A. Khan. 1992. The adhesive strength of the palettes of males of a boreal water beetle, *Dytiscus alaskanus* Browne, J. Balfour (Coleoptera: Dytiscidae). *Canadian Journal of Zoology* 70: 1321–1324.

Alaghbari, A., H. N. Engel, and D. P. Froman. 1992. Analysis of seminal plasma from roosters carrying the sd (sperm degeneration) allele. *Biology of Reproduction* 47: 1059–1063.

Alatalo, R. V., A. Carlson, A. Lundberg, and S. Ulfstrand. 1981. The conflict between male polygamy and female monogamy: The case of the pied flycatcher *Ficedula hypoleuca*. *American Naturalist* 117: 738–753.

Alatalo, R. V., and A. Lundberg. 1986. The sexy son hypothesis: Data from the pied flycatcher *Ficedula hypoleuca*. *Animal Behaviour* 34: 1454–1462.

Alatalo, R. V., and O. Ratti. 1995. Sexy son hypothesis: Controversial once more. *Trends in Ecology & Evolution* 10: 52–53.

Albone, E. S., and S. G. Shirley. 1984. *Mammalian semiochemistry: The investigation of chemical signals between mammals.* Wiley, Chichester.

Alcock, J. 1994. Postinsemination associations between males and females in insects— the mate-guarding hypothesis. *Annual Review of Entomology* 39: 1–21.

Alcock, J., and A. Forsyth. 1988. Post-copulatory aggression toward their mates by males of the rove beetle *Leistotrophus versicolor* (Coleoptera, Staphylinidae). *Behavioral Ecology and Sociobiology* 22: 303–308.

Alcock, J., C. E. Jones, and S. L. Buchmann. 1977. Male mating strategies in bee *Centris pallida* Fox (Anthophoridae: Hymenoptera). *American Naturalist* 111: 145–155.

Alexander, R. D. 1962. The evolution of mating behavior in arthropods. *American Zoologist* 2: 501–502.

Alleman, M., and J. Doctor. 2000. Genomic imprinting in plants: observations and evolutionary implications. *Plant Molecular Biology* 43: 147–161.

Allen, G. R., and L. W. Simmons. 1996. Coercive mating, fluctuating asymmetry and male mating success in the dung fly *Sepsis cynipsea*. *Animal Behaviour* 52: 737–741.

Aloia, R. C., and R. L. Moretti. 1973. Mating behavior and ultrastructural aspects of copulation in the rotifer *Asplanchna brightwelli*. *Transactions of the American Microscopical Society* 92: 371–380.

Alonzo, S. H., and R. R. Warner. 2000a. Dynamic games and field experiments examining intra- and intersexual conflict: Explaining counterintuitive mating behavior in a Mediterranean wrasse, *Symphodus ocellatus*. *Behavioral Ecology* 11: 56–70.

Alonzo, S. H., and R. R. Warner. 2000b. Female choice, conflict between the sexes and the evolution of male alternative reproductive behaviours. *Evolutionary Ecology Research* 2: 149–170.

Andersen, N. M. 1997. A phylogenetic analysis of the evolution of sexual dimorphism and mating systems in water striders (Hemiptera: Gerridae). *Biological Journal of the Linnean Society* 61: 345–368.

Andersen, N. M., and P. P. Chen. 1995. A taxonomic revision of the Prilomerine genus *Rhyacobates* Esaki (Hemipter: Gerridae), with five new species from China and adjacent countries. *Tijdschrift voor Entomologie* 138: 51–67.

Andersen, N. M., and T. A. Weir. 2001. New genera of Veliidae (Hemiptera: Heteroptera) from Australia, with notes on the generic classification of the subfamily Microveliinae. *Invertebrate Taxonomy* 15: 217–258.

Andersson, J., A. K. Borg-Karlson, and C. Wiklund. 2000. Sexual cooperation and conflict in butterflies: A male-transferred anti-aphrodisiac reduces harassment of recently mated females. *Proceedings of the Royal Society of London Series B—Biological Sciences* 267: 1271–1275.

Andersson, M. 1982. Sexual selection, natural selection and quality advertisement. *Biological Journal of the Linnean Society* 17: 375–393.

Andersson M. 1986. Evolution of condition-dependent sex ornaments and mating preferences: Sexual selection based on viability differences. *Evolution* 40: 804–816.

Andersson, M. 2004. Social polyandry, parental investment, sexual selection, and evolution of reduced female gamete size. *Evolution* 58: 24–34.

Andersson, M. B. 1994. *Sexual Selection*. Princeton University Press, Princeton, N.J.

Andersson, S., S. R. Pryke, J. Ornborg, M. J. Lawes, and M. Andersson. 2002. Multiple receivers, multiple ornaments, and a trade-off between agonistic and epigamic signaling in a widowbird. *American Naturalist* 160: 683–691.

Andrade, M.C.B. 1996. Sexual selection for male sacrifice in the Australian redback spider. *Science* 271: 70–72.

Andrade, M.C.B. 1998. Female hunger can explain variation in cannibalistic behavior despite male sacrifice in redback spiders. *Behavioral Ecology* 9: 33–42.

Andrade, M.C.B., and E. M. Banta. 2002. Value of male remating and functional sterility in redback spiders. *Animal Behaviour* 63: 857–870.

Andrés, J. A., and G. Arnqvist. 2001. Genetic divergence of the seminal signal-receptor system in houseflies: The footprints of sexually antagonistic coevolution? *Proceedings of the Royal Society of London Series B—Biological Sciences* 268: 399–405.

Andrés, J. A., and A. Cordero. 1998. Effects of water mites on the damselfly *Ceriagrion tenellum*. *Ecological Entomology* 23: 103–109.

Andrés, J. A., and A. Cordero. 1999. The inheritance of female colour morphs in the damselfly *Ceriagrion tenellum* (Odonata, Coenagrionidae). *Heredity* 82: 328–335.

Andrés, J. A., and E. H. Morrow. 2003. The origin of interlocus sexual conflict: Is sex-linkage important? *Journal of Evolutionary Biology* 16: 219–223.

Andrés, J. A., and A. C. Rivera. 2001. Survival rates in a natural population of the damselfly *Ceriagrion tenellum*: Effects of sex and female phenotype. *Ecological Entomology* 26: 341–346.

Andrés, J. A., R. A. Sanchez-Guillen, and A. C. Rivera. 2000. Molecular evidence for selection on female color polymorphism in the damselfly *Ischnura graellsii*. *Evolution* 54: 2156–2161.

Andrés, J. A., R. A. Sanchez-Guillen, and A. C. Rivera. 2002. Evolution of female colour polymorphism in damselflies: Testing the hypotheses. *Animal Behaviour* 63: 677–685.

Arak, A. 1983. Male-male competition and mate choice in anuran amphibians. In P. Bateson (ed.), *Mate Choice*, Cambridge University Press, Cambridge, pp. 181–210.

Arak, A., and M. Enquist. 1993. Hidden preferences and the evolution of signals. *Philosophical Transactions of the Royal Society of London Series B—Biological Sciences* 340: 207–213.

Arak, A., and M. Enquist. 1995. Conflict, receiver bias and the evolution of signal form. *Philosophical Transactions of the Royal Society of London Series B—Biological Sciences* 349: 337–344.

Arnold, S. J. 1977. The courtship behavior of North American salamanders with some omments of Old World salamandrids. In D. H. Taylor and S. I. Guttman (eds.), *The Reproductive Biology of Amphibians*, Plenum, New York, pp. 141–183.

Arnold, S. J. 1983. Sexual selection the interface of theory and empiricism. In P. Bateson (ed.), *Mate Choice*, Cambridge University Press, Cambridge, pp. 67–108.

Arnold, S. J., and D. Duvall. 1994. Animal mating systems: A synthesis based on selection theory. *American Naturalist* 143: 317–348.

Arnold, S. J., and L. D. Houck. 1982. Courtship pheromones: Evolution by natural and sexual selection. In M. Nitecke (ed.), *Biochemical Aspects of Evolutionary Biology*, University of Chicago Press, Chicago, pp. 173–211.

Arnold, S. J., and M. J. Wade. 1984. On the measurement of natural and sexual selection: Theory. *Evolution* 38: 709–719.

Arnqvist, G. 1989a. Multiple mating in a water strider: Mutual benefits or intersexual conflict? *Animal Behaviour* 38: 749–756.

Arnqvist, G. 1989b. Sexual selection in a water strider: The function, mechanism of selection and heritability of a male grasping apparatus. *Oikos* 56: 344–350.

Arnqvist, G. 1992. Precopulatory fighting in a water strider: Intersexual conflict or mate assessment? *Animal Behaviour* 43: 559–567.

Arnqvist, G. 1997a. The evolution of water strider mating systems: Causes and consequences of sexual conflicts. In J. C. Choe and B. J. Crespi (eds.), *The Evolution of Mating Systems in Insects and Arachnids*, Cambridge University Press, Cambridge, pp. 146–163.

Arnqvist, G. 1997b. The evolution of animal genitalia: Distinguishing between hypotheses by single species studies. *Biological Journal of the Linnean Society* 60: 365–379.

Arnqvist, G. 1998. Comparative evidence for the evolution of genitalia by sexual selection. *Nature* 393: 784–786.

Arnqvist, G. 2004. Sexual conflict and sexual selection: Lost in the chase. *Evolution* 58: 1383–1388.

Arnqvist, G., and I. Danielsson. 1999a. Copulatory behavior, genital morphology, and male fertilization success in water striders. *Evolution* 53: 147–156.

Arnqvist, G., and I. Danielsson. 1999b. Postmating sexual selection: The effects of male body size and recovery period on paternity and egg production rate in a water strider. *Behavioral Ecology* 10: 358–365.

Arnqvist, G., M. Edvardsson, U. Friberg, and T. Nilsson. 2000. Sexual conflict promotes speciation in insects. *Proceedings of the National Academy of Sciences of the United States of America* 97: 10460–10464.

Arnqvist, G., and S. Henriksson. 1997. Sexual cannibalism in the fishing spider and a model for the evolution of sexual cannibalism based on genetic constraints. *Evolutionary Ecology* 11: 255–273.

Arnqvist, G., T. M. Jones, and M. A. Elgar. 2003. Reversal of sex roles in nuptial feeding. *Nature* 424: 387.

Arnqvist, G., and T. Nilsson. 2000. The evolution of polyandry: Multiple mating and female fitness in insects. *Animal Behaviour* 60: 145–164.

Arnqvist, G., T. Nilsson, and M. Katvala. 2004. Mating rate and fitness in female bean weevils. *Behavioral Ecology* (in press).

Arnqvist, G., and L. Rowe. 1995. Sexual conflict and arms races between the sexes: A morphological adaptation for control of mating in a female insect. *Proceedings of the Royal Society of London Series B—Biological Sciences* 261: 123–127.

Arnqvist, G., and L. Rowe. 2002a. Antagonistic coevolution between the sexes in a group of insects. *Nature* 415: 787–789.

Arnqvist, G., and L. Rowe. 2002b. Correlated evolution of male and female morphologies in water striders. *Evolution* 56: 936–947.

Arnqvist, G., R. Thornhill, and L. Rowe. 1997. Evolution of animal genitalia: morphological correlates of fitness components in a water strider. *Journal of Evolutionary Biology* 10: 613–640.

Asch, R., C. Simerly, T. Ord, V. A. Ord, and G. Schatten. 1995. The stages at which human fertilization arrests—microtubule and chromosome configurations in inseminated oocytes which failed to complete fertilization and development in humans. *Human Reproduction* 10: 1897–1906.

Aumüller, G., and W. Krause (eds.). 1990. International Workshop on Seminal and Sperm-Specific Proteins. *Andrologia* 22 supplement 1.

Aumüller, G., and A. Riva. 1992. Morphology and functions of the human seminal vesicle. *Andrologia* 24: 183–196.

Aureli, F., and F.B.M. de Waal (eds.) 2000. *Natural Conflict Resolution.* University of California Press, Berkeley.

Austin, C. R. 1956. Effects of hypothermia and hyperthermia on fertilization in rat eggs. *Journal of Experimental Biology* 33: 348–357.

Austin, C. R. 1957. Fertilization, early cleavage and associated phenomena in the field vole (*Microtus agrestis*). *Journal of Anatomy* 91: 1–11.

Austin, C. R. 1965. *Fertilization.* Prentice-Hall, Englewood Cliffs, N.J.

Austin, C. R., and A.W.H. Braden. 1953. An investigation of polyspermy in the rat and rabbit. *Australian Journal of Biological Sciences* 6: 674–692.

Axelrod, R., and W. D. Hamilton. 1981. The evolution of coorperation. *Science* 211: 1390–1396.

Badyaev, A. V. 2002. Growing apart: An ontogenetic perspective on the evolution of sexual size dimorphism. *Trends in Ecology & Evolution* 17: 369–378.

Baer, B., R. Maile, P. Schmid-Hempel, E. D. Morgan, and G. R. Jones. 2000. Chemistry of a mating plug in bumblebees. *Journal of Chemical Ecology* 26: 1869–1875.

Baer, B., E. D. Morgan, and P. Schmid-Hempel. 2001. A nonspecific fatty acid within the bumblebee mating plug prevents females from remating. *Proceedings of the National Academy of Sciences of the United States of America* 98: 3926–3928.

Baer, B., and P. Schmid-Hempel. 1999. Experimental variation in polyandry affects parasite loads and fitness in a bumble-bee. *Nature* 397: 151–154.

Baker, R. H., R.I.S. Ashwell, T. A. Richards, K. Fowler, T. Chapman, and A. Pomiankowski. 2001. Effects of multiple mating and male eye span on female reproductive output in the stalk-eyed fly, *Cyrtodiopsis dalmanni*. *Behavioral Ecology* 12: 732–739.

Bali, G., A. K. Raina, T. G. Kingan, and J. D. Lopez. 1996. Ovipositional behavior of newly colonized corn earworm (Lepidoptera: Noctuidae) females and evidence for an oviposition stimulating factor of male origin. *Annals of the Entomological Society of America* 89: 475–480.

Ball, M. A., and G. A. Parker. 2003. Sperm competition games: Sperm selection by females. *Journal of Theoretical Biology* 224: 27–42.

Balzer, A. L., and T. D. Williams. 1998. Do female zebra finches vary primary reproductive effort in relation to mate attractiveness? *Behaviour* 135: 297–309.

Banerjee, B. N. 1968. Effect of oestrogen on fertility of rat when transferred through semen during mating. *Journal of Reproduction and Fertility* 17: 157–159.

Barnes, H., M. Barnes, and W. Klepal. 1977. Studies on reproduction of cirripedes. I. Introduction—copulation, release of oocytes, and formation of egg lamellae. *Journal of Experimental Marine Biology and Ecology* 27: 195–218.

Barnett, M., S. R. Telford, and B. J. Tibbles. 1995. Female mediation of sperm competition in the millipede *Alloporus uncinatus* (Diplopoda: Spirostreptidae). *Behavioral Ecology and Sociobiology* 36: 413–419.

Barraclough, T. G., P. H. Harvey, and S. Nee. 1995. Sexual selection and taxonomic diversity in passerine birds. *Proceedings of the Royal Society of London Series B—Biological Sciences* 259: 211–215.

Barrett, S.C.H. 2002. Sexual interference of the floral kind. *Heredity* 88: 154–159.

Barrett, S.C.H. 2003. Mating strategies in flowering plants: The outcrossing-selfing paradigm and beyond. *Philosophical Transactions of the Royal Society of London Series B—Biological Sciences* 358: 991–1004.

Barta, Z. N., A. I. Houston, J. M. Mcnamara, and T. Szekely. 2002. Sexual conflict about parental care: The role of reserves. *American Naturalist* 159: 687–705.

Barth, F. G., and A. Schmitt. 1991. Species recognition and species isolation in wandering spiders (*Cuppemius* spp.; Ctenidae). *Behavioral Ecology and Sociobiology* 29: 333–339.

Bartolomei, M. S., and S. M. Tilghman. 1997. Genomic imprinting in mammals. *Annual Review of Genetics* 31: 493–525.

Basolo A. L. 1990. Female preference predates the evolution of the sword in swordtail fish. *Science* 250: 808–810.

Basolo, A. L., and J. A. Endler. 1995. Sensory biases and the evolution of sensory systems. *Trends in Ecology & Evolution* 10: 489.

Bateman, A. J. 1948. Intra-sexual selection in *Drosophila*. *Heredity* 2: 349–368.

Bateson, P.P.G. 1983. Optimal outbreeding. In P.P.G. Bateson (ed.), *Mate Choice*, Cambridge University Press, Cambridge, pp. 257–277.

Baur, B. 1994. Parental care in terrestrial gastropods. *Experientia* 50: 5–14.

Baur, B. 1998. Sperm competition in Molluscs. In T. R. Birkhead and A. P. Møller (eds.), *Sperm Competition and Sexual Selection*, Academic, London, pp. 255–305.

Baur, B., and A. Baur. 1992. Effect of courtship and repeated copulation on egg production in the simultaneously hermaphroditic land snail *Arianta arbustorum*. *Invertebrate Reproduction & Development* 21: 201–206.

Beck, C. W. 1998. Mode of fertilization and parental care in anurans. *Animal Behaviour* 55: 439–449.

Beck, M. W., and E. F. Connor. 1992. Factors affecting the reproductive success of the crab spider *Misumenoides formosipes*: The covariance between juvenile and adult traits. *Oecologia* 92: 287–295.

Bedwal, R. S., and A. Bahuguna. 1994. Zinc, copper and selenium in reproduction. *Experientia* 50: 626–640.

Beekman, M., J. Komdeur, and F.L.W. Ratnieks. 2003. Reproductive conflicts in social animals: Who has power? *Trends in Ecology & Evolution* 18: 277–282.

Begun, D. J., P. Whitley, B. L. Todd, H. M. Waldrip-Dail, and A. G. Clark. 2000. Molecular population genetics of male accessory gland proteins in *Drosophila*. *Genetics* 156: 1879–1888.

Bell, G. 1978. Handicap principle in sexual selection. *Evolution* 32: 872–885.

Benavides, E., J. C. Ortiz, and J. R. Formas. 2002. A new species of Telmatobius (Anura: Leptodactylidae) from northern Chile. *Herpetologica* 58: 210–220.

Bensch, S. 1997. The cost of polygyny—Definitions and applications. *Journal of Avian Biology* 28: 345–352.

Benton, T. 2001. Reproductive ecology. In P. Brownell and G. A. Polis (eds.), Scorpion Biology and Research, Oxford University Press, New York, pp. 278–301.

Bergh, J. C., M. O. Harris, and S. Rose. 1992. Factors inducing mated behavior in female Hessian flies (Diptera: Cecidomyiidae). *Annals of the Entomological Society of America* 85: 224–233.

Berglund, A. 1993. Risky sex: Male pipefishes mate at random in the presence of a predator. *Animal Behaviour* 46: 169–175.

Berglund, A., A. Bisazza, and A. Pilastro. 1996. Armaments and ornaments: an evolutionary explanation of traits of dual utility. *Biological Journal of the Linnean Society* 58: 385–399.

Bergsten, J., A. Toyra, and A. N. Nilsson. 2001. Intraspecific variation and intersexual correlation in secondary sexual characters of three diving beetles (Coleoptera: Dytiscidae). *Biological Journal of the Linnean Society* 73: 221–232.

Bernasconi, G., and L. Keller. 2001. Female polyandry affects their sons' reproductive success in the red flour beetle *Tribolium castaneum*. *Journal of Evolutionary Biology* 14: 186–193.

Berrill, M., and M. Arsenault. 1984. The breeding behavior of a northern temperate orconectid crayfish, *Orconectes rusticus*. *Animal Behaviour* 32: 333–339.

Birkhead, T., and A. Møller. 1993. Female control of paternity. *Trends in Ecology & Evolution* 8: 100–104.

Birkhead, T. R. 1997. Darwin on sex. *Biologist* 44: 397–399.

Birkhead, T. R. 1998. Cryptic female choice: Criteria for establishing female sperm choice. *Evolution* 52: 1212–1218.

Birkhead, T. R. 2000. Defining and demonstrating postcopulatory female choice—again. *Evolution* 54: 1057–1060.

Birkhead, T. R., L. Atkin, and A. P. Møller. 1987. Copulation behavior of birds. *Behaviour* 101: 101–138.

Birkhead, T. R., K. E. Lee, and P. Young. 1988. Sexual cannibalism in the praying mantis *Hieroula membranacea*. *Behaviour* 106: 112–118.

Birkhead, T. R., and A. P. Møller (eds.). 1992. *Sperm Competition in Birds—Evolutionary Causes and Consequences*. Academic, London.

Birkhead, T. R., and A. P. Møller (eds.). 1998. *Sperm Competition and Sexual Selection*. Academic, London.

Birkhead, T. R., A. P. Møller, and W. J. Sutherland. 1993. Why do females make it so difficult for males to fertilize their eggs? *Journal of Theoretical Biology* 161: 51–60.

Birkhead, T. R., B. C. Sheldon, and F. Fletcher. 1994. A comparative study of sperm-egg interactions in birds. *Journal of Reproduction and Fertility* 101: 353–361.

Bisazza, A., S. Manfredi, and A. Pilastro. 2000. Sexual competition, coercive mating and mate assessment in the one-sided livebearer, *Jenynsia multidentata*: Are they predictive of sexual dimorphism? *Ethology* 106: 961–978.

Bisazza, A., G. Vaccari, and A. Pilastro. 2001. Female mate choice in a mating system dominated by male sexual coercion. *Behavioral Ecology* 12: 59–64.

Bishop, J.D.D. 1996. Female control of paternity in the internally fertilizing compound ascidian *Diplosoma listerianum*. 1. Autoradiographic investigation of sperm movements in the female reproductive tract. *Proceedings of the Royal Society of London Series B—Biological Sciences* 263: 369–376.

Bishop, J.D.D., P. H. Manriquez, and R. N. Hughes. 2000. Water-borne sperm trigger vitellogenic egg growth in two sessile marine invertebrates. *Proceedings of the Royal Society of London Series B—Biological Sciences* 267: 1165–1169.

Bissoondath, C. J., and C. Wiklund. 1995. Protein-content of spermatophores in relation to monandry/polyandry in butterflies. *Behavioral Ecology and Sociobiology* 37: 365–371.

Björklund, M., and J. C. Senar. 2001. Sex differences in survival selection in the serin, *Serinus serinus*. *Journal of Evolutionary Biology* 14: 841–849.

Blanckenhorn, W. U., D. J. Hosken, O. Y. Martin, C. Reim, Y. Teuschl, and P. I. Ward. 2002. The costs of copulating in the dung fly *Sepsis cynipsea*. *Behavioral Ecology* 13: 353–358.

Blanckenhorn, W. U., C. Morf, C. Mühlhäuser, and T. Reusch. 1999. Spatiotemporal variation in selection on body size in the dung fly *Sepsis cynipsea*. *Journal of Evolutionary Biology* 12: 563–576.

Blandy, J. P., and B. Lytton (eds.). 1986. *The Prostate*. Butterworths, London.

Bloch-Qazi, M. C., J. T. Herbeck, and S. M. Lewis. 1996. Mechanisms of sperm transfer and storage in the red flour beetle (Coleoptera: Tenebrionidae). *Annals of the Entomological Society of America* 89: 892–897.

Bluhm, C. K., R. E. Phillips, W. H. Burke, and G. N. Gupta. 1984. Effects of male courtship and gonadal steroids on pair formation, egg-laying, and serum lutenizing hormone in Canvasback ducks (*Aythya valisineria*). *Journal of Zoology* 204: 185–200.

Bluhm, C. K., I. Rozenboim, J. Silsby, and M. El Halawani. 2000. Sex-related differences in the effects of late winter pairing activity and seasonal influences on neuroendocrinology and gonadal development of mallards. *General and Comparative Endocrinology* 118: 310–321.

Blumer, L. S. 1986. Parental care sex differences in the brown bullhead (*Ictalurus nebulosus*, Pisces, Ictaluridae). *Behavioral Ecology and Sociobiology* 19: 97–104.

Boake, C.R.B. 2000. Flying apart: Mating behavior and speciation. *Bioscience* 50: 501–508.

Boiteau, G. 1988. Sperm utilization and post-copulatory female-guarding in the Colorado potato beetle, *Leptinotarsa decemlineata*. *Entomologia Experimentalis et Applicata* 47: 183–187.

Bond16riansky, R. 2001. The evolution of male mate choice in insects: A synthesis of ideas and evidence. *Biological Reviews* 76: 305–339.

Bonduriansky, R. 2003. Layered sexual selection: A comparative analysis of sexual behaviour within an assemblage of piophilid flies. *Canadian Journal of Zoology* 81: 479–491.

Bonduriansky, R., and L. Rowe. 2003. Interactions among mechanisms of sexual selection on male body size and head shape in a sexually dimorphic fly *Prochyliza xanthostoma*. *Evolution* 57: 2046–2053.

Bonduriansky, R., and L. Rowe. 2004. Sexual selection, genetic architecture, and the condition dependence of body size and shape in the sexually dimorphic fly *Prochyliz xanthostoma* (Diptera: Piophilidae). *Evolution* (in press).

Bonduriansky, R., J. Wheeler, and L. Rowe. 2005. Ejaculate feeding expedites oviposition and increases female fecundity in the "waltzing fly" *Prochyliza xanthostoma* (Diptera: Piophilidae). *Animal Behavior* 69: 489–497.

Boness, D. J., W. D. Bowen, and S. J. Iverson. 1995. Does male harassment of females contribute to reproductive synchrony in the grey seal by affecting maternal performance? *Behavioral Ecology and Sociobiology* 36: 1–10.

Boomsma, J. J. 1996. Split sex ratios and queen-male conflict over sperm allocation. *Proceedings of the Royal Society of London Series B—Biological Sciences* 263: 697–704.

Borgia, G. 1979. Sexual selection and the evolution of mating systems. In M. S. Blum and N. A. Blum (eds.), *Sexual Selection and Reproductive Competition in Insects*, Academic, New York, pp. 19–80.

Borgia, G. 1981. Mate selection in the fly *Scatophaga stercoraria*: Female choice in a male-controlled system. *Animal Behaviour* 29: 71–80.

Borgia, G. 1987. A critical review of sexual selection models. In J. W. Bradbury and M. B. Andersson (eds.), *Sexual Selection: Testing the Alternatives*, Wiley, Chichester, pp. 55–66.

Borovsky, D., D. A. Carlson, R. G. Hancock, H. Rembold, and E. Vanhandel. 1994. De novo biosynthesis of juvenile hormone III and hormone I by the accessory glands of the male mosquito. *Insect Biochemistry and Molecular Biology* 24: 437–444.

Borovsky, D., and F. Mahmood. 1995. Feeding the mosquito *Aedes aegypti* with TMOF and its analogs: Effect of trypsin biosynthesis and egg development. *Regulatory Peptides* 57: 273–281.

Borowsky, B. 1980. Reproductive patterns of 3 inter-tidal salt marsh gammaridean amphipods. *Marine Biology* 55: 327–334.

Borowsky, B. 1984. The use of the males' gnathopods during precopulation in some gammaridean amphipods. *Crustaceana* 47: 245–250.

Borries, C., K. Launhardt, C. Epplen, J. T. Epplen, and P. Winkler. 1999. DNA analyses support the hypothesis that infanticide is adaptive in langur monkeys. *Proceedings of the Royal Society of London Series B—Biological Sciences* 266: 901–904.

Boucher, D. H. 1977. Wasting parental investment. *American Naturalist* 111: 786–788.

Boucher, L., and J. Huignard. 1987. Transfer of male secretions from the spermatophore to the female insect in *Caryedon serratus* (OL): Analysis of the possible trophic role of these secretions. *Journal of Insect Physiology* 33: 949–957.

Boughman, J. W. 2002. How sensory drive can promote speciation. *Trends in Ecology & Evolution* 17: 571–577.

Bourne, G. R. 1998. Amphisexual parental behavior of a terrestrial breeding frog *Eleutherodactylus johnstonei* in Guyana. *Behavioral Ecology* 9: 1–7.

Bradbury, J. W., and M. B. Andersson (eds.). 1987. *Sexual selection: Testing the alternatives: Report of the Dahlem Workshop, Berlin 1986, August 31—September 5.* Wiley, Chichester.

Braden, A.W.H. 1953. Distribution of sperms in the genital tract of the female rabbit after coitus. *Australian Journal of Biological Sciences* 6: 693–705.

Braden, A.W.H. 1957. Variation between strains in the incidence of various abnormalities of egg maturation and fertilization in the mouse. *Journal of Genetics* 55: 478–486.

Braden, A.W.H., C. R. Austin, and H. A. David. 1954. The reaction of the zona pellucida to sperm penetration. *Australian Journal of Biological Sciences* 7: 391–409.

Breed, W. G., and C. M. Leigh. 1990. Morphological changes in the oocyte and its surrounding vestments during in vivo fertilization in the dasyurid marsupial *Sminthopsis crassicaudata*. *Journal of Morphology* 204: 177–196.

Breene, R. G., and M. H. Sweet. 1985. Evidence of insemination of multiple females by the male black widow spider, *Latrodectus mactans* (Araneae, Theridiidae). *Journal of Arachnology* 13: 331–335.

Briceno, D. 1987. How spiders determine clutch size. *Revista De Biologia Tropical* 35: 25–29.

Bride, J., and L. Gomot. 1991a. Asynchronous development of the genital tract during growth and reproduction in the snail, *Helix aspersa*. *Reproduction Nutrition Development* 31: 81–96.

Bride, J., and L. Gomot. 1991b. In vitro demonstration of direct stimulation of the *Helix aspersa* albumin glad galactogen synthesis by extracts from the male accessory glands of the genital tract. *Comptes Rendus de L'Academie des Sciences, Serie Iii: Sciences de la Vie* 313: 565–571.

Briskie, J. V., and R. Montgomerie. 1992. Sperm size and sperm competition in birds. *Proceedings of the Royal Society of London Series B—Biological Sciences* 247: 89–95.

Briskie, J. V., and R. Montgomerie. 1993. Patterns of sperm storage in relation to sperm competition in passerine birds. *Condor* 95: 442–454.

Briskie, J. V., R. Montgomerie, and T. R. Birkhead. 1997. The evolution of sperm size in birds. *Evolution* 51: 937–945.

Brockes, J. P. 1999. Topics in prion cell biology. *Current Opinion in Neurobiology* 9: 571–577.

Brockmann, H. J., and A. Grafen. 1989. Mate conflict and male behavior in a solitary wasp, *Trypoxylon* (Trypargilum) *politum* (Hymenoptera, Sphecidae). *Animal Behaviour* 37: 232–255.

Brooks, R. 2000. Negative genetic correlation between male sexual attractiveness and survival. *Nature* 406: 67–70.

Brower, A.V.Z. 1997. The evolution of ecologically important characters in Heliconius butterflies (Lepidoptera: Nymphalidae): A cladistic review. *Zoological Journal of the Linnean Society* 119: 457–472.

Brown, D. V., and P. E. Eady. 2001. Functional incompatibility between the fertilization systems of two allopatric populations of Callosobruchus maculatus (Coleoptera: Bruchidae). *Evolution* 55: 2257–2262.

Brown, M.J.F., and P. Schmid-Hempel. 2003. The evolution of female multiple mating in social hymenoptera. *Evolution* 57: 2067–2081.

Brun, L. O., J. Stuart, V. Gaudichon, K. Aronstein, and R. H. Ffrenchconstant. 1995. Functional haplodiploidy: A mechanism for the spread of insecticide resistance in an important international insect pest. *Proceedings of the National Academy of Sciences of the United States of America* 92: 9861–9865.

Bryant, E. H. 1979. Inbreeding and hertogamic mating: Alternative to Averhoff and Richardson. *Behavior Genetics* 9: 249–256.

Bulmer M. 1989. Structural instability of models of sexual selection. *Theoretical Population Biology* 35: 195–206.

Bulmer, M. G., and G. A. Parker. 2002. The evolution of anisogamy: A game-theoretic approach. *Proceedings of the Royal Society of London Series B—Biological Sciences* 269: 2603.

Buntin, J. D. 1996. Neural and hormonal control of parental behaviour in birds. In J. S. Rosenblatt and C. T. Snowdon (eds.), *Parental Care, Evolution Mechanisms, and Adaptive Significance*. Advances in the Study of Behavior, Vol. 25, Academic, London, pp. 161–213.

Burd, M. 1994. Bateman principle and plant reproduction: The role of pollen limitation in fruit and seed set. *Botanical Review* 60: 83–139.

Burd, M. 1998. "Excess" flower production and selective fruit abortion: A model of potential benefits. *Ecology* 79: 2123–2132.

Burford, F.R.L., P. K. Mcgregor, and R. F. Oliveira. 2001. Male-like mudballing behavior of some female fiddler crabs (*Uca tangeri*). *Journal of Ethology* 19: 97–103.

Burley, N. 1981. The evolution of sexual indistinguishability. In R. D. Alexander and D. W. Tinkle (eds.), *Natural Selection and Social Behavior: Recent Research and New Theory*, Chiron, New York / distributed outside North America by Blackwell Scientific, Oxford, pp. 121–137.

Burley, N. 1986. Sexual selection for aesthetic traits in species with biparental care. *American Naturalist* 127: 415–445.

Burley, N. 1988. The differential allocation hypothesis: An experimental test. *American Naturalist* 132: 611–628.

Burns, J. R., and S. H. Weitzman. 2005. Insemination in ostariophysan fishes. In *Proceedings of the Second International Symposium on Livebearing Fishes*, New Life Publications, Homestead, Fla. (in press).

Burns, J. R., S. H. Weitzman, H. J. Grier, and N. A. Menezes. 1995. Internal fertilization, testis and sperm morphology in glandulocaudine fishes (Teleostei: Characidae: Glandulocaudinae). *Journal of Morphology* 224: 131–145.

Burt, A. 1995. Perspective: The evolution of fitness. *Evolution* 49: 1–8.

Burt, A. 2000. Natural selection in the wild. *Trends in Ecology & Evolution* 15: 306–307.

Bushmann, P. J., J. R. Burns, and S. H. Weitzman. 2002. Gill-derived glands in glandulocaudine fishes (Teleostei: Characidae: Glandulocaudinae). *Journal of Morphology* 253: 187–195.

Buskirk, R. E., C. Frohlich, and K. G. Ross. 1984. The natural selection of sexual cannibalism. *American Naturalist* 123: 612–625.

Bussing, W. A. 1974. *Pterobrycon myrnae*, a remarkable new glandulocaudinae characid fis from Costa Rica. *Revista de Biologica Tropical* 22: 135–159.

Bussing, W. A., and T. R. Roberts. 1971. Rediscovery of glandulocaudine fish *Pterobrycon* and hypothetical significance of its spectacular humeral scales (Pisces: Characidae). *Copeia* 1971: 179–181.

Butlin, R. K., and M. G. Ritchie. 1994. Mating behaviour and speciation. In P.J.B. Slater and T. R. Halliday (eds.), *Behaviour and Evolution*, Cambridge University Press, Cambridge, pp. 43–79.

Butt, K. R., and V. Nuutinen. 1998. Reproduction of the earthworm *Lumbricus terrestris* Linne after the first mating. *Canadian Journal of Zoology* 76: 104–109.

Byrd, E. W., and F. D. Collins. 1975. Absence of fast block to polyspermy in eggs of sea urchin *Strongylocentrotus purpuratus*. *Nature* 257: 675–677.

Byrne, P. G., and J. D. Roberts. 1999. Simultaneous mating with multiple males reduces fertilization success in the myobatrachid frog *Crinia georgiana*. *Proceedings of the Royal Society of London Series B—Biological Sciences* 266: 717–721.

Callaini, G., and M. G. Riparbelli. 1996. Fertilization in *Drosophila melanogaster*: Centrosome inheritance and organization of the first mitotic spindle. *Developmental Biology* 176: 199–208.

Cameron, E., T. Day, and L. Rowe. 2003. Sexual conflict and indirect benefits. *Journal of Evolutionary Biology* 16: 1055–1060.

Campagna, C., B. J. Leboeuf, and H. L. Cappozzo. 1988. Group raids: A mating strategy of male southern sea lions. *Behaviour* 105: 224–249.

Carayon, J. 1966. Tramatic insemination and paragenital system. In R. L. Usinger (ed.), *Monograph of Cimicidae (Hemiptera, Heteroptera)*, Entomological Society of America, College Park, Md., pp. 81–166.

Castro, J. I., P. M. Bubucis, and N. A. Overstrom. 1988. The reproductive biology of the chain dogfish, *Scyliorhinus retifer*. *Copeia* 740–746.

Catchpole, C., B. Leisler, and H. Winkler. 1985. Poygyny in the great reed warbler, *Acrocephalus arundinaceus*: A possible case of deception. *Behavioral Ecology and Sociobiology* 16: 285–291.

Cerolini, S., P. Surai, A. Maldjian, T. Gliozzi, and R. Noble. 1997. Lipid composition of semen in different fowl breeders. *Poultry and Avian Biology Reviews* 8: 141–148.

Chandra, H. S., and S. W. Brown. 1975. Chromosome imprinting and the mammalian X-chromosome. *Nature* 253: 165–168.

Chang, M. C., and L. Fernandez-Cano. 1958. Effects of delayed fertilization on the development of pronucleus and the segmentation of hamster ova. *Anatomical Record* 132: 307–319.

Chapman, R. F. 1998. *The Insects: Structure and Function*, 4th Ed. Cambridge University Press, Cambridge.

Chapman, T. 2001. Seminal fluid-mediated fitness traits in *Drosophila*. *Heredity* 87: 511–521.

Chapman, T., G. Arnqvist, J. Bangham, and L. Rowe. 2003a. Sexual conflict. *Trends in Ecology & Evolution* 18: 41–47.

Chapman, T., G. Arnqvist, J. Bangham, and L. Rowe. 2003b. Response to Eberhard and Cordero, and Córdoba-Aguilar and Contreras-Garduño: Sexual conflict and female choice. *Trends in Ecology & Evolution* 18: 440–441.

Chapman, T., J. Bangham, G. Vinti, B. Seifried, O. Lung, M. F. Wolfner, H. K. Smith, and L. Partridge. 2003c. The sex peptide of *Drosophila melanogaster*: Female postmating responses analyzed by using RNA interference. *Proceedings of the National Academy of Sciences of the United States of America* 100: 9923–9928.

Chapman, T., L. A. Herndon, Y. Heifetz, L. Partridge, and M. F. Wolfner. 2001. The Acp26Aa seminal fluid protein is a modulator of early egg hatchability in *Drosophila melanogaster*. *Proceedings of the Royal Society of London Series B—Biological Sciences* 268: 1647–1654.

Chapman, T., J. Hutchings, and L. Partridge. 1993. No reduction in the cost of mating for *Drosophila melanogaster* females mating with spermless males. *Proceedings of the Royal Society of London Series B—Biological Sciences* 253: 211–217.

Chapman, T., L. F. Liddle, J. M. Kalb, M. F. Wolfner, and L. Partridge. 1995. Cost of mating in *Drosophila melanogaster* females is mediated by male accessory gland products. *Nature* 373: 241–244.

Chapman, T., T. Miyatake, H. K. Smith, and L. Partridge. 1998. Interactions of mating, egg production and death rates in females of the Mediterranean fruit fly, *Ceratitis capitata*. *Proceedings of the Royal Society of London Series B—Biological Sciences* 265: 1879–1894.

Chapman, T., D. M. Neubaum, M. F. Wolfner, and L. Partridge. 2000. The role of male accessory gland protein Acp36DE in sperm competition in *Drosophila melanogaster*. *Proceedings of the Royal Society of London Series B—Biological Sciences* 267: 1097–1105.

Chapman, T., and L. Partridge. 1996a. Sexual conflict as fuel for evolution. *Nature* 381: 189–190.

Chapman, T., and L. Partridge. 1996b. Female fitness in *Drosophila melanogaster*: An interaction between the effect of nutrition and of encounter rate with males. *Proceedings of the Royal Society of London Series B—Biological Sciences* 263: 755–759.

Chapman, T., S. Trevitt, and L. Partridge. 1994. Remating and male-derived nutrients in *Drosophila melanogaster*. *Journal of Evolutionary Biology* 7: 51–69.

Charlesworth, B. 1987. The heritability of fitness. In J. W. Bradbury and M. B. Andersson (eds.), *Sexual Selection: Testing the Alternatives*, Wiley, Chichester, pp. 21–40.

Charlesworth D., D. W. Schemske, and V. L. Sork. 1987. The evolution of plant reproductive characters: Sexual versus natural selection. In S. C. Stearns (ed.), *The Evolution of Sex and Its Consequences*, Birkhäuser, Basel, pp. 317–336.

Charnov, E. L. 1979. Simultaneous hermaphroditism and sexual selection. *Proceedings of the National Academy of Sciences of the United States of America* 76: 2480–2484.

Charnov, E. L. 1982. *The Theory of Sex Allocation*, Princeton University Press, Princeton, N.J.

Charnov, E. L. 1984. Behavioural ecology of plants. In J. R. Krebs and N. B. Davies (eds.), *Behavioural Ecology: An Evolutionary Approach*, 2nd Ed., Sinauer Associates, Sunderland, Mass., pp. 362–379.

Charnov, E. L. 1996. Sperm competition and sex allocation in simultaneous hermaphrodites. *Evolutionary Ecology* 10: 457–462.

Chase I. D. 1980. Cooperative and noncooperative behavior in animals. *American Naturalist* 115: 827–857.

Chen, P. S. 1984. The functional morphology and biochemistry of insect male accessory glands and their secretions. *Annual Review of Entomology* 29: 233–255.

Chen, P. S. 1996. The accessory gland proteins in male *Drosophila*: Structural, reproductive, and evolutionary aspects. *Experientia* 52: 503–510.

Chippindale, A. K., J. R. Gibson, and W. R. Rice. 2001. Negative genetic correlation for adult fitness between sexes reveals ontogenetic conflict in *Drosophila*. *Proceedings of the National Academy of Sciences of the United States of America* 98: 1671–1675.

Chippindale, A. K., and W. R. Rice. 2001. Y chromosome polymorphism is a strong determinant of male fitness in *Drosophila melanogaster*. *Proceedings of the National Academy of Sciences of the United States of America* 98: 5677–5682.

Choe, J. C., and B. J. Crespi (eds.). 1997a. *The Evolution of Mating Systems in Insects and Arachnids*. Cambridge University Press, Cambridge.

Choe, J. C., and B. J. Crespi (eds.). 1997b. *The Evolution of Social Behavior in Insects and Arachnids*. Cambridge University Press, Cambridge.

Chung, D.J.D. 1986. Stimulation of genital eversion in the land snail *Helix aspersa* by extracts of the glands of dart apparatus. *Journal of Experimental Zoology* 238: 129–139.

Chung, D.J.D. 1987. Courtship and dart shooting behavior of the land snail *Helix aspersa*. *Veliger* 30: 24–39.

Civetta, A., and A. G. Clark. 2000. Correlated effects of sperm competition and post-mating female mortality. *Proceedings of the National Academy of Sciences of the United States of America* 97: 13162–13165.

Civetta, A., and R. S. Singh. 1995. High divergence of reproductive tract proteins and their association with postzygotic reproductive isolation in *Drosophila melanogaster* and *Drosophila virilis* group species. *Journal of Molecular Evolution* 41: 1085–1095.

Civetta, A., and R. S. Singh. 1998. Sex and speciation: Genetic architecture and evolutionary potential of sexual versus nonsexual traits in the sibling species of the *Drosophila melanogaster* complex. *Evolution* 52: 1080–1092.

Clark, A. G., M. Aguade, T. Prout, L. G. Harshman, and C. H. Langley. 1995. Variation in sperm displacement and its association with accessory gland protein loci in *Drosophila melanogaster*. *Genetics* 139: 189–201.

Clark, A. G., and D. J. Begun. 1998. Female genotypes affect sperm displacement in *Drosophila*. *Genetics* 149: 1487–1493.

Clark, A. G., D. J. Begun, and T. Prout. 1999. Female × male interactions in *Drosophila* sperm competition. *Science* 283: 217–220.

Clark, S. J. 1988. The effects of operational sex ratio and food deprivation on copulation duration in the water strider (*Gerris remigis* Say). *Behavioral Ecology and Sociobiology* 23: 317–322.

Clavert, A., C. Cranz, and C. Bollack. 1990. Functions of the seminal vesicle. *Andrologia* 22: 185–192.

Clutton-Brock, T. H. 1991. *The Evolution of Parental Care*. Princeton University Press, Princeton, N.J.

Clutton-Brock, T. H., S. D. Albon, and F. E. Guinness. 1981. Parental investment in male and female offspring in polygynous mammals. *Nature* 289: 487–489.

Clutton-Brock, T. H., and G. R. Iason. 1986. Sex-ratio variation in mammals. *Quarterly Review of Biology* 61: 339–374.

Clutton-Brock, T. H., and G. A. Parker. 1992. Potential reproductive rates and the operation of sexual selection. *Quarterly Review of Biology* 67: 437–456.

Clutton-Brock, T. H., and G. A. Parker. 1995. Sexual coercion in animal societies. *Animal Behaviour* 49: 1345–1365.

Cockburn, A. 1998. Evolution of helping behavior in cooperatively breeding birds. *Annual Review of Ecology and Systematics* 29: 141–177.

Cockburn, A., S. Legge, and M. C. Double. 2002. Sex ratios in birds and mammals: Can the hypotheses be disentangled? In I.C.W. Hardy (ed.), *Sex Ratios: Concepts and Research Methods*, Cambridge University Press, New York.

Colegrave, N. 2001. Differential allocation and "good genes"—Comment from Colegrave. *Trends in Ecology & Evolution* 16: 22–23.

Conlan, K. E. 1991. Precopulatory mating behavior and sexual dimorphism in the amphipod Crustacea. *Hydrobiologia* 223: 255–282.

Constantz, G. D. 1989. Reproductive biology of Poeciliid fishes. In G. K. Meffe and F. F. Snelson (eds.), *Ecology and Evolution of Livebearing Fishes (Poeciliidae)*, Prentice Hall, Englewood Cliffs, N.J., pp. 33–50.

Cook, D. F. 1992. The effect of male size on receptivity in female *Lucilia cuprina* (Diptera, Calliphoridae). *Journal of Insect Behavior* 5: 365–374.

Cook, P. A., and N. Wedell. 1999. Non-fertile sperm delay female remating. *Nature* 397: 486.

Cook, S. E., J. G. Vernon, M. Bateson, and T. Guilford. 1994. Mate choice in the polymorphic Africa swallowtail butterfly, *Papilio dardanus*: Male-like females may avoid sexual harassment. *Animal Behaviour* 47: 389–397.

Cordero, A. 1990. The inheritance of female polymorphism in the damselfly *Ischnura graellsii* (Rambur) (Odonata, Coenagrionidae). *Heredity* 64: 341–346.

Cordero, A. 1992. Density-dependent mating success and color polymorphism in females of the damselfly *Ischnura graellsii* (Odonata, Coenagrionidae). *Journal of Animal Ecology* 61: 769–780.

Cordero, A., S. S. Carbone, and C. Utzeri. 1998. Mating opportunities and mating costs are reduced in androchrome female damselflies, *Ischnura elegans* (Odonata). *Animal Behaviour* 55: 185–197.

Cordero, A., and P. L. Miller. 1992. Sperm transfer, displacement and precedence in *Ischnura graellsii* (Odonata, Coenagrionidae). *Behavioral Ecology and Sociobiology* 30: 261–267.

Cordero, C. 1995. Ejaculate substances that affect female insect reproductive physiology and behavior: Honest or arbitrary traits? *Journal of Theoretical Biology* 174: 453–461.

Cordero, C., and W. G. Eberhard. 2003. Female choice of sexually antagonistic male adaptations: A critical review of some current research. *Journal of Evolutionary Biology* 16: 1–6.

Córdoba-Aguilar, A., and J. Contreras-Garduno. 2003. Sexual conflict. *Trends in Ecology & Evolution* 18: 439–440.

Cotton, S., K. Fowler, and A. Pomiankowski. 2004. Do sexual ornaments demonstrate heightened condition-dependent expression as predicted by the handicap hypothesis? *Proceedings of the Royal Society of London Series B—Biological Sciences* 271: 771–783.

Crean, C. S., D. W. Dunn, T. H. Day, and A. S. Gilburn. 2000. Female mate choice for large males in several species of seaweed fly (Diptera: Coelopidae). *Animal Behaviour* 59: 121–126.

Crean, C. S., and A. S. Gilburn. 1998. Sexual selection as a side-effect of sexual conflict in the seaweed fly, *Coelopa ursina* (Diptera: Coelopidae). *Animal Behaviour* 56: 1405–1410.

Crouse, H. V. 1960. The controlling element in sex chromosome behavior in *Sciara*. *Genetics* 45: 1429–1443.

Crowley, P. H., T. Cottrell, T. Garcia, M. Hatch, R. C. Sargent, B. J. Stokes, and J. M. White. 1998. Solving the complementarity dilemma: Evolving strategies for simultaneous hermaphroditism. *Journal of Theoretical Biology* 195: 13–26.

Crudgington, H. S., and M. T. Siva-Jothy. 2000. Genital damage, kicking and early death—the battle of the sexes takes a sinister turn in the bean weevil. *Nature* 407: 855–856.

Crump, M. L. 1988. Aggression in harlequin frogs: Male-male competition and a possible conflict of interest between the sexes. *Animal Behaviour* 36: 1064–1077.

Cumming, J. M. 1994. Sexual selection and the evolution of dance fly mating systems (Diptera: Empididae: Empidinae). *Canadian Entomologist* 126: 907–920.

Cunningham, E., and T. Birkhead. 1997. Female roles in perspective. *Trends in Ecology & Evolution* 12: 337–338.

Cunningham, E.J.A., and A. F. Russell. 2000. Egg investment is influenced by male attractiveness in the mallard. *Nature* 404: 74–77.

Cunningham, E.J.A., and A. F. Russell. 2001. Differential allocation and "good genes"—Comment from Cunningham & Russell. *Trends in Ecology & Evolution* 16: 21.

Curtsinger, J. W. 1991. Sperm competition and the evolution of multiple mating. *American Naturalist* 138: 93–102.

Cusson, M., J. Delisle, and D. Miller. 1999. Juvenile hormone titers in virgin and mated *Choristoneura fumiferana* and *C-rosaceana* females: Assessment of the capacity of males to produce and transfer JH to the female during copulation. *Journal of Insect Physiology* 45: 637–646.

Dale, S., and T. Slagsvold. 1994. Polygyny and deception in the pied flycatcher: Can females determine male mating status? *Animal Behaviour* 48: 1207–1217.

Daly, M. 1978. The cost of mating. *American Naturalist* 112: 771–774.

Darwin, C. 1871. *The Descent of Man and Selection in Relation to Sex*. J. Murray, London.

Darwin, C. 1859. *On the Origin of Species by Means of Natural Selection*. J. Murray, London.

Darwin, E. 1794. *Zoonomia; or, The Laws of Organic Life*. P. Byrne and W. Jones, Dublin.

Das, A. K., J. Huignard, M. Barbier, and A. Quesneau-Thierry. 1980. Isolation of the 2 paragonial substance deposited into the spermatophores of *Acanthoscelides obtectus* (Coleoptera, Bruchidae). *Experientia* 36: 918–920.

Davies, N. B. 1989. Sexual conflict and the polygamy threshold. *Animal Behaviour* 38: 226–234.

Davies, N. B. 1992. *Dunnock Behaviour and Social Evolution*. Oxford University Press, Oxford.

Davies, N. B., and T. R. Halliday. 1979. Competitive mate searching in male common toads, *Bufo bufo*. *Animal Behaviour* 27: 1253–1267.

Davies, N. B., and I. R. Hartley. 1996. Food patchiness, territory overlap and social systems: An experiment with dunnocks *Prunella modularis*. *Journal of Animal Ecology* 65: 837–846.

Davies, N. B., I. R. Hartley, B. J. Hatchwell, A. Desrochers, J. Skeer, and D. Nebel. 1995. The polysynandrous mating system of the alpine accentor, *Prunella collaris*. I. Ecological causes and reproductive conflicts. *Animal Behaviour* 49: 769–788.

Davies, N. B., I. R. Hartley, B. J. Hatchwell, and N. E. Langmore. 1996. Female control of copulations to maximize male help: A comparison of polygynandrous alpine accentors, *Prunella collaris*, and dunnocks, *P-modularis*. *Animal Behaviour* 51: 27–47.

Davies, N. B., and A. I. Houston. 1986. Reproductive success of dunnocks, *Prunella modularis*, in a variable mating system. II. Conflicts of interest among breeding adults. *Journal of Animal Ecology* 55: 139–154.

Davis, N. T. 1965a. Studies of the reproductive physiology of the Cimicidae (Hemiptera). II. Artificial insemination and the function of the seminal fluid. *Journal of Insect Physiology* 11: 355–366.

Davis, N. T. 1965b. Studies of the reproductive physiology of the Cimicidae (Hemiptera). III. The seminal stimulus. *Journal of Insect Physiology* 11: 1199–1211.

Dawkins, M. S., and T. Guilford. 1991. The corruption of honest signaling. *Animal Behaviour* 41: 865–873.

Dawkins, R. 1976. *The Selfish Gene*. Oxford University Press, Oxford.

Dawkins, R., and T. R. Carlisle. 1976. Parental investment, mate desertion and a fallacy. *Nature* 262: 131–133.

Dawkins, R., and J. R. Krebs. 1978. Animal signals: Information or manipulation? In J. R. Krebs and N. B. Davies (eds.), *Behavioural Ecology: An Evolutionary Approach*, Blackwell Scientific, Oxford, pp. 282–309.

Dawkins, R., and J. R. Krebs. 1979. Arms races between and within species. *Proceedings of the Royal Society of London Series B—Biological Sciences* 205: 489–511.

Day, T. 2000. Sexual selection and the evolution of costly female preferences: Spatial effects. *Evolution* 54: 715–730.

Day, T. H., and A. S. Gilburn. 1997. Sexual selection in seaweed flies. *Advances in the Study of Behavior* 26: 1–57.

De La Riva, I. 1994. A new aquatic frog of the genus Telmatobius (Anura: Leptodactylidae) from Bolivian cloud forests. *Herpetologica* 50: 38–45.

De Lope, F., and A. P. Møller. 1993. Female reproductive effort depends on the degree of ornamentation of their mates. *Evolution* 47: 1152–1160.

De Visser, J.A.G.M., A. Ter Maat, and C. Zonneveld. 1994. Energy budgets and reproductive allocation in the simultaneous hermaphrodite pond snail, *Lymnaea stagnalis* (L): A trade-off between male and female function. *American Naturalist* 144: 861–867.

Deinert, E. I., J. T. Longino, and L. E. Gilbert. 1994. Mate competition in butterflies. *Nature* 370: 23–24.

Delasena, C. A., N. S. Fechheimer, and K. E. Nestor. 1992. Evidence for genetic etiology of hereroploidy in embryois of the Japanese quail (*Coturnix coturnix japonica*). *Cytogenetics and Cell Genetics* 60: 140–145.

Delph, L. F., and K. Havens. 1998. Pollen competition in flowering plants. In T. R. Birkhead and A. P. Moller (eds.), *Sperm Competition and Sexual Selection*, Academic, London, pp. 149–174.

Dennis, D. S., and R. J. Lavigne. 1976. Ethology of *Efferia varipes* with comments on species coexistence (Diptera: Asilidae). *Journal of the Kansas Entomological Society* 49: 48–62.

Dennis, D. S., R. J. Lavigne, and S. W. Bullington. 1986. Ethology of *Efferia cressoni* with a review of the comparative ethology of the genus (Diptera: Asilidae). *Proceedings of the Entomological Society of Washington* 88: 42–55.

Destephano, D. B., and U. E. Brady. 1977. Prostaglandin and prostaglandin synthetase in cricket, *Acheta domesticus. Journal of Insect Physiology* 23: 905–911.

Dewsbury, D. A. 1972. Patterns of copluatory behavior in male mammals. *Quarterly Review of Biology* 47: 1–33.

Dewsbury, D. A. 1982. Ejaculate cost and male choice. *American Naturalist* 119: 601–610.

Dickinson, J. L. 1992. Scramble competition polygyny in the milkweed leaf beetle: Combat, mobility, and the importance of being there. *Behavioral Ecology* 3: 32–41.

Dickinson, J. L. 1997. Multiple mating, sperm competition, and cryptic female choice in the leaf beetles (Coleoptera: Chrysomelidae). In J. C. Choe and B. J. Crespi (eds.), *Social Competition and Cooperation in Insects and Arachnids: Evolution of Mating Systems*, Princeton University Press, Princeton, N.J.

Diesel, R. 1989. Structure and function of the reproductive system of the symbiotic spider crab *Inachus phalangium* (Decapoda: Majidae): Observations on sperm transfer, sperm storage, and spawning. *Journal of Crustacean Biology* 9: 266–277.

Dixson, A. F. 1995. Baculum length and copulatory behavior in carnivores and pinnipeds (Grand Order Ferae). *Journal of Zoology* 235: 67–76.

Donaldson, W. E. 1987. An ethogram and description of mating behavior for the tanner crab, *Chionoecetes bairdi. American Zoologist* 27: A35.

Donaldson, W. E., and A. E. Adams. 1989. Ethogram of behavior with emphasis on mating for the tanner crab *Chionoecetes bairdi* Rathbun. *Journal of Crustacean Biology* 9: 37–53.

Doyle, L. R., and L. N. Gleason. 1991. Suckers and other bursal structures of *Pomphorhynchus bulbocolli* and *Acanthocephalus dirus* (Acanthocephala). *Journal of Parasitology* 77: 437–440.

Dressler, R. L. 1982. Biology of the orchid bees (Euglossini). *Annual Review of Ecology and Systematics* 13: 373–394.

Droney, D. C. 2003. Females lay fewer eggs for males with greater courtship success in a lekking *Drosophila. Animal Behaviour* 65: 371–378.

Dubois, A., and M. Matsui. 1983. A new species of frog (genus Rana, subgenus Paa) from Western Nepal (Amphibia: Anura). *Copeia* 895–901.

Duellman, W. E., and O. Ochoa. 1991. A new species of *Bufo* (Anura: Bufonidae) from the Andes of southern Peru. *Copeia* 137–141.

Dunham, P., T. Alexander, and A. Hurshman. 1986. Precopulatory mate guarding in an amphipod, *Gammarus lawrencianus* Bousfield. *Animal Behaviour* 34: 1680–1686.

Dunn, D. W., C. S. Crean, C. L. Wilson, and A. S. Gilburn. 1999. Male choice, willingness to mate and body size in seaweed flies (Diptera: Coelopidae). *Animal Behaviour* 57: 847–853.

Dunn, P. O., L. A. Whittingham, and T. E. Pitcher. 2001. Mating systems, sperm competition, and the evolution of sexual dimorphism in birds. *Evolution* 55: 161–175.

Duvoisin, N., B. Baer, and P. Schmid-Hempel. 1999. Sperm transfer and male competition in a bumblebee. *Animal Behaviour* 58: 743–749.

Dybas, L. K., and H. S. Dybas. 1981. Coadaptation and taxonomic differentiation of sperm and spermathecae in featherwing beetles. *Evolution* 35: 168–174.

Eady, P. 1994. Intraspecific variation in sperm precedence in the bruchid beetle *Callosobruchus maculatus*. *Ecological Entomology* 19: 11–16.

Eady, P. E. 1995. Why do male *Callosobruchus maculatus* beetles inseminate so many sperm? *Behavioral Ecology and Sociobiology* 36: 25–32.

Eady, P. E. 2001. Postcopulatory, prezygotic reproductive isolation. *Journal of Zoology* 253: 47–52.

East, M. L., H. Hofer, and W. Wickler. 1993. The erect "penis" is a flag of submission in a female-dominated society: Greetings in Serengeti spotted hyenas. *Behavioral Ecology and Sociobiology* 33: 355–370.

Eberhard, W. G. 1985. *Sexual Selection and Animal Genitalia*. Harvard University Press, Cambridge, Mass.

Eberhard, W. G. 1993a. Copulatory courtship and genital mechanics of 3 species of *Macrodactylus* (Coleoptera Scarabaeidae Melolonthinae). *Ethology Ecology & Evolution* 5: 19–63.

Eberhard, W. G. 1993b. Copulatory courtship and morhpology of genitalic coupling in 7 *Phyllophaga* species (Coleoptera, Melolonthidae). *Journal of Natural History* 27: 683–717.

Eberhard, W. G. 1994. Evidence for widespread courtship during copulation in 131 species of insects and spiders, and implications for crytptic female choice. *Evolution* 48: 711–733.

Eberhard, W. G. 1996. *Female Control:Sexual Selection by Cryptic Female Choice*. Princeton University Press, Princeton, N.J.

Eberhard, W. G. 1998. Female roles in sperm competition. In T. R. Birkhead and A. P. Møller (eds.), *Sperm Competition and Sexual Selection*, Academic, London, pp. 91–116.

Eberhard, W. G. 2002. The function of female resistance behavior: Intromission by male coercion vs. female cooperation in sepsid flies (Diptera: Sepsidae). *Revista de Biologia Tropical* 50: 485–505.

Eberhard, W. G., and C. Cordero. 2003. Sexual conflict and female choice. *Trends in Ecology & Evolution* 18: 438–439.

Eberhard, W. G., and F. Pereira. 1993. Functions of the male genitalis surstyli in the mediterranean fruit fly, *Ceratitis capitata* (Diptera: Tephritidae). *Journal of the Kansas Entomological Society* 66: 427–433.

Eberhard, W. G., and F. Pereira. 1995. The process of intromission in the mediterranean fruit fly, *Ceratitis capitata* (Diptera: Tephritidae). *Psyche* 102: 99–120.

Eens, M., and R. Pinxten. 1995. Intersexual conflicts over copulations in the European starling: evidence for the female mate guarding hypothesis. *Behavioral Ecology and Sociobiology* 36: 71–81.

Eens, M., and R. Pinxten. 1996. Female European starlings increase their copulation solicitation rate when faced with the risk of polygyny. *Animal Behaviour* 51: 1141–1147.

Eggert, A. K., and S. K. Sakaluk. 1995. Female-coerced monogamy in burying beetles. *Behavioral Ecology and Sociobiology* 37: 147–153.

Eibl-Eibesfeldt, I. 1970. *Ethology, the Biology of Behavior.* Holt, Rinehart and Winston, New York.

Eisner, T., S. R. Smedley, D. K. Young, M. Eisner, B. Roach, and J. Meinwald. 1996. Chemical basis of courtship in a beetle (*Neopyrochroa flabellata*): Cantharidin as precopulatory "enticing" agent. *Proceedings of the National Academy of Sciences of the United States of America* 93: 6494–6498.

Ekman, J., and B. Sklepkovych. 1994. Conflict of interest between the sexes in Siberian jay winter flocks. *Animal Behaviour* 48: 485–487.

Elgar, M. 1992. Sexual cannibalism in spiders and other invertebrates. In M. A. Elgar and B. J. Crespi (eds.), *Cannibalism: Ecology and Evolution Among Diverse Taxa,* Oxford University Press, Oxford, pp. 128–155.

Elgar, M. A. 1991. Sexual cannibalism, size dimorphism, and courtship behavior in orb-weaving spiders (Araneidae). *Evolution* 45: 444–448.

Elgar, M. A., and B. F. Fahey. 1995. Sexual cannibalism, competition and size dimorphism in the orb-weaving spider *Nephila plumipes* Latreille (Araneae: Araneoidea). Manuscript.

Elgar, M. A., and B. F. Fahey. 1996. Sexual cannibalism, competition, and size dimorphism in the orb-weaving spider *Nephila plumipes* Latreille (Araneae: Araneoidea). *Behavioral Ecology* 7: 195–198.

Elgar, M. A., N. Ghaffar, and A. F. Read. 1990. Sexual dimorphism in leg length among orb-weaving spiders: A possible role for sexual cannibalism. *Journal of Zoology* 222: 455–470.

Elinson, R. P. 1987. Fertilization and aqueous development of the Puerto Rican terrestrial breeding frog, *Eleutherodactylus coqui. Journal of Morphology* 193: 217–224.

Ellegren, H. 2001. Hens, cocks and avian sex determination—a quest for genes on Z or W? *Embo Reports* 2: 192–196.

Eltz, T., W. M. Whitten, D. W. Roubik, and K. E. Linsenmair. 1999. Fragrance collection, storage, and accumulation by individual male orchid bees. *Journal of Chemical Ecology* 25: 157–176.

Emlen, S. T., and L. W. Oring. 1977. Ecology, sexual selection, and evolution of mating systems. *Science* 197: 215–223.

Endler, J. A. 1980. Natural selection on color patterns in *Poecilia reticulata. Evolution* 34: 76–91.

Endler, J. A. 1986. *Natural Selection in the Wild.* Princeton University Press, Princeton, N.J.

Endler, J. A. 1987. Predation, light intensity and courtship behavior in *Poecilia reticulata* (Pisces: Poecilidae). *Animal Behaviour* 35: 1376–1385.

Endler, J. A. 1992. Signals, signal conditions, and the direction of evolution. *American Naturalist* 139: S125–S153.

Endler, J. A., and A. L. Basolo. 1998. Sensory ecology, receiver biases and sexual selection. *Trends in Ecology & Evolution* 13: 415–420.

Engelhard, G., S. P. Foster, and T. H. Day. 1989. Genetic differences in mating success and female choice in seaweed flies (*Coelopa frigida*). *Heredity* 62: 123–131.

Enquist, M., and A. Arak. 1993. Selection of exaggerated male traits by female aesthetic senses. *Nature* 361: 446–448.

Enquist, M., A. Arak, S. Ghirlanda, and C. A. Wachtmeister. 2002. Spectacular phenomena and limits to rationality in genetic and cultural evolution. *Philosophical*

Transactions of the Royal Society of London Series B—Biological Sciences 357: 1585–1594.

Epstein, M. S., and D. G. Blackburn. 1997. Histology and histochemistry of androgen-stimulated nuptial pads in the leopard frog, *Rana pipiens*, with notes on nuptial gland evolution. *Canadian Journal of Zoology* 75: 472–477.

Ericsson, R. J., and V. F. Baker. 1966. Transport of oestrogens in semen to female rat during maitng and its effect on fertility. *Journal of Reproduction and Fertility* 12: 381–384.

Eroglu, A., M. Toner, L. Leykin, and T. L. Toth. 1998. Cytoskeleton and polyploidy after maturation and fertilization of cryopreserved germinal vesicle stage mouse oocytes. *Journal of Assisted Reproduction and Genetics* 15: 447–454.

Estoup, A., A. Scholl, A. Pouvreau, and M. Solignac. 1995. Monandry and polyandry in bumble bees (Hymenoptera, Bombinae) as evidenced by highly variable microsatellites. *Molecular Ecology* 4: 89–93.

Etges, W. J. 2002. Divergence in mate choice systems: Does evolution play by rules? *Genetica* 116: 151–166.

Etges, W. J., and W. B. Heed. 1992. Remating effects of the genetic structure of female life histories in populations of *Drosophila mojavensis*. *Heredity* 68: 515–528.

Evans, J. P. 2000. Getting sperm and egg together: Things conserved and things diverged. *Biology of Reproduction* 63: 355–360.

Evans, J. P., J. L. Kelley, I. W. Ramnarine, and A. Pilastro. 2002. Female behaviour mediates male courtship under predation risk in the guppy (*Poecilia reticulata*). *Behavioral Ecology and Sociobiology* 52: 496–502.

Evans, J. P., A. Pilastro, and I. W. Ramnarine. 2003. Sperm transfer through forced matings and its evolutionary implications in natural guppy (*Poecilia reticulata*) populations. *Biological Journal of the Linnean Society* 78: 605–612.

Fairbairn, D. J. 1993. Costs of loading associated with mate carrying in the water strider, *Aquarius remigis*. *Behavioral Ecology* 4: 224–231.

Fairbairn, D. J., R. Vermette, N. N. Kapoor, and N. Zahiri. 2003. Functional morphology of sexually selected gentalia in the water strider *Aquarius remigis*. *Canadian Journal of Zoology* 81: 400–413.

Fairfull, R. 1990. Heterosis. In R. D. Crawford (ed.), *Poultry Breeding and Genetics*, Elsevier, Amsterdam, pp. 913–933.

Fan, Y. L., A. Rafaeli, C. Gileadi, E. Kubli, and S. W. Applebaum. 1999. *Drosophila melanogaster* sex peptide stimulates juvenile hormone synthesis and depresses sex pheromone production in *Helicoverpa armigera*. *Journal of Insect Physiology* 45: 127–133.

Fan, Y. L., A. Rafaeli, P. Moshitzky, E. Kubli, Y. Choffat, and S. W. Applebaum. 2000. Common functional elements of *Drosophila melanogaster* seminal peptides involved in reproduction of *Drosophila melanogaster* and *Helicoverpa armigera* females. *Insect Biochemistry and Molecular Biology* 30: 805–812.

Farr, J. A. 1975. Role of predation in evolution of social behavior of natural populations of guppy, *Poecilia reticulata* (Pisces: Poeciliidae). *Evolution* 29: 151–158.

Festa-Bianchet, M. 1996. Offspring sex ratio studies of mammals: Does publication depend upon the quality of the research or the direction of the results? *Ecoscience* 3: 42–44.

Field, S. A., and B. Yuval. 1999. Nutritional status affects copula duration in the Mediterranean fruit fly, *Ceratitis capitata* (Insecta Tephritidae). *Ethology Ecology & Evolution* 11: 61–70.

Fielding, K., and C. B. Knisley. 1995. Mating behavior in 2 tiger beetles, *Cicindela dorsalis* and *C-puritana* (Coleoptera: Cicindelidae). *Entomological News* 106: 61–67.

Fincke, O. M. 1984. Sperm competition in the damselflly *Enallagma hageni* Walsh (Odonata: Coenagrionidae): Benefits of multiple mating to males and females. *Behavioral Ecology and Sociobiology* 14: 235–240.

Fincke, O. M. 1994a. On the difficulty of detecting density dependent selection on polymorphic females of the damselfly *Ischnura graellsii*: Failure to reject the null. *Evolutionary Ecology* 8: 328–329.

Fincke, O. M. 1994b. Female color polymorphism in damselflies: Failure to reject the null hypothesis. *Animal Behaviour* 47: 1249–1266.

Fischer, E. A. 1980. The relationship between mating system and simultaneous hermaphroditism in the coral reef fish, *Hypoplectrus nigricans* (Serranidae). *Animal Behaviour* 28: 620–633.

Fisher, R. A. 1915. The evolution of sexual preference. *Eugenics Review* 7: 184–192.

Fisher, R. A. 1930. *The Genetical Theory of Natural Selection*. Clarendon, Oxford.

Flores, G. 1985. A new Centrolenella (Anura) from Ecuador, with comments on nuptial pads and prepollical spines in Centrolenella. *Journal of Herpetology* 19: 313–320.

Forbes, M.R.L. 1991. Female morphs of the damselfly *Enallagma boreale selys* (Odonata: Coenagrionidae): A benefit for androchromatypes. *Canadian Journal of Zoology* 69: 1969–1970.

Forbes, M.R.L., J.M.L. Richarson, and R. L. Baker. 1995. Frequency of female morphs is related to an index of male density in the damselfly, *Nehalennia irene* (Hagen). *Ecoscience* 2: 28–33.

Forsberg, J., and C. Wiklund. 1989. Mating in the afternoon: Time saving in courtship and remating by females of a polyandrous butterfly *Pieris napi*. *Behavioral Ecology and Sociobiology* 25: 349–356.

Forster, L. M. 1992. The stereotyped behavior of sexual cannibalism in *Latrodectus hasselti* Thorell (Araneae: Theridiidae): The Autralian redback spider. *Australian Journal of Zoology* 40: 1–11.

Foster, S. P., and R. H. Ayers. 1996. Multiple mating and its effects in the light-brown apple moth, *Epiphyas postvittana* (Walker). *Journal of Insect Physiology* 42: 657–667.

Fowler, K., and L. Partridge. 1989. A cost of mating in female fruit flies. *Nature* 338: 760–761.

Fox, C. W. 1993. Multiple mating, lifetime fecundity and female mortality of the bruchid beetle, *Callosobruchus maculatus* (Coleoptera: Bruchidae). *Functional Ecology* 7: 203–208.

Frank, S. A. 2000. Sperm competition and female avoidance of polyspermy mediated by sperm-egg biochemistry. *Evolutionary Ecology Research* 2: 613–625.

Frank, S. A., and L. D. Hurst. 1996. Mitochondria and male disease. *Nature* 383: 224.

Franke, E. S., R. C. Babcock, and C. A. Styan. 2002. Sexual conflict and polyspermy under sperm-limited conditions: In situ evidence from field simulations with the free-spawning marine echinoid *Evechinus chloroticus*. *American Naturalist* 160: 485–496.

Franke, E. S., R. C. Babcock, and C. A. Styan. 2003. Correction: Sexual conflict and polyspermy under sperm-limited conditions: In situ evidence from field simulations with the free-spawning marine echinoid *Evechinus chloroticus* (vol 160, pg 485, 2002). *American Naturalist* 161: 169.

Franks, L. 1983. Origins of benign prostatic hypertrophy. In F. Hinman (ed.), *Benign Prostatic Hypertrophy*, Springer, New York, pp. 141–167.

Franz, D. 1991. Mating system and strategy of reproduction in penduline tit *Remiz pendulinus*. *Journal für Ornithologie* 132: 241–266.

Friberg, U., and G. Arnqvist. 2003. Fitness effects of female mate choice: Preferred males are detrimental for *Drosophila melanogaster* females. *Journal of Evolutionary Biology* 16: 797–811.

Friedlander, M. 1980. Monospermic fertilization in *Chrysopa carnea* (Neurotera: Chrsopidae): Behavior of the fertilizing spermatozoa prior to syngamy. *International Journal of Insect Morphology & Embryology* 9: 53–57.

Friedlander, M. 1997. Control of the eupyrene-apyrene sperm dimorphism in lepidoptera. *Journal of Insect Physiology* 43: 1085–1092.

Friedman, M., and D. S. Lehrman. 1968. Physiological conditions for stimulation of prolactin secretion by external stimuli in male ring dove. *Animal Behaviour* 16: 233–237.

Froman, D. P., T. Pizzari, A. J. Feltmann, H. Castillo-Juarez, and T. R. Birkhead. 2002. Sperm mobility: Mechanisms of fertilizing efficiency, genetic variation and phenotypic relationship with male status in the domestic fowl, *Gallus gallus domesticus*. *Proceedings of the Royal Society of London Series B—Biological Sciences* 269: 607–612.

Fugo, H., and N. Arisawa. 1992. Oviposition behaviour of the moths which mated with males sterilized by high temperature in the silkworm, *Bombyx mori*. *Journal of Sericultural Science of Japan* 61: 110–115.

Fujihara, N. 1992. Accessory reproductive fluids and organs in male domestic birds. *World's Poultry Science Journal* 48: 39–56.

Fuller, R., and A. Berglund. 1996. Behavioral responses of a sex-role reversed pipefish to a gradient of perceived predation risk. *Behavioral Ecology* 7: 69–75.

Funahashi, H., and B. N. Day. 1997. Advances in in vitro production of pig embryos. *Journal of Reproduction and Fertility* 271–283.

Funk, D. H., and D. W. Tallamy. 2000. Courtship role reversal and deceptive signals in the long-tailed dance fly, *Rhamphomyia longicuada*. *Animal Behaviour* 59: 411–421.

Fuyama, Y. 1983. Species-specificity of paragonial substances as an isolating mechanism in *Drosophila*. *Experientia* 39: 190–192.

Gage, M.J.G. 1994. Associations between body size, mating pattern, testis size and sperm lengths across butterflies. *Proceedings of the Royal Society of London Series B—Biological Sciences* 258: 247–254.

Gage, M.J.G. 1998. Mammalian sperm morphometry. *Proceedings of the Royal Society of London Series B—Biological Sciences* 265: 97–103.

Gage, M.J.G., G. A. Parker, S. Nylin, and C. Wiklund. 2002. Sexual selection and speciation in mammals, butterflies and spiders. *Proceedings of the Royal Society of London Series B—Biological Sciences* 269: 2309–2316.

Galan, P. 2000. Females that imitate males: Dorsal coloration varies with reproductive stage in female *Podarcis bocagei* (Lacertidae). *Copeia* 819–825.

Galimberti, F., L. Boitani, and I. Marzetti. 2000a. Female strategies of harassment reduction in southern elephant seals. *Ethology Ecology & Evolution* 12: 367–388.

Galimberti, F., L. Boitani, and I. Marzetti. 2000b. The frequency and costs of harassment in southern elephant seals. *Ethology Ecology & Evolution* 12: 345–365.

Galimberti, F., L. Boitani, and I. Marzetti. 2000c. Harassment during arrival on land and departure to sea in southern elephant seals. *Ethology Ecology & Evolution* 12: 389–404.

Galindo, B. E., V. D. Vacquier, and W. J. Swanson. 2003. Positive selection in the egg receptor for abalone sperm lysin. *Proceedings of the National Academy of Sciences of the United States of America* 100: 4639–4643.

Galtung, J., C. G. Jacobsen, and K. F. Brand-Jacobsen. 2002. *Searching for Peace: The Road to Transcend*, 2nd Ed. Pluto, London.

Galvani, A., and R. Johnstone. 1998. Sperm allocation in an uncertain world. *Behavioral Ecology and Sociobiology* 44: 161–168.

Gavrilets, S. 1997. Coevolutionary chase in exploiter-victim systems with polygenic characters. *Journal of Theoretical Biology* 186: 527–534.

Gavrilets, S. 2000. Rapid evolution of reproductive barriers driven by sexual conflict. *Nature* 403: 886–889.

Gavrilets, S., G. Arnqvist, and U. Friberg. 2001. The evolution of female mate choice by sexual conflict. *Proceedings of the Royal Society of London Series B—Biological Sciences* 268: 531–539.

Gavrilets, S., and A. Hastings. 1998. Coevolutionary chase in two-species systems with applications to mimicry. *Journal of Theoretical Biology* 191: 415–427.

Gavrilets, S., and D. Waxman. 2002. Sympatric speciation by sexual conflict. *Proceedings of the National Academy of Sciences of the United States of America* 99: 10533–10538.

Gebhard, J. 1995. Observations on the mating behaviour of *Nyctalus noctula* (Schreber, 1774) in the hibernaculum. *Myotis* 32–33: 123–129.

Gemmell, N. J., and F. W. Allendorf. 2001. Mitochondrial mutations may decrease population viability. *Trends in Ecology & Evolution* 16: 115–117.

Gems, D., and D. L. Riddle. 1996. Longevity in *Caenorhabditis elegans* reduced by mating but not gamete production. *Nature* 379: 723–725.

Getty, T. 1998. Handicap signalling: When fecundity and viability do not add up. *Animal Behaviour* 56: 127–130.

Getty, T. 1999. Chase-away sexual selection as noisy reliable signaling. *Evolution* 53: 299–302.

Gianaroli, L., M. C. Magli, A. P. Ferraretti, A. Fiorentino, E. Tosti, S. Panzella, and B. Dale. 1996. Reducing the time of sperm-oocyte interaction in human in-vitro fertilization improves the implantation rate. *Human Reproduction* 11: 166–171.

Gibson, A. R., and J. B. Falls. 1975. Evidence for multiple insemination in common garter snake, *Thamnophis sirtalis. Canadian Journal of Zoology* 53: 1362–1368.

Gibson, J. R., A. K. Chippindale, and W. R. Rice. 2002. The X chromosome is a hot spot for sexually antagonistic fitness variation. *Proceedings of the Royal Society of London Series B—Biological Sciences* 269: 499–505.

Gil, D., and J. Graves. 2001. Differential allocation and "good genes"—Comment from Gil & Graves. *Trends in Ecology & Evolution* 16: 21–22.

Gil, D., J. Graves, N. Hazon, and A. Wells. 1999. Male attractiveness and differential testosterone investment in zebra finch eggs. *Science* 286: 126–128.

Gilburn, A. S., and T. H. Day. 1999. Female mating behaviour, sexual selection and chromosome I inversion karyotype in the seaweed fly, *Coelopa frigida. Heredity* 82: 276–281.

Gilchrist, A. S., and L. Partridge. 1997. Heritability of pre-adult viability differences can explain apparent heritability of sperm displacement ability in *Drosophila melanogaster. Proceedings of the Royal Society of London Series B—Biological Sciences* 264: 1271–1275.

Gillott, C. 1988. Arthropoda—Insecta. In R. G. Adiyodi and K. G. Adiyodi (eds.), *Reproductive Biology of Invertebrates. Vol. IV P.B. Fertilization, Development and Parental Care*, Wiley, Chichester, pp. 319–471.

Gillott, C. 1996. Male insect accessory glands: Functions and control of secretory activity. *Invertebrate Reproduction & Development* 30: 199–205.

Gillott, C. 2003. Male accessory gland secretions: Modulators of female reproductive physiology and behavior. *Annual Review of Entomology* 48: 163–184.

Ginzberg, A. S. 1972. *Fertilization in Fishes and the Problem of Polyspermy*, Israel Program for Scientific Translations, Jerusalem.

Giusti, F., and A. Lepri. 1980. Aspetti morfologici ed etologici dell'accoppiamento in alcune specie della famiglia Helicidae (Gas-tropoda, Pulmonata). *Atti della Accademia dei Fisiocritici in Siena* 11–71.

Godfray, H. C. J. 1995. Evolutionary theory of parent-offspring conflict. *Nature* 376: 133–138.

Gomendio, M., A. H. Harcourt, and E.R.S. Roldan. 1998. Sperm competition in mammals. In T. R. Birkhead and A. P. Moller (eds.), *Sperm Competition and Sexual Selection*, Academic, London, pp. 667–756.

Gomendio, M., and E.R.S. Roldan. 1991. Sperm competition influences sperm size in mammals. *Proceedings of the Royal Society of London Series B—Biological Sciences* 243: 181–185.

Gomez, M. I., and M. O. Cabada. 1994. Amphibian cross-fertilization and polyspermy. *Journal of Experimental Zoology* 269: 560–565.

Gordon, I. 1993. Pre-copulatory behavior of captive sandtiger sharks, *Carcharias taurus. Environmental Biology of Fishes* 38: 159–164.

Gould, S. J. 1984. Only his wings remained. *Natural History* 93: 10–18.

Gowaty, P. A. 1994. Architects of sperm competition. *Trends in Ecology & Evolution* 9: 160–162.

Gowaty, P. A. (ed.). 1997a. *Feminism and Evolutionary Biology: Boundaries, Intersection, and Frontiers*. Chapman & Hall, New York.

Gowaty, P. A. 1997b. Sexual dialectics, sexual selection, and variation in mating behavior. In P. A. Gowaty (ed.), *Feminism and Evolutionary Biology: Boundaries, Intersection, and Frontiers*, Chapman & Hall, New York, pp. 351–384.

Gowaty, P. A. 2003. Sexual natures: How feminism changed evolutionary biology. *Signs* 28: 901–922.

Gowaty, P. A., and N. Buschhaus. 1998. Ultimate causation of aggressive and forced copulation in birds: female resistance, the CODE hypothesis, and social monogamy. *American Zoologist* 38: 207–225.

Grafen, A. 1990a. Biological signals as handicaps. *Journal of Theoretical Biology* 144: 517–546.

Grafen, A. 1990b. Sexual selection unhandicapped by the Fisher process. *Journal of Theoretical Biology* 144: 473–516.

Gray, D. A., and W. H. Cade. 2000. Sexual selection and speciation in field crickets. *Proceedings of the National Academy of Sciences of the United States of America* 97: 14449–14454.

Grbic, M., P. J. Ode, and M. R. Strand. 1992. Sibling rivalry and brood sex-ratios in polyembryonic wasps. *Nature* 360: 254–256.

Greeff, J. M., and N. K. Michiels. 1999a. Sperm digestion and reciprocal sperm transfer can drive hermaphrodite sex allocation to equality. *American Naturalist* 153: 421–430.

Greeff, J. M., and N. K. Michiels. 1999b. Low potential for sexual selection in simultaneously hermaphroditic animals. *Proceedings of the Royal Society of London Series B—Biological Sciences* 266: 1671–1676.

Greeff, J. M., and G. A. Parker. 2000. Spermicide by females: What should males do? *Proceedings of the Royal Society of London Series B—Biological Sciences* 267: 1759–1763.

Gross, M. R., and R. C. Sargent. 1985. The evloution of male and female parental care in fishes. *American Zoologist* 25: 807–822.

Gross, M. R., and R. Shine. 1981. Parental care and mode of fertilization in ectothermic vertebrates. *Evolution* 35: 775–793.

Gubernick, D. J., and P. H. Klopfer. 1981. *Parental Care in Mammals*. Plenum, New York.

Gwynne, D. T. 1984. Male mating effort, confidence of paternity, and insect sperm competition. In R. L. Smith (ed.), *Sperm Competition and the Evolution of Animal Mating Systems*, Academic, Orlando, Fla., pp. 117–149.

Gwynne, D. T. 1986. Courtship feeding in katydids (Orthoptera: Tettigoniidae): Investment in offspring or in obtaining fertilizations? *American Naturalist* 128: 342–352.

Gwynne, D. T. 1989. Does copulation increase the risk of predation? *Trends in Ecology & Evolution* 4: 54–56.

Gwynne, D. T. 1991. Sexual competition among females: What causes courtship role reversal? *Trends in Ecology & Evolution* 6: 118–121.

Gwynne, D. T. 1993. Food quality controls sexual selection in mormon crickets by altering male mating investment. *Ecology* 74: 1406–1413.

Gwynne, D. T. 1997. The evolution of edible sperm sacs and other forms of courtship feeding in crickets, katydids and their kin. In J. C. Choe and B. J. Crespi (eds.), *The Evolution of Mating Systems in Insects and Arachnids*, Cambridge University Press, Cambridge.

Gwynne, D. T., and L. W. Simmons. 1990. Experimental reversal of courtship roles in an insect. *Nature* 346: 172–174.

Habib, F. K., S. Q. Maddy, and S. R. Stitch. 1980. Zinc induced changes in the progesterone binding properties of the human endometrium. *Acta Endocrinologica* 94: 99–106.

Haig, D. 1992. Genomic imprinting and the theory of parent-offspring conflict. *Seminars in Developmental Biology* 3: 153–160.

Haig, D. 1997. Parental antagonism, relatedness asymmetries, and genomic imprinting. *Proceedings of the Royal Society of London Series B—Biological Sciences* 264: 1657–1662.

Haig, D. 2000. The kinship theory of genomic imprinting. *Annual Review of Ecology and Systematics* 31: 9–32.

Haig, D., and M. Westoby. 1989. Parent-specific gene expression and the triploid endosperm. *American Naturalist* 134: 147–155.

Haig, D., and M. Westoby. 1991. Genomic imprinting in endosperm: Its effect of seed development in crosses between species, and between different ploidies of the same species, and its implications for the evolution of apomixis. *Philosophical Transactions of the Royal Society of London Series B—Biological Sciences* 333: 1–13.

Hall, D. W., M. Kirkpatrick, and B. West. 2000. Runaway sexual selection when female preferences are directly selected. *Evolution* 54: 1862–1869.

Halliday, T., and S. J. Arnold. 1987. Multiple mating by females: A perspective from quantitative genetics. *Animal Behaviour* 35: 939–941.

Halliday, T. R. 1983. The study of mate choice. In P. Bateson (ed.), *Mate Choice*, Cambridge University Press, Cambridge, pp. 3–32.

Hamilton, W. D. 1993. Haploid dynamic polymorphism in a host with matching parasites: Effects of mutation subdivision, linkage, and patterns of selection. *Journal of Heredity* 84: 328–338.

Hamilton, W. D., and M. Zuk. 1982. Heritable true fitness and bright birds: A role for parasites? *Science* 218: 384–387.

Hammerstein, P., and G. A. Parker. 1987. Sexual selection: Games between the sexes. In J. W. Bradbury and M. B. Andersson (eds.), *Sexual Selection: Testing the Alternatives*, Wiley, Chichester, pp. 119–142.

Han, Y. M., W. H. Wang, L. R. Abeydeera, A. L. Petersen, J. H. Kim, C. Murphy, B. N. Day, and R. S. Prather. 1999. Pronuclear location before the first cell division determines ploidy of polyspermic pig embryos. *Biology of Reproduction* 61: 1340–1346.

Hancock, J. L. 1959. Polyspermy of pig ova. *Animal Production* 1: 103–106.

Harcourt, A. H., P. H. Harvey, S. G. Larson, and R. V. Short. 1981. Testis weight, body weight and breeding system in primates. *Nature* 293: 55–57.

Härdling, R. 1999. Arms races, conflict costs and evolutionary dynamics. *Journal of Theoretical Biology* 196: 163–167.

Härdling, R., and P. Nilsson. 2001. A model of triploid endosperm evolution driven by parent-offspring conflict. *Oikos* 92: 417–423.

Härdling, R., H. G. Smith, V. Jormalainen, and J. Tuomi. 2001. Resolution of evolutionary conflicts: Costly behaviours enforce the evolution of cost-free competition. *Evolutionary Ecology Research* 3: 829–844.

Harmsen, R., and D. R. Clark. 1987. The effects of inbreeding and ventilation on mating behavior in *Drosophila pseudoobscura*. *Behavioural Processes* 15: 181–190.

Harper, A. B. 1988. *The Banana Slug*. Bay Leaves Press, Aptos, Calif.

Harshman, L. G., and T. Prout. 1994. Sperm displacement without sperm transfer in *Drosophila melanogaster*. *Evolution* 48: 758–766.

Hartley, I. R., N. B. Davies, B. J. Hatchwell, A. Desrochers, D. Nebel, and T. Burke. 1995. The polysynandrous mating system of the alpine accentor, *Prunell collaris*. II. Multiple paternity and parental effort. *Animal Behaviour* 49: 789–803.

Hartmann, R., and W. Loher. 1974. Control of sexual behavior pattern secondary defence in female grasshopper, *Chorthippus curtipennis*. *Journal of Insect Physiology* 20: 1713–1728.

Hartmann, R., and W. Loher. 1996. Control mechanisms of the behavior "secondary defense" in the grasshopper *Gomphocerus rufus* L (Gomphocerinae: Orthoptera). *Journal of Comparative Physiology A—Sensory, Neural, and Behavioral Physiology* 178: 329–336.

Hastings, I. M. 1994. Manifestations of sexual selection may depend on the genetic basis of sex determination. *Proceedings of the Royal Society of London Series B—Biological Sciences* 258: 83–87.

Haupt, H., and I. Todte. 1992. Contribution to the breeding biology of the pendulin tit (*Remiz pendulinus*). *Beitrage zur Vogelkunde* 38: 231–248.

Hausfater, G., and S. B. Hrdy. 1984. *Infanticide: Comparative and Evolutionary Perspectives*, Aldine, New York.

Hawkes, K., A. R. Rogers, and E. L. Charnov. 1995. The males' dilemma: Increased offspring production is more paternity to steal. *Evolutionary Ecology* 9: 662–677.

Hawkes, P. G. 1992. Sex-ratio stability and male-female conflict over sex-ratio control in hymenopteran parasitoids. *South African Journal of Science* 88: 423–430.

Hayashi, F. 1998. Multiple mating and lifetime reproductive output in female dobsonflies that receive nuptial gifts. *Ecological Research* 13: 283–289.

Hayashi, F. 1999. Rapid evacuation of spermatophore contents and male post-mating behaviour in aldersflies (Megaloptera: Sialidae). *Entomological Science* 2: 49–56.

Hébert, P. N., and S. G. Sealy. 1993. Hatching asynchrony and feeding rates in yellow warblers: A test of the sexual conflict hypothesis. *American Naturalist* 142: 881–892.

Hedrick, A. V., and L. M. Dill. 1993. Mate choice by female crickets is influenced by predation risk. *Animal Behaviour* 46: 193–196.

Heifetz, Y., O. Lung, E. A. Frongillo, and M. F. Wolfner. 2000. The *Drosophila* seminal fluid protein Acp26Aa stimulates release of oocytes by the ovary. *Current Biology* 10: 99–102.

Heller, K. G., S. Faltin, P. Fleischmann, and O. Von Helversen. 1998. The chemical composition of the spermatophore in some species of phanteropterid bushcrickets (Orthoptera: Tettigonioidea). *Journal of Insect Physiology* 44: 1001–1008.

Heller, K. G., P. Fleischmann, and A. Lutz-Roder. 2000. Carotenoids in the spermatophores of bushcrickets (Orthoptera: Ephippigerinae). *Proceedings of the Royal Society of London Series B—Biological Sciences* 267: 1905–1908.

Heming–van Battum, K. F., and B. S. Heming. 1986. Structure, function and evolution of the reproductive system in females of *Hebrus pusillus* and *H. ruficeps* (Hemiptera, Gerromorpha, Hebridae). *Journal of Morphology* 190: 121–167.

Henriksson, S. 1993. Sexuell kannibalism hos kärrspindeln. M.Sc. thesis, University of Umeå, Sweden.

Henry, C. S., M.L.M. Wells, and K. E. Holsinger. 2002. The inheritance of mating songs in two cryptic, sibling lacewing species (Neuroptera: Chrysopidae: Chrysoperla). *Genetica* 116: 269–289.

Herberstein, M. E., J. M. Schneider, and M. A. Elgar. 2002. Costs of courtship and mating in a sexually cannibalistic orb-web spider: Female mating strategies and their consequences for males. *Behavioral Ecology and Sociobiology* 51: 440–446.

Heywood, J. S. 1989. Sexual selection by the handicap mechanism. *Evolution* 43: 1387–1397.

Hildreth, P. E., and J. C. Lucchesi. 1963. Fertilization in *Drosophila*. I. Evidence for regular occurrence of monospermy. *Developmental Biology* 6: 262–278.

Hill, G. E. 1994. Trait elaboration via adaptive mate choice: Sexual conflict in the evolution of signals of male quality. *Ethology Ecology & Evolution* 6: 351–370.

Hinde, R. A., and E. Steel. 1978. The influence of daylength and male vocalizations on the estrogen-dependent behaviour of female canaries and budgerigars, with discussion of data from other species. *Advances in the Study of Behaviour* 8: 39–73.

Hinnekint, B.O.N. 1987. Population dynamics of *Ischnura e. elegans* (Vander Linden (Insecta: Odonata) with special referecne to morphological color changes, female polymorphism, multiannual cycles and their influence on behavior. *Hydrobiologia* 146: 3–31.

Hinton, H. E. 1964. Sperm transfer in insects and the evolution of haemocoelic insemination. In K. C. Highnam (ed.), *Insect Reproduction*, Royal Entomological Society, London, pp. 95–107.

Hiraiwa-Hasegawa, M. 1988. Adaptive significance of infanticide in primates. *Trends in Ecology & Evolution* 3: 102–105.

Hirschenhauser, K., E. Mostl, and K. Kotrschal. 1999. Within-pair testosterone covariation and reproductive output in Greylag Geese *Anser anser*. *Ibis* 141: 577–586.

Ho, P. C., W.S.B. Yeung, Y. F. Chan, W.W.K. So, and S.T.H. Chan. 1994. Factors affecting the incidence of polyploidy in a human in-vitro fertilization program. *International Journal of Fertility and Menopausal Studies* 39: 14–19.

Hocini, H., A. Barra, L. Belec, S. Iscaki, J. L. Preudhomme, J. Pillot, and J. P. Bouvet. 1995. Systemic and secretory humoral immunity in the normal human vaginal tract. *Scandinavian Journal of Immunology* 42: 269–274.

Hockham, L. R., and M. G. Ritchie. 2000. Female secondary sexual characteristics: Appearances might be deceptive. *Trends in Ecology & Evolution* 15: 436–438.

Hoekstra, H. E., J. M. Hoekstra, D. Berrigan, S. N. Vignieri, A. Hoang, C. E. Hill, P. Beerli, and J. G. Kingsolver. 2001. Strength and tempo of directional selection in the wild. *Proceedings of the National Academy of Sciences of the United States of America* 98: 9157–9160.

Hogg, J. T. 1988. Copulatory tactics in relation to sperm competition in Rocky Mountain bighorn sheep. *Behavioral Ecology and Sociobiology* 22: 49–59.

Hoi, H., B. Schleicher, and F. Valera. 1994. Female mate choice and nest desertion in penduline tits, *Remiz pendulinus*: The importance of nest quality. *Animal Behaviour* 48: 743–746.

Holland, B. 2002. Sexual selection fails to promote adaptation to a new environment. *Evolution* 56: 721–730.

Holland, B., and W. R. Rice. 1998. Perspective: Chase-away sexual selection: Antagonistic seduction versus resistance. *Evolution* 52: 1–7.

Holland, B., and W. R. Rice. 1999. Experimental removal of sexual selection reverses intersexual antagonistic coevolution and removes a reproductive load. *Proceedings of the National Academy of Sciences of the United States of America* 96: 5083–5088.

Holmgren, N.M.A., and M. Enquist. 1999. Dynamics of mimicry evolution. *Biological Journal of the Linnean Society* 66: 145–158.

Hosken, D. J. 1997. Sperm competition in bats. *Proceedings of the Royal Society of London Series B—Biological Sciences* 264: 385–392.

Hosken, D. J., W. U. Blanckenhorn, and T.W.J. Garner. 2002. Heteropopulation males have a fertilization advantage during sperm competition in the yellow dung fly (*Scathophaga stercoraria*). *Proceedings of the Royal Society of London Series B—Biological Sciences* 269: 1701–1707.

Hosken, D. J., T.W.J. Garner, T. Tregenza, N. Wedell, and P. I. Ward. 2003a. Superior sperm competitors sire higher-quality young. *Proceedings of the Royal Society of London Series B—Biological Sciences* 270: 1933–1938.

Hosken, D. J., T. W. J. Garner, and P. I. Ward. 2001. Sexual conflict selects for male and female reproductive characters. *Current Biology* 11: 489–493.

Hosken, D. J., O. Y. Martin, J. Born, and F. Huber. 2003b. Sexual conflict in *Sepsis cynipsea*: Female reluctance, fertility and mate choice. *Journal of Evolutionary Biology* 16: 485–490.

Houde, A. E. 1997. *Sex, Color, and Mate Choice in Guppies*. Princeton University Press, Princeton, N.J.

Houde, A. E., and J. A. Endler. 1990. Correlated evolution of female mating preferences and male color patterns in the guppy *Poecilia reticulata*. *Science* 248: 1405–1408.

Houle, D. 1992. Comparing evolvability and variability of quantitative traits. *Genetics* 130: 195–204.

Houle, D. 1998. How should we explain variation in the genetic variance of traits? *Genetica* 103: 241–253.

Houle, D., and A. S. Kondrashov. 2002. Coevolution of costly mate choice and condition-dependent display of good genes. *Proceedings of the Royal Society of London Series B—Biological Sciences* 269: 97–104.

Houston, A. I., and N. B. Davies. 1985. The evolution of cooperation and life history in the dunnock *Prunell modularis*. In R. M. Sibly and R. H. Smith (eds.), *Behavioural Ecology: Ecological Consequences of Adaptive Behaviour*, Blackwell Science, Oxford, pp. 471–487.

Howard, D. J. 1999. Conspecific sperm and pollen precedence and speciation. *Annual Review of Ecology and Systematics* 30: 109–132.

Howard, D. J., and S. H. Berlocher. 1998. *Endless Forms: Species and Speciation*, Oxford University Press, New York.

Howard, D. J., M. Reece, P. G. Gregory, J. Chu, and M. L. Cain. 1998. The evolution of barriers to fertilization between closely related organisms. In D. J. Howard and S. H. Berlocher (eds.), *Endless Forms: Species and Speciation*, Oxford University Press, New York, pp. 279–288.

Howard, R. D., and A. G. Kluge. 1985. Proximate mechanisma of sexual selection in wood frogs. *Evolution* 39: 260–277.

Hrdy, S. B. 1979. Infanticide among animals: A review, classification, and examination of the implications for the reproductive strategies of females. *Ethology and Sociobiology* 1: 13–40.

Hrdy, S. B. 1981. *The Woman that Never Evolved*. Harvard University Press, Cambridge, Mass.

Hughes, K. A. 1997. Quantitative genetics of sperm precedence in *Drosophila melanogaster*. *Genetics* 145: 139–151.

Hugot, J. P. 1984. Traumatic insemination in pinworms of dermoptera and rabbits: Morphological study: Considerations about phylogenesis. *Annales de Parasitologie Humaine et Comparée* 59: 379–385.

Hunter, F. M., M. Petrie, M. Otronen, T. Birkhead, and A. P. Moller. 1993. Why do females copulate repeatedly with one male? *Trends in Ecology & Evolution* 8: 21–26.

Hunter, R.H.F. 1996. Ovarian control of very low sperm/egg ratios at the commencement of mammalian fertilisation to avoid polyspermy. *Molecular Reproduction and Development* 44: 417–422.

Hunter, R.H.F., and T. Greve. 1998. Deep uterine insemination of cattle: A fruitful way forward with smaller numbers of spermatozoa. *Acta Veterinaria Scandinavica* 39: 149–163.

Hunter, R.H.F., H. H. Petersen, and T. Greve. 1999. Ovarian follicular fluid, progesterone and Ca^{2+} ion influences on sperm release from the Fallopian tube reservoir. *Molecular Reproduction and Development* 54: 283–291.

Hurd, L. E., R. M. Eisenberg, W. F. Fagan, K. J. Tilmon, W. E. Snyder, K. S. Vandersall, S. G. Datz, and J. D. Welch. 1994. Cannibalism reverses male-biased sex-ratio in adult mantids: Female strategy against food limitation? *Oikos* 69: 193–198.

Hurst, G.D.D., R. G. Sharpe, A. H. Broomfield, L. E. Walker, T.M.O. Majerus, I. A. Zakharov, and M.E.N. Majerus. 1995. Sexually-transmitted disease in a promiscuous insect, *Adalia bipunctata*. *Ecological Entomology* 20: 230–236.

Hurst, G.D.D., and J. H. Werren. 2001. The role of selfish genetic elements in eukaryotic evolution. *Nature Reviews Genetics* 2: 597–606.

Hurst, L. D. 1993. scat+ is a selfish gene analogous to Medea of *Tribolium castaneum*. *Cell* 75: 407–408.

Hurst, L. D. 1997. Evolutionary theories of genomic imprinting. In W. Reik and A. Surani (eds.), *Genomic Imprinting*, IRL, Oxford.

Hurst, L. D. 1999. Is multiple paternity necessary for the evolution of genomic imprinting? *Genetics* 153: 509–512.

Hurst, L. D., and G. T. McVean. 1998. Do we understand the evolution of genomic imprinting? *Current Opinion in Genetics & Development* 8: 701–708.

Hurst, L. D., and J. P. Randerson. 1999. An eXceptional chromosome. *Trends in Genetics* 15: 383–385.

Ilango, K., and R. P. Lane. 2000. Coadaptation of male aedeagal filaments and female spermathecal ducts of the old world Phlebotomine sand flies (Diptera: Psychodidae). *Journal of Medical Entomology* 37: 653–659.

Inceoglu, B., J. Lango, J. Jing, L. L. Chen, F. Doymaz, I. N. Pessah, and B. D. Hammock. 2003. One scorpion, two venoms: Prevenom of *Parabuthus transvaalicus* acts as an alternative type of venom with distinct mechanism of action. *Proceedings of the National Academy of Sciences of the United States of America* 100: 922–927.

Inglehart, R. 1997. *Modernization and Postmodernization: Cultural, Economic, and Political Change in 43 Societies*. Princeton University Press, Princeton, N.J.

Iwasa, Y., and A. Pomiankowski. 1991. The evolution of costly mate preferences. II. The handicap principle. *Evolution* 45: 1431–1442.

Iwasa, Y., and A. Pomiankowski. 1994. The evolution of mate preferences for multiple sexual ornaments. *Evolution* 48: 853–867.

Iyengar, V. K., H. K. Reeve, and T. Eisner. 2002. Paternal inheritance of a female moth's mating preference. *Nature* 419: 830–832.

Jaenike, J. 2001. Sex chromosome meiotic drive. *Annual Review of Ecology and Systematics* 32: 25–49.

Jaffe, L. A., and M. Gould. 1985. Polyspermy-preventing mechanisms. In C. B. Metz and A. Monroy (eds.), *Biology of Fertilization*, Academic, New York, pp. 223–250.

Janetos, A. C. 1980. Strategies of female mate choice: A theoretical analysis. *Behavioral Ecology and Sociobiology* 7: 107–112.

Jennions, M. D. 1997. Female promiscuity and genetic incompatibility. *Trends in Ecology & Evolution* 12: 251–253.

Jennions, M. D., and D. W. Macdonald. 1994. Cooperative breeding in mammals. *Trends in Ecology & Evolution* 9: 89–93.

Jennions, M. D., A. P. Møller, and M. Petrie. 2001. Sexually selected traits and adult survival: A meta-analysis. *Quarterly Review of Biology* 76: 3–36.

Jennions, M. D., and M. Petrie. 2000. Why do females mate multiply? A review of the genetic benefits. *Biological Reviews* 75: 21–64.

Johnson, C. 1964. The inheritance of female dimorphism in damselfly *Ischnura damula*. *Genetics* 49: 513–519.

Johnson, C. 1966. Genetics of female dimorphism in *Ischnura demorsa*. *Heredity* 21: 453–459.

Johnson, C. 1975. Polymorphism and natrual selection ischnuran damselflies. *Evolution Theory* 1: 81–90.

Johnson, J. C. 2001. Sexual cannibalism in fishing spiders (*Dolomedes triton*): An evaluation of two explanations for female aggression towards potential mates. *Animal Behaviour* 61: 905–914.

Johnson, M. A., and D. Preuss. 2002. Plotting a course: Multiple signals guide pollen tubes to their targets. *Developmental Cell* 2: 273–281.

Johnstone, R. A. 1995. Honest advertisement of multiple qualities using multiple signals. *Journal of Theoretical Biology* 177: 87–94.

Johnstone, R. A., and A. Grafen. 1992. Error-prone signaling. *Proceedings of the Royal Society of London Series B—Biological Sciences* 248: 229–233.

Johnstone, R. A., and L. Keller. 2000. How males can gain by harming their mates: Sexual conflict, seminal toxins, and the cost of mating. *American Naturalist* 156: 368–377.

Jones, K. T. 1998. Protein kinase C action at fertilization: Overstated or undervalued? *Reviews of Reproduction* 3: 7–12.

Jormalainen, V. 1998. Precopulatory mate guarding in crustaceans: Male competitive strategy and intersexual conflict. *Quarterly Review of Biology* 73: 275–304.

Jormalainen, V., and S. Merilaita. 1993. Female resistance and precopulatory guarding in the isopod *Idotea baltica* (Pallas). *Behaviour* 125: 219–231.

Jormalainen, V., and S. Merilaita. 1995. Female resistance and duration of mate-guarding in 3 aquatic peracarids (Crustacea). *Behavioral Ecology and Sociobiology* 36: 43–48.

Jormalainen, V., S. Merilaita, and R. Härdling. 2000. Dynamics of intersexual conflict over precopulatory mate guarding in two populations of the isopod *Idotea baltica*. *Animal Behaviour* 60: 85–93.

Jormalainen, V., S. Merilaita, and J. Riihimaki. 2001. Costs of intersexual conflict in the isopod *Idotea baltica*. *Journal of Evolutionary Biology* 14: 763–772.

Jormalainen, V., J. Tuomi, and S. Merilaita. 1994a. Effect of female resistance on size-dependent precopula duration in mate-guarding Crustacea. *Animal Behaviour* 47: 1471–1474.

Jormalainen, V., J. Tuomi, and N. Yamamura. 1994b. Intersexual conflict over precopula duration in mate guarding Crustacea. *Behavioural Processes* 32: 265–283.

Jungfer, K. H., and W. Hodl. 2002. A new species of Osteocephalus from Ecuador and a redescription of *O-leprieurii* (Dumeril & Bibron, 1841) (Anura: Hylidae). *Amphibia-Reptilia* 23: 21–46.

Kaitala, A., and H. Dingle. 1993. Wing dimorphism, territoriality and mating frequency of the water strider *Aquarius remigis* (Say). *Annales Zoologici Fennici* 30: 163–168.

Kaitala, A., and C. Wiklund. 1994. Polyandrous female butterflies forage for matings. *Behavioral Ecology and Sociobiology* 35: 385–388.

Kajiura, S. M., A. P. Sebastian, and T. C. Tricas. 2000. Dermal bite wounds as indicators of reproductive seasonality and behaviour in the Atlantic stingray, *Dasyatis sabina*. *Environmental Biology of Fishes* 58: 23–31.

Kajiura, S. M., and T. C. Tricas. 1996. Seasonal dynamics of dental sexual dimorphism in the Atlantic stingray *Dasyatis sabina*. *Journal of Experimental Biology* 199: 2297–2306.

Karlsson, B. 1995. Resource allocation and mating systems in butterflies. *Evolution* 49: 955–961.

Karlsson, B. 1996. Male reproductive reserves in relation to mating system in butterflies: A comparative study. *Proceedings of the Royal Society of London Series B—Biological Sciences* 263: 187–192.

Karlsson, B. 1998. Nuptial gifts, resource budgets, and reproductive output in a polyandrous butterfly. *Ecology* 79: 2931–2940.

Karube, F., and M. Kobayashi. 1999a. Combinative stimulation inactivates sex pheromone production in the silkworm moth, *Bombyx mori*. *Archives of Insect Biochemistry and Physiology* 42: 111–118.

Karube, F., and M. Kobayashi. 1999b. Presence of eupyrene spermatozoa in vestibulum accelerates oviposition in the silkworm moth, *Bombyx mori*. *Journal of Insect Physiology* 45: 947–957.

Kasuga, H., T. Aigaki, and M. Osanai. 1987. System for supply of free arginine in the spermatophore of *Bombyx mori:* Arginine liberating activities of contents of male reproductive glands. *Insect Biochemistry* 17: 317–322.

Katzourakis, A., A. Purvis, S. Azmeh, G. Rotheray, and F. Gilbert. 2001. Macroevolution of hoverflies (Diptera: Syrphidae): The effect of using higher-level taxa in studies of biodiversity, and correlates of species richness. *Journal of Evolutionary Biology* 14: 219–227.

Keenleyside, M.H.A. 1985. Bigamy and mate choice in the biparental cichlid fish *Cichlasoma nigrofasciatum*. *Behavioral Ecology and Sociobiology* 17: 285–290.

Keenleyside, M.H.A., and R. W. Mackereth. 1992. Effects of loss of male parent on brood survival in a biparental cichlid fish. *Environmental Biology of Fishes* 34: 207–212.

Keller, L. 1995. Evolutionary biology: All's fair when love is war. *Nature* 373: 190–191.

Keller, L., and H. K. Reeve. 1995. Why do females mate with multiple males? The sexually selected sperm hypothesis. *Advances in the Study of Behavior* 24: 291–315.

Kelly, C. D., J.G.J. Godin, and G. Abdallah. 2000. Geographical variation in the male intromittent organ of the Trinidadian guppy (*Poecilia reticulata*). *Canadian Journal of Zoology* 78: 1674–1680.

Kempenaers, B. 1995. Polygyny in the blue tit: Intra-sexual and inter-sexual conflicts. *Animal Behaviour* 49: 1047–1064.

Kenagy, G. J., and S. C. Trombulak. 1986. Size and function of mammalian testes in relation to body size. *Journal of Mammalogy* 67: 1–22.

Kence, A., and E. H. Bryant. 1978. Model of mating behavior in flies. *American Naturalist* 112: 1047–1062.

Kennedy, J. S. 1992. *The New Anthropomorphism*. Cambridge University Press, Cambridge.

Ketterson, E. D., and V. Nolan. 1994. Male parental behavior in birds. *Annual Review of Ecology and Systematics* 25: 601–628.

Khan, H. R., M. L. Ashton, S. T. Mukai, and A. S. M. Saleuddin. 1990. The effect of mating on the fine-structure of neurosecretory caudodorsal cells in *Helisoma duryi* (Mollusca). *Canadian Journal of Zoology* 68: 1233–1240.

Killian, J. K., C. M. Nolan, N. Stewart, B. L. Munday, N. A. Andersen, S. Nicol, and R. L. Jirtle. 2001. Monotreme IGF2 expression and ancestral origin of genomic imprinting. *Journal of Experimental Zoology* 291: 205–212.

Kim, H. U., R. Cotter, S. Johnson, M. Senda, P. Dodds, R. Kulikauskas, W. H. Tang, I. Ezcurra, P. Herzmark, and S. Mccormick. 2002. New pollen-specific receptor kinases identified in tomato, maize and *Arabidopsis*: The tomato kinases show overlapping but distinct localization patterns on pollen tubes. *Plant Molecular Biology* 50: 1–16.

Kingan, T. G., W. M. Bodnar, A. K. Raina, J. Shabanowitz, and D. F. Hunt. 1995. The loss of female sex-pheromone after mating in the corn-earworm moth *Helicoverpa zea*: Identification of a male pheromonostatic peptide. *Proceedings of the National Academy of Sciences of the United States of America* 92: 5082–5086.

Kingsolver, J. G., H. E. Hoekstra, J. M. Hoekstra, D. Berrigan, S. N. Vignieri, C. E. Hill, A. Hoang, P. Gibert, and P. Beerli. 2001. The strength of phenotypic selection in natural populations. *American Naturalist* 157: 245–261.

Kinoshita, T., R. Yadegari, J. J. Harada, R. B. Goldberg, and R. L. Fischer. 1999. Imprinting of the MEDEA polycomb gene in the *Arabidopsis* endosperm. *Plant Cell* 11: 1945–1952.

Kirkpatrick, M. 1982. Sexual selection and the evolution of female choice. *Evolution* 36: 1–12.

Kirkpatrick, M. 1985. Evolution of female choice and male parental investment in polygynous species: The demise of the "sexy son." *American Naturalist* 125: 788–810.

Kirkpatrick, M. 1986. The handicap mechanism of sexual selection does not work. *American Naturalist* 127: 222–240.

Kirkpatrick, M. 1987a. The evolutionary forces acting on female mating preferences in polygynous animals. In J. W. Bradbury and M. B. Andersson (eds.), *Sexual Selection: Testing the Alternatives*, Wiley, Chichester, pp. 67–82.

Kirkpatrick, M. 1987b. Sexual selection by female choice in polygynous animals. *Annual Review of Ecology and Systematics* 18: 43–70.

Kirkpatrick, M. 1996. Good genes and direct selection in evolution of mating preferences. *Evolution* 50: 2125–2140.

Kirkpatrick, M., and N. H. Barton. 1997. The strength of indirect selection on female mating preferences. *Proceedings of the National Academy of Sciences of the United States of America* 94: 1282–1286.

Kirkpatrick, M., T. Price, and S. J. Arnold. 1990. The Darwin-Fisher theory of sexual selection in monogamous birds. *Evolution* 44: 180–193.

Kirkpatrick, M., and V. Ravigné. 2002. Speciation by natural and sexual selection: Models and experiments. *American Naturalist* 159: S22–S35.

Kirkpatrick, M., and M. J. Ryan. 1991. The evolution of mating preferences and the paradox of the lek. *Nature* 350: 33–38.

Klepal, W., H. Barnes, and M. Barnes. 1977. Studies on reproduction of cirripedes. VI. Passage of spermatozoa into oviductal sac and closure of pores. *Journal of Experimental Marine Biology and Ecology* 27: 289–304.

Klimley, A. P. 1980. Observations of courtship and copulation in the nurse shark, *Ginglymostoma cirratum*. *Copeia* 878–882.

Klowden, M. J. 1999. The check is in the male: Male mosquitoes affect female physiology and behavior. *Journal of the American Mosquito Control Association* 15: 213–220.

Knowles, L. L., and T. A. Markow. 2001. Sexually antagonistic coevolution of a postmating-prezygotic reproductive character in desert *Drosophila*. *Proceedings of the National Academy of Sciences of the United States of America* 98: 8692–8696.

Koene, J. M., and R. Chase. 1998a. The love dart of *Helix aspersa* Müller is not a gift of calcium. *Journal of Molluscan Studies* 64: 75–80.

Koene, J. M., and R. Chase. 1998b. Changes in the reproductive system of the snail *Helix aspersa* caused by mucus from the love dart. *Journal of Experimental Biology* 201: 2313–2319.

Koene, J. M., and H. Schulenburg. 2003. Shooting darts: Sexual conflict drives a coevolutionary arms race in hermaphroditic land snails. Manuscript.

Koene, J. M., G. Sundermann, and N. K. Michiels. 2002. On the function of body piercing during copulation in earthworms. *Invertebrate Reproduction & Development* 41: 35–40.

Koene, J. M., and A. Ter Maat. 2001. "Allohormones": a class of bioactive substances favoured by sexual selection. *Journal of Comparative Physiology, A: Neuroethology, Sensory, Neural, and Behavioral Physiology* 187: 323–326.

Koga, T., and K. Hayashi. 1993. Territorial behavior of both sexes in the water strider *Metrocoris histrio* (Hemiptera: Gerridae) during the mating season. *Journal of Insect Behavior* 6: 65–77.

Kokita, T. 2002. The role of female behavior in maintaining monogamy of a coral-reef filefish. *Ethology* 108: 157–168.

Kokita, T., and A. Nakazono. 2001. Sexual conflict over mating system: The case of a pair-territorial filefish without parental care. *Animal Behaviour* 62: 147–155.

Kokko, H. 1997. Evolutionarily stable strategies of age-dependent sexual advertisement. *Behavioral Ecology and Sociobiology* 41: 99–107.

Kokko, H. 1998. Should advertising parental care be honest? *Proceedings of the Royal Society of London Series B—Biological Sciences* 265: 1871–1878.

Kokko, H. 2001. Fisherian and "good genes" benefits of mate choice: how (not) to distinguish between them. *Ecology Letters* 4: 322–326.

Kokko, H., and R. Brooks. 2003. Sexy to die for? Sexual selection and the risk of extinction. *Annales Zoologici Fennici* 40: 207–219.

Kokko, H., R. Brooks, M. D. Jennions, and J. Morley. 2003. The evolution of mate choice and mating biases. *Proceedings of the Royal Society of London Series B—Biological Sciences* 270: 653–664.

Kokko, H., and M. Jennions. 2003. It takes two to tango. *Trends in Ecology & Evolution* 18: 103–104.

Komdeur, J., P. Wiersma, and M. Magrath. 2002. Paternal care and male mate-attraction effort in the European starling is adjusted to clutch size. *Proceedings of the Royal Society of London Series B—Biological Sciences* 269: 1253–1261.

Kondoh, M., and M. Higashi. 2000. Reproductive isolation mechanism resulting from resolution of intragenomic conflict. *American Naturalist* 156: 511–518.

Koprowski, J. L. 1992. Removal of copulatory plugs by female tree squirrels. *Journal of Mammalogy* 73: 572–576.

Korpimäki, E. 1994. Nest predation may not explain poor reproductive success of polysynously mated female tengmalms owls. *Journal of Avian Biology* 25: 161–164.

Kraus, B., and R. C. Lederhouse. 1983. Contact guarding during courtship in the tiger beetle *Cincindela marutha* Dow (Coeoptera: Cicindelidae). *American Midland Naturalist* 110: 207–211.

Kreiter, N. A., and D. H. Wise. 2001. Prey availability limits fecundity and influences the movement pattern of female fishing spiders. *Oecologia* 127: 417–424.

Kritsky, G., and S. Simon. 1995. Mandibular sexual dimorphism in *Cicindela* Linnaeus (Coleoptera: Cicindelidae). *Coleopterists Bulletin* 49: 143–148.

Krupa, J. J., W. R. Leopold, and A. Sih. 1990. Avoidance of male giant water striders by females. *Behaviour* 115: 247–253.

Krupa, J. J., and A. Sih. 1993. Experimental studies on water strider mating dynamics: Spatial variation in density and sex-ratio. *Behavioral Ecology and Sociobiology* 33: 107–120.

Krupa, J. J., and A. Sih. 1998. Fishing spiders, green sunfish, and a stream-dwelling water strider: Male-female conflict and prey responses to single versus multiple predator environments. *Oecologia* 117: 258–265.

Kuhn, T. S. 1970. *The Structure of Scientific Revolutions*, 2nd Ed. University of Chicago Press, Chicago.

Kulkarni, C. V. 1940. On the systematic position, structural modifications, bionomics and development of a remarkable new family of cyprinodont fishes from the province of Bombay. *Records of the Indian Museum, Delhi* 42: 379–423.

Kurpisz, M., and N. Fernandez (eds.). 1995. *Immunology of Human Reproduction*. BIOS Science, Oxford.

Kvarnemo, C., and I. Ahnesjö. 1996. The dynamics of operational sex ratios and competition for mates. *Trends in Ecology & Evolution* 11: 404–408.

Kvarnemo, C., and I. Ahnesjö. 2002. Operational sex ratios and mating competition. In I.C.W. Hardy (ed.), *Sex Ratios: Concepts and Research Methods*, Cambridge University Press, New York.

Kwet, A. 2001. *Frösche im brasilianischen Araukarienwald—Anurengemeinschaft des Araukarienwaldes von Rio Grande do Sul: Diversität, Reproducktion und Ressourcenaufteilung*. Natur und Tier Verlag, Münster.

Kynaston, S. E., P. Mcerlainward, and P. J. Mill. 1994. Courtship, mating behavior and sexual cannibalism in the praying mantis, *Sphodromantis lineoa*. *Animal Behaviour* 47: 739–741.

Lack, D. L. 1968. *Ecological Adaptations for Breeding in Birds*. Methuen, London.

Lambert, C. C. 2000. Germ-cell warfare in ascidians: Sperm from one species can interfere with the fertilization of a second species. *Biological Bulletin* 198: 22–25.

Lams, H. 1913. Etude de l'oeuf de Cobaye aux premiers stades d'embryogenese. *Archives de Biologie* 28: 229–323.

LaMunyon, C. W., and T. Eisner. 1994. Spermatophore size as determinant of paternity in an arctiid moth (*Utetheisa ornatrix*). *Proceedings of the National Academy of Sciences of the United States of America* 91: 7081–7084.

Lande, R. 1980. Sexual dimorphism, sexual selection, and adaptation in polygenic characters. *Evolution* 34: 292–305.

Lande, R. 1981. Models of speciation by sexual selection on polygenic traits. *Proceedings of the National Academy of Sciences of the United States of America* 78: 3721–3725.

Lande, R. 1987. Genetic correlations between the sexes in the evolution of sexual dimorphism and mating preferences. In J. W. Bradbury and M. B. Andersson (eds.), *Sexual Selection: Testing the Alternatives*, Wiley, Chichester, pp. 83–94.

Lande, R., and S. J. Arnold. 1983. The measurement of selection on correlated characters. *Evolution* 37: 1210–1226.

Landolfa, M. A. 2002. On the adaptive function of the love dart of *Helix aspersa*. *Veliger* 45: 231–249.

Landolfa, M. A., D. M. Green, and R. Chase. 2001. Dart shooting influences paternal reproductive success in the snail *Helix aspersa* (Pulmonata, Stylommatophora). *Behavioral Ecology* 12: 773–777.

Lange, A. B. 1984. The transfer of prostaglandin-synthesizing activity during mating in *Locusta migratoria*. *Insect Biochemistry* 14: 551–556.

Langmore, N. E., and N. B. Davies. 1997. Female dunnocks use vocalizations to compete for males. *Animal Behaviour* 53: 881–890.

Langmore, N. E., N. B. Davies, B. J. Hatchwell, and I. R. Hartley. 1996. Female song attracts males in the alpine accentor *Prunella collaris*. *Proceedings of the Royal Society of London Series B—Biological Sciences* 263: 141–146.

Larson, B.M.H., and S.C.H. Barrett. 2000. A comparative analysis of pollen limitation in flowering plants. *Biological Journal of the Linnean Society* 69: 503–520.

Lauer, M. J. 1996. Effect of sperm depletion and starvation on female mating behavior in the water strider, *Aquarius remigis*. *Behavioral Ecology and Sociobiology* 38: 89–96.

Lauer, M. J., A. Sih, and J. J. Krupa. 1996. Male density, female density and intersexual conflict in a stream-dwelling insect. *Animal Behaviour* 52: 929–939.

Lavilla, E. O., and P. E. Sandoval. 1999. A new Bolivian species of the genus Telmatobius (Anura: Leptodactylidae) with a humeral spine. *Amphibia-Reptilia* 20: 55–64.

Lawrence, S. E. 1992. Sexual cannibalism in the praying mantid, *Mantis religiosa*: A field study. *Animal Behaviour* 43: 569–583.

Lazarus, J. 1990. The logic of mate desertion. *Animal Behaviour* 39: 672–684.

LeBas, N. R., L. R. Hockham, and M. G. Ritchie. 2003. Nonlinear and correlational sexual selection on "honest" female ornamentation. *Proceedings of the Royal Society of London Series B—Biological Sciences* 270: 2159–2165.

Le Boeuf, B. J. 1972. Sexual behavior in northern elephant seal *Mirounga angustirostris*. *Behaviour* 41: 1–26.

Le Boeuf, B. J., and S. Mesnick. 1991. Sexual behavior of male northern elephant seals. I. Lethal injuries to adult females. *Behaviour* 116: 143–162.

Lee, K. H., N. S. Fechheimer, and H. Abplanalp. 1990. Euploid chicken embryos from eggs containing one, two or several yolks. *Journal of Reproduction and Fertility* 89: 85–90.

Lee, P.L.M., P. F. Brain, D. W. Forman, R. B. Bradbury, and R. Griffiths. 2002. Sex and death: Chdiz associated with high mortality in moorhens. *Evolution* 56: 2548–2553.

Lee, Y. H., T. Ota, and V. D. Vacquier. 1995. Positive selection is a general phenomenon in the evolution of abalone sperm lysin. *Molecular Biology and Evolution* 12: 231–238.

Leonard, J. L. 1990. The hermaphrodite's dilemma. *Journal of Theoretical Biology* 147: 361–372.

Leonard, J. L. 1992. The love dart in Helicid snails: A gift of calcium or a firm commitment? *Journal of Theoretical Biology* 159: 513–521.

Leonard, J. L. 1993. Sexual conflict in simultaneous hermaphrodites: Evidence from serranid fishes. *Environmental Biology of Fishes* 36: 135–148.

Leonard, J. L., and K. Lukowiak. 1985. Courtship, copulation, and sperm trading in the sea slug, *Navanax inermis* (Opisthobranchia, Cephalaspidea). *Canadian Journal of Zoology* 63: 2719–2729.

Leonard, J. L., and K. Lukowiak. 1991. Sex and the simultaneous hermaphrodite: Testing models of male-female conflict in a sea slug, *Navanax inermis* (Opisthobranchia). *Animal Behaviour* 41: 255–266.

Leong, K.L.H., D. Frey, D. Hamaoka, and K. Honma. 1993. Wing damage in overwintering populations of monarch butterfly at two California sites. *Annals of the Entomological Society of America* 86: 728–733.

Leonhard-Marek, S. 2000. Why do trace elements have an influence on fertility? *Tierarztliche Praxis Ausgabe Grobtiere Nutztiere* 28: 60–65.

Leopold, R. A. 1970. Cytological and cytochemical studies on ejaculatory duct and accessory secretion in *Musca domestica*. *Journal of Insect Physiology* 16: 1859–1872.

Leopold, R. A., A. C. Terranov, B. J. Thorson, and M. E. DeGrugillier. 1971. Biosynthesis of male housefly accessory secretion and its fate in the mated female. *Journal of Insect Physiology* 17: 987–1003.

Lessells, C. M. 1999. Sexual conflict in animals. In L. Keller (ed.), *Levels of Selection in Evolution*, Princeton University Press, Princeton, N.J., pp. 75–99.

Lewis, C. T., and J. N. Pollock. 1975. Engagement of phallosome in blowflies. *Journal of Entomology Series A—Physiology & Behaviour* 49: 137–147.

Lewis, S. M., and S. N. Austad. 1990. Sources of intraspecific variation in sperm precedence in red flour beetles. *American Naturalist* 135: 351–359.

Lewis, S. M., and S. N. Austad. 1994. Sexual selection in flour beetles: The relationship between sperm precedence and male olfactory attractiveness. *Behavioral Ecology* 5: 219–224.

Li, L. J., R. W. Garden, P. D. Floyd, T. P. Moroz, J. M. Gleeson, J. V. Sweedler, L. Pasa-Tolic, and R. D. Smith. 1999. Egg-laying hormone peptides in the Aplysiidae family. *Journal of Experimental Biology* 202: 2961–2973.

Liersch, S., and P. Schmid-Hempel. 1998. Genetic variation within social insect colonies reduces parasite load. *Proceedings of the Royal Society of London Series B—Biological Sciences* 265: 221–225.

Lifjeld, J. T., P. O. Dunn, and D. F. Westneat. 1994. Sexual selection by sperm competition in birds: Male-male competition or female choice. *Journal of Avian Biology* 25: 244–250.

Lifjeld, J. T., and T. Slagsvold. 1991. Sexual conflict among polygynous pied flycatchers feeding young. *Behavioral Ecology* 2: 106–115.

Ligon, J. D. 1999. *The Evolution of Avian Breeding Systems*. Oxford University Press, Oxford.

Liker, A., and T. Szekely. 1997. Aggression among female lapwings, *Vanellus vanellus*. *Animal Behaviour* 54: 797–802.

Lind, H. 1973. Functional significance of spermatophore and fate of spermatozoa in genital tract of *Helix pomatia* (Gastropoda: Stylommatophora). *Journal of Zoology* 169: 39–64.

Lindenfors, P. 2002. Sexually antagonistic selection on primate size. *Journal of Evolutionary Biology* 15: 595–607.

Lindenfors, P., and B. S. Tullberg. 1998. Phylogenetic analyses of primate size evolution: The consequences of sexual selection. *Biological Journal of the Linnean Society* 64: 413–447.

Lindfors, V. 1998. Butterfly life history and mating system. Ph.D. thesis, University of Stockholm.

Linley, J. R., and G. M. Adams. 1975. Sexual receptivity in *Culicoides melleus* (Diptera: Ceratopogonidae). *Transactions of the Royal Entomological Society of London* 126: 279–303.

Linley, J. R., and M. J. Hinds. 1975. Quantity of male ejaculate influenced by female unreceptivity in fly, *Culicoides melleus*. *Journal of Insect Physiology* 21: 281–285.

Linley, J. R., and M. S. Mook. 1975. Behavioural interaction between sexually experienced *Culicoides melleus* (Coquillett) (Diptera: Ceratopogonidae). *Behaviour* 54: 97–110.

Liske, E., and W. J. Davis. 1987. Courtship and mating behavior of the Chinese praying mantis, *Tenodera aridifolia sinensis*. *Animal Behaviour* 35: 1524–1537.

Liu, H. F., and E. Kubli. 2003. Sex-peptide is the molecular basis of the sperm effect in *Drosophila melanogaster*. *Proceedings of the National Academy of Sciences of the United States of America* 100: 9929–9933.

Lloyd, J. E. 1979. Mating behavior and natural selection. *Florida Entomologist* 62: 17–34.

Locher, R., and B. Baur. 2000. Mating frequency and resource allocation to male and female function in the simultaneous hermaphrodite land snail *Arianta arbustorum*. *Journal of Evolutionary Biology* 13: 607–614.

Loehrl, H. 1990. Observations on penduline tits (*Remiz pendulinus*) in an aviary. *Vogelwarte* 35: 331–320.

Loher, W., I. Ganjian, I. Kubo, D. Stanley-Samuelson, and S. S. Tobe. 1981. Prostaglandins: Their role in egg laying of the cricket *Teleogryllus commodus*. *Proceedings of the National Academy of Sciences of the United States of America* 78: 7835–7838.

Lorch, P. D., S. Proulx, L. Rowe, and T. Day . 2003. Condition-dependent sexual selection can accelerate adaptation. *Evolutionary Ecology Research* 5: 867–881.

Lorch, P. D., G. S. Wilkinson, and P. R. Reillo. 1993. Copulation duration and sperm precedence in the stalk-eyed fly *Cyrtodiopsis whitei* (Diptera: Diopsidae). *Behavioral Ecology and Sociobiology* 32: 303–311.

Low, B. S., and W. T. Wcislo. 1992. Male foretibial plates and mating in *Crabro cribrellifer* (Packard) (Hymenoptera: Sphecidae), with a survery of expanded male forelegs in Apoidea. *Annals of the Entomological Society of America* 85: 219–223.

Luce, R. D., and H. Raiffa. 1957. *Games and Decisions: Introduction and Critical Survey*. John Wiley and Sons, New York.

Lung, O., U. Tram, C. M. Finnerty, M. A. Eipper-Mains, J. M. Kalb, and M. F. Wolfner. 2002. The *Drosophila melanogaster* seminal fluid protein Acp62F is a protease inhibitor that is toxic upon ectopic expression. *Genetics* 160: 211–224.

Lung, O., and M. F. Wolfner. 1999. *Drosophila* seminal fluid proteins enter the circulatory system of the mated female fly by crossing the posterior vaginal wall. *Insect Biochemistry and Molecular Biology* 29: 1043–1052.

Lynch, J. D., and P. M. Ruiz-Carranza. 1996. A remarkable new centrolenid frog from Colombia with a review of nuptial excrescences in the family. *Herpetologica* 52: 525–535.

Mackay, T.F.C. 2002. The nature of quantitative genetic variation for *Drosophila* longevity. *Mechanisms of Ageing and Development* 123: 95–104.

Magnhagen, C. 1991. Predation risk as a cost of reproduction. *Trends in Ecology & Evolution* 6: 183–185.

Magurran, A. E. 1998. Population differentiation without speciation. *Philosophical Transactions of the Royal Society of London Series B—Biological Sciences* 353: 275–286.

Magurran, A. E. 2001. Sexual conflict and evolution in Trinidadian guppies. *Genetica* 112: 463–474.

Magurran, A. E., and R. M. May (eds.). 1999. *Evolution of Biological Diversity*. Oxford University Press, Oxford.

Magurran, A. E., and M. A. Nowak. 1991. Another battle of the sexes: the consequences of sexual asymmetry in mating costs and predation risk in the guppy, *Poecilia reticulata*. *Proceedings of the Royal Society of London Series B—Biological Sciences* 246: 31–38.

Magurran, A. E., and B. H. Seghers. 1994a. Sexual conflict as a consequence of ecology: Evidence from guppy, *Poecilia reticulata*, populations in Trinidad. *Proceedings of the Royal Society of London Series B—Biological Sciences* 255: 31–36.

Magurran, A. E., and B. H. Seghers. 1994b. A cost of sexual harassment in the guppy, *Poecilia reticulata*. *Proceedings of the Royal Society of London Series B—Biological Sciences* 258: 89–92.

Maheshwari, P. (ed.). 1950. *An Introduction to the Embryology of Angiosperms*, 1st Ed. McGraw-Hill, New York.

Maier, G., I. Berger, W. Burghard, and B. Nassal. 2000. Is mating of copepods associated with increased risk of predation? *Journal of Plankton Research* 22: 1977–1987.

Mair, J., and A. Blackwell. 1996. Mating behavior of *Culicoides nubeculosus* (Diptera: Ceratopogonidae). *Journal of Medical Entomology* 33: 856–858.

Majerus, M.E.N. 2003. *Sex Wars: Genes, Bacteria, and Biased Sex Ratios*. Princeton University Press, Princeton, N.J.

Mallet, J. 1999. Causes and consequences of a lack of coevolution in Müllerian mimicry. *Evolutionary Ecology* 13: 777–806.

Mann, T. 1964. *The Biochemistry of Semen and of the Male Reproductive Tract*. Methuen, London.

Mann, T., and C. Lutwak-Mann. 1981. *Male Reproductive Function and Semen: Themes and Trends in Physiology, Biochemistry, and Investigative Andrology*. Springer, Berlin.

Markow, T. A. 2000. Forced matings in natural populations of *Drosophila*. *American Naturalist* 156: 100–103.

Markow, T. A., and P. F. Ankney. 1984. *Drosophila* males contribute to oogenesis in a multiple mating species. *Science* 224: 302–303.

Markow, T. A., A. Coppola, and T. D. Watts. 2001. How *Drosophila* males make eggs: It is elemental. *Proceedings of the Royal Society of London Series B—Biological Sciences* 268: 1527–1532.

Markow, T. A., P. D. Gallagher, and R. A. Krebs. 1990. Ejaculate-derived nutritional contribution and female reproductive success in *Drosophila mojavensis* (Patterson and Crow). *Functional Ecology* 4: 67–73.

Marshall, D. L., M. W. Folsom, C. Hatfield, and T. Bennett. 1996. Does interference competition among pollen grains occur in wild radish? *Evolution* 50: 1842–1848.

Martens, A., and G. Rehfeldt. 1989. Female aggregation in *Platycypha caligata* (Odonata: Chlorocyphidae): A tactic to evade male interference during oviposition. *Animal Behaviour* 38: 369–374.

Martin, O. Y., and D. J. Hosken. 2003a. The evolution of reproductive isolation through sexual conflict. *Nature* 423: 979–982.

Martin, O. Y., and D. J. Hosken. 2003b. Costs and benefits of evolving under experimentally enforced polyandry or monogamy. *Evolution* 57: 2765–2772.

Martin, O. Y., R. R. Leugger, N. Zeltner, and D. J. Hosken. 2003. Male age, mating probability and mating costs in the fly Sepsis cynipsea. *Evolutionary Ecology Research* 5: 119–129.

Martin-Alganza, A., M. D. Lopez-Leon, J. Cabrero, and J.P.M. Camacho. 1997. Somatic condition determines female mating frequency in a field population of the grasshopper *Eyprepocnemis plorans*. *Heredity* 79: 524–530.

Martin-Vivaldi, M., J. J. Palomino, M. Soler, and J. G. Martinez. 1999. Song strophelength and reproductive success in a non-passerine bird, the Hoopoe Upupa epops. *Ibis* 141: 670–679.

Maruska, K. P., E. G. Cowie, and T. C. Tricas. 1996. Periodic gonadal activity and protracted mating in elasmobranch fishes. *Journal of Experimental Zoology* 276: 219–232.

Massa, R., V. Galanti, and L. Bottoni. 1996. Mate choice and reproductive success in the domesticated budgerigar, *Melopsittacus undulatus*. *Italian Journal of Zoology* 63: 243–246.

Masta, S. E., and W. P. Maddison. 2002. Sexual selection driving diversification in jumping spiders. *Proceedings of the National Academy of Sciences of the United States of America* 99: 4442–4447.

Mathis, A., and W. W. Hoback. 1997. The influence of chemical stimuli from predators on precopulatory pairing by the amphipod, *Gammarus pseudolimnaeus*. *Ethology* 103: 33–40.

Maxwell, M. R. 2000. Does a single meal affect female reproductive output in the sexually cannibalistic praying mantid *Iris oratoria*? *Ecological Entomology* 25: 54–62.

Mayfield, J. A., A. Fiebig, S. E. Johnstone, and D. Preuss. 2001. Gene families from the Arabidopsis thaliana pollen coat proteome. *Science* 292: 2482–2485.

Maynard-Smith, J. 1956. Fertility, mating behaviour and sexual selection in *Drosophila subobscura*. *Journal of Genetics* 54: 261–279.

Maynard-Smith, J. 1977. Parental investment: prospective analysis. *Animal Behaviour* 25: 1–9.

Maynard-Smith, J. 1978. *The Evolution of Sex*. Cambridge University Press, Cambridge.

Maynard-Smith, J. 1985. Sexual selection, handicaps and true fitness. *Journal of Theoretical Biology* 115: 1–8.

Maynard-Smith, J. 1987. Sexual selection: A classification of models. In J. W. Bradbury and M. B. Andersson (eds.), *Sexual Selection: Testing the Alternatives*. Wiley, Chichester, pp. 9–20.

Maynard-Smith, J. 1991. Honest signaling: The Philip Sidney game. *Animal Behaviour* 42: 1034–1035.

Mayr, E. 1940. Speciation phenomena in birds. *American Naturalist* 74: 249–278.

Mayr, E. 1963. *Animal Species and Evolution*. Belknap Press of Harvard University Press, Cambridge, Mass.

Mazuc, J., O. Chastel, and G. Sorci. 2003. No evidence for differential maternal allocation to offspring in the house sparrow (*Passer domesticus*). *Behavioral Ecology* 14: 340–346.

McFadden, D. E., and S. Langlois. 2000. Parental and meiotic origin of triploidy in the embryonic and fetal periods. *Clinical Genetics* 58: 192–200.

McKinney, F., S. R. Derrickson, and P. Mineau. 1983. Forced copulation in waterfowl. *Behaviour* 86: 250–294.

McLain, D. K. 1998. Non-genetic benefits of mate choice: Fecundity enhancement and sexy sons. *Animal Behaviour* 55: 1191–1201.

McLain, D. K., D. L. Lanier, and N. B. Marsh. 1990. Effects of female size, mate size, and number of copulations on fecundity, fertility, and longevity of *Nezara viridula* (Hemiptera: Pentatomidae). *Annals of the Entomological Society of America* 83: 1130–1136.

McLain, D. K., M. P. Moulton, and T. P. Redfearn. 1995. Sexual selection and the risk of extinction of introduced birds on oceanic islands. *Oikos* 74: 27–34.

McLain, D. K., M. P. Moulton, and J. G. Sanderson. 1999. Sexual selection and extinction: The fate of plumage-dimorphic and plumage-monomorphic birds introduced onto islands. *Evolutionary Ecology Research* 1: 549–565.

McLain, D. K., and A. E. Pratt. 1999. The cost of sexual coercion and heterospecific sexual harassment on the fecundity of a host-specific, seed-eating insect (*Neacoryphus bicrucis*). *Behavioral Ecology and Sociobiology* 46: 164–170.

McNamara, J. M., A. I. Houston, Z. Barta, and J. L. Osorno. 2003. Should young ever be better off with one parent than with two? *Behavioral Ecology* 14: 301–310.

McVean, G. T., and L. D. Hurst. 1997. Molecular evolution of imprinted genes: No evidence for antagonistic coevolution. *Proceedings of the Royal Society of London Series B—Biological Sciences* 264: 739–746.

Mead, A. R. 1943. Revision of the giant west coast land slugs of the genus *Ariolimax* Moerch (Pulmonata: Arionidae). *American Midland Naturalist* 30: 675–717.

Melser, C., and P.G.L. Klinkhamer. 2001. Selective seed abortion increases offspring survival in *Cynoglossum officinale* (Boraginaceae). *American Journal of Botany* 88: 1033–1040.

Merilä, J., B. C. Sheldon, and H. Ellegren. 1997. Antagonistic natural selection revealed by molecular sex identification of nestling collared flycatchers. *Molecular Ecology* 6: 1167–1175.

Merritt, D. J. 1989. The morphology of the phallosome and accessory gland material transfer during copulation in the blowfly, *Lucilia cuprina* (Insecta: Diptera). *Zoomorphology* 108: 359–366.

Mesnick, S. L. 1997. Sexual alliances: Evidence and evolutionary implications. In P. A. Gowaty (ed.), *Feminism and Evolutionary Biology: Boundaries, Intersection, and Frontiers*, Chapman & Hall, New York, pp. 207–260.

Mesnick, S. L., and B. J. Le Boeuf. 1991. Sexual behavior of male northern elephant seals. II. Female response to potentially injurious encounters. *Behaviour* 117: 262–280.

Mestel, R. 1994. Seamy side of sea otter life. *New Scientist* 141: 5.

Metz, E. C., R. E. Kane, H. Yanagimachi, and S. R. Palumbi. 1994. Specificity of gamete binding and early stages of fusion in closely related sea urchins (genus *Echinometra*). *Biological Bulletin* 187: 23–34.

Metz, E. C., and S. R. Palumbi. 1996. Positive selection and sequence rearrangements generate extensive polymorphism in the gamete recognition protein bindin. *Molecular Biology and Evolution* 13: 397–406.

Michiels, N. K. 1998. Mating conflicts and sperm competition in simultaneous hermaphrodites. In T. R. Birkhead and A. P. Møller (eds.), *Sperm Competition and Sexual Selection*, Academic, London, pp. 219–255.

Michiels, N. K., and B. Bakovski. 2000. Sperm trading in a hermaphroditic flatworm: reluctant fathers and sexy mothers. *Animal Behaviour* 59: 319–325.

Michiels, N. K., and J. M. Koene. 2004. Love hurts when genders merge: Harmful mating tactics escalate more in hermaphrodites than in gonochorists. Manuscript.

Michiels, N. K., and A. Kuhl. 2003. Altruistic sperm donation in a sperm-dependent parthenogenetic hermaphrodite is stabilized by reciprocal sperm exchange. *Journal of Zoology* 259: 77–82.

Michiels, N. K., and L. J. Newman. 1998. Sex and violence in hermaphrodites. *Nature* 391: 647.

Michiels, N. K., A. Raven-Yoo-Heufes, and K. K. Brockmann. 2003. Sperm trading and sex roles in the hermaphroditic opisthobranch sea slug *Navanax inermis*: Eager females or opportunistic males? *Biological Journal of the Linnean Society* 78: 105–116.

Michiels, N. K., and A. Streng. 1998. Sperm exchange in a simultaneous hermaphrodite. *Behavioral Ecology and Sociobiology* 42: 171–178.

Miller, J. R., J. L. Spencer, A. J. Lentz, E. D. Walker, and J. F. Leycam. 1994. Sexpeptides: potentially important and useful regulators of insect reproduction. In P. A. Hedin, J. J. Menn, and R. M. Hollingworth (eds.), *Natural and Engineered Pest Management Agents*, American Chemical Society, Washington, D.C., pp. 292–318.

Miller, K. B. 2003. The phylogeny of diving beetles (Coleoptera: Dytiscidae) and the evolution of sexual conflict. *Biological Journal of the Linnean Society* 79: 359–388.

Miller, M. N., and O. M. Fincke. 1999. Cues for mate recognition and the effect of prior experience on mate recognition in *Enallagma* damselflies. *Journal of Insect Behavior* 12: 801–814.

Misof, B. 2002. Diversity of Anisoptera (Odonata): Inferring speciation processes from patterns of morphological diversity. *Zoology* 105: 355–365.

Mitra, S., H. Landel, and S. Pruett-Jones. 1996. Species richness covaries with mating system in birds. *Auk* 113: 544–551.

Mochizuki, A., Y. Takeda, and Y. Iwasa. 1996. The evolution of genomic imprinting. *Genetics* 144: 1283–1295.

Mock, D. W., and G. A. Parker. 1997. *The Evolution of Sibling Rivalry*. Oxford University Press, Oxford.

Møller, A. P. 1988. Testes size, ejaculate quality and sperm competition in birds. *Biological Journal of the Linnean Society* 33: 273–283.

Møller, A. P. 1990. Deceptive use of alarm calls by male swallows *Hirundo rustica*: A new paternity guard. *Behavioral Ecology* 1: 1–6.

Møller, A. P., and R. V. Alatalo. 1999. Good-genes effects in sexual selection. *Proceedings of the Royal Society of London Series B—Biological Sciences* 266: 85–91.

Møller, A. P., and J. J. Cuervo. 1998. Speciation and feather ornamentation in birds. *Evolution* 52: 859–869.

Møller, A. P., and M. D. Jennions. 2001. How important are direct fitness benefits of sexual selection? *Naturwissenschaften* 88: 401–415.

Møller, A. P., and M. Petrie. 2002. Condition dependence, multiple sexual signals, and immunocompetence in peacocks. *Behavioral Ecology* 13: 248–253.

Møller, A. P., and R. Thornhill. 1998. Male parental care, differential parental investment by females and sexual selection. *Animal Behaviour* 55: 1507–1515.

Moore, A. J. 1989. The behavioral ecology of *Libellula luctuosa* (Burmeister) (Odonata: Libellulidae). III. Male density, osr, and male and female mating behavior. *Ethology* 80: 120–136.

Moore, A. J., P. A. Gowaty, and P. J. Moore. 2003. Females avoid manipulative males and live longer. *Journal of Evolutionary Biology* 16: 523–530.

Moore, A. J., P. A. Gowaty, W. G. Wallin, and P. J. Moore. 2001. Sexual conflict and the evolution of female mate choice and male social dominance. *Proceedings of the Royal Society of London Series B—Biological Sciences* 268: 517–523.

Moore, A. J., and P. J. Moore. 1999. Balancing sexual selection through opposing mate choice and male competition. *Proceedings of the Royal Society of London Series B— Biological Sciences* 266: 711–716.

Moore, S. L., and K. Wilson. 2002. Parasites as a viability cost of sexual selection in natural populations of mammals. *Science* 297: 2015–2018.

Moore, T. 2001. Genetic conflict, genomic imprinting and establishment of the epigenotype in relation to growth. *Reproduction* 122: 185–193.

Moore, T., and D. Haig. 1991. Genomic imprinting in mammalian development: A parental tug-of-war. *Trends in Genetics* 7: 45–49.

Moreno, J., M. Soler, A. P. Møller, and M. Linden. 1994. The function of stone carrying in the black wheatear, *Oenanthe leucura*. *Animal Behaviour* 47: 1297–1309.

Morgan, M. T. 1994. Models of sexual selection in hermaphrodites, especially plants. *American Naturalist* 144: S100–S125.

Morris, M. R., W. E. Wagner, and M. J. Ryan. 1996. A negative correlation between trait and mate preference in *Xiphophorus pygmaeus*. *Animal Behaviour* 52: 1193–1203.

Morrow, E. H., and G. Arnqvist. 2003. Costly traumatic insemination and a female counter-adaptation in bed bugs. *Proceedings of the Royal Society of London Series B—Biological Sciences* 270: 2377–2381.

Morrow, E. H., G. Arnqvist, and T. E. Pitcher. 2002. The evolution of infertility: Does hatching rate in birds coevolve with female polyandry? *Journal of Evolutionary Biology* 15: 702–709.

Morrow, E. H., G. Arnqvist, and S. Pitnick. 2003a. Adaptation versus pleiotropy: Why do males harm their mates? *Behavioral Ecology* 14: 802–806.

Morrow, E. H., and M. J. G. Gage. 2000. The evolution of sperm length in moths. *Proceedings of the Royal Society of London Series B—Biological Sciences* 267: 307–313.

Morrow, E. H., and T. E. Pitcher. 2003. Sexual selection and the risk of extinction in birds. *Proceedings of the Royal Society of London Series B—Biological Sciences* 270: 1793–1799.

Morrow, E. H., T. E. Pitcher, and G. Arnqvist. 2003b. No evidence that sexual selection is an "engine of speciation" in birds. *Ecology Letters* 6: 228–234.

Muhlhauser, C., and W. U. Blanckenhorn. 2002. The costs of avoiding matings in the dung fly *Sepsis cynipsea*. *Behavioral Ecology* 13: 359–365.

Murray, A. M., and P. S. Giller. 1990. The life-history of *Aquarius naja* De Geer (Hemipter: Gerridae) in Southern Ireland. *The Entomologist* 109: 53–64.

Murtaugh, M. P., and D. L. Denlinger. 1985. Physiological regulation of long-term oviposition in the house cricket, *Acheta domesticus*. *Journal of Insect Physiology* 31: 611–617.

Myers, H. S., B. D. Barry, J. A. Burnside, and R. H. Rhode. 1976. Sperm precedence in female apple maggots (Coleoptera: Curculionidae) alternately mated to normal and irradiated males. *Annals of the Entomological Society of America* 69: 39–41.

Nagai, T. 1994. Current status and perspectived in IVM-IVF of porcine oocytes. *Theriogenology* 41: 73–78.

Nasrallah, J. B. 2002. Recognition and rejection of self in plant reproduction. *Science* 296: 305–308.

Nei, M., T. Maruyama, and C. I. Wu. 1983. Models of evolution of reproductive isolation. *Genetics* 103: 557–579.

Nelson, K. 1964. Behaviour and morphology in the glandulocaudin fishes (Ostariophysi: Characidae). *University of California Publications in Zoology* 75: 59–152.

Newberry, K. 1989. The effects on domestic infestation of *Cimex lectularius* bedbugs of interspecific mating with *Cimex hemipterus*. *Medical and Veterinary Entomology* 3: 407–414.

Newman, J. A., and M. A. Elgar. 1991. Sexual cannibalism in orb-weaving spiders: An economic model. *American Naturalist* 138: 1372–1395.

Nijhout, H. F. 1994. *Insect Hormones*. Princeton University Press, Princeton, N.J.

Nilakhe, S. S. 1977. Longevity and fecundity of female boll weevil (Coleoptera: Curculionidae) placed with varying numbers of males. *Annals of the Entomological Society of America* 70: 673–674.

Nilsson, T., C. Fricke, and G. Arnqvist. 2002. Patterns of divergence in the effects of mating on female reproductive performance in flour beetles. *Evolution* 56: 111–120.

Nilsson, T., C. Fricke, and G. Arnqvist. 2003. The effects of male and female genotype on variance in male fertilization success in the red flour beetle (*Tribolium castaneum*). *Behavioral Ecology and Sociobiology* 53: 227–233.

Nolan, C. M., J. K. Killian, J. N. Petitte, and R. L. Jirtle. 2001. Imprint status of M6P/IGF2R and IGF2 in chickens. *Development Genes and Evolution* 211: 179–183.

Nordell, S. E. 1994. Observation of the mating behavior and dentition of the round stingray, *Urolophus halleri*. *Environmental Biology of Fishes* 39: 219–229.

Norman, M. D., and C. C. Lu. 1997. Sex in giant squid. *Nature* 389: 683–684.

Norris, K. J. 1990a. Female choice and the evolution of the conspicuous plumage coloration of monogamous male great tits. *Behavioral Ecology and Sociobiology* 26: 129–138.

Norris, K. J. 1990b. Female choice and the quality of parental care in the great tit *Parus major*. *Behavioral Ecology and Sociobiology* 27: 275–281.

Nuismer, S. L., J. N. Thompson, and R. Gomulkiewicz. 2000. Coevolutionary clines across selection mosaics. *Evolution* 54: 1102–1115.

Nur, N., and O. Hasson. 1984. Phenotypic plasiticity and the handicap principle. *Journal of Theoretical Biology* 110: 275–297.

Nur, U. 1962. Sperms, sperm bundles and fertilization in a mealy bug, *Pseudococcus obscurus* Essig: (Homoptera: Coccoidea). *Journal of Morphology* 111: 173–199.

Nur, U., J. H. Werren, D. G. Eickbush, W. D. Burke, and T. H. Eickbush. 1988. A "selfish" B-chromosome that enhances its transmission by eliminating the paternal genome. *Science* 240: 512–514.

Oberhauser, K. S. 1989. Effects of spermatophores on male and female monarch butterfly reproductive success. *Behavioral Ecology and Sociobiology* 25: 237–246.

O'Donald, P. 1962. Theory of sexual selection. *Heredity* 17: 541–552.

Odor, D. L., and R. J. Blandau. 1956. Incidence of polyspermy in normal and delayed matings in rats of the Wistar strain. *Fertility and Sterility* 7: 456–467.

Ohlsson, R. (ed.). 1999. *Genomic Imprinting: An Interdisciplinary Approach*. Springer, Berlin.

Ohlsson, R., K. Hall, and M. Ritzén (eds.). 1995. *Genomic Imprinting: Causes and Consequences*. Cambridge University Press, Cambridge.

Oliver, J. H. 1986. Induction of oogenesis and oviposition in ticks. In J. R. Sauer and J. A. Hair (eds.), *Morphology, Physiology, and Behavioral Biology of Ticks*, E. Horwood Halsted, Chichester, pp. 233–247.

Olsson, M. 1995. Forced copulation and costly female resistance behavior in the Lake Eyre dragon, *Ctenophorus maculosus*. *Herpetologica* 51: 19–24.

Omu, A. E., T. Fatinikun, N. Mannazhath, and S. Abraham. 1999. Significance of simultaneous determination of serum and seminal plasma alpha-tocopherol and retinol in infertile men by high-performance liquid chromatography. *Andrologia* 31: 347–354.

O'Neill, M. J., R. S. Ingram, P. B. Vrana, and S. M. Tilghman. 2000. Allelic expression of IGF2 in marsupials and birds. *Development Genes and Evolution* 210: 18–20.

Opp, S. B., and R. J. Prokopy. 1986. Variation in laboratory oviposition by *Rhagoletis pomonella* (Diptera: Tephritidae) in relation to mating status. *Annals of the Entomological Society of America* 79: 705–710.

Orians, G. H. 1969. On evolution of mating systems in birds and mammals. *American Naturalist* 103: 589–603.

Orsetti, D. M., and R. L. Rutowski. 2001. No material benefits, and a fertilzation cost, for multiple mating by femalt Colorado potato beetles (*Leptinotarsa decemlineata*). Manuscript.

Ortigosa, A., and L. Rowe. 2002. The effect of hunger on mating behaviour and sexual selection for male body size in *Gerris buenoi*. *Animal Behaviour* 64: 369–375.

Ortigosa, A., and L. Rowe. 2003. The role of mating history and male size in determining mating behaviours and sexual conflict in a water strider. *Animal Behaviour* 65: 851–858.

Osanai, M., T. Aigaki, and H. Kasuga. 1987. Energy metabolism in the spermatophor of the silkmoth, *Bombyx mori*, associated with accumulation of alanine drived from argenine. *Insect Biochemistry* 17: 71–75.

Osanai, M., T. Aigaki, H. Kasuga, and Y. Yonezawa. 1986. Role of arginase transferred from the vesicula seminalis during mating and changes in amino acid pools of the spermatophore after ejaculation in the silkworm, *Bombyx mori*. *Insect Biochemistry* 16: 879–885.

Osanai, M., and P. S. Chen. 1993. A comparative study on the arginine degradation cascade for sperm maturation of *Bombyx mori* and *Drosophila melanogaster*. *Amino Acids* 5: 341–350.

Osanai, M., H. Kasuga, and T. Aigaki. 1987a. Physiological role of apyrene spermatozoa of *Bombyx mori*. *Experientia* 43: 593–596.

Osanai, M., H. Kasuga, and T. Aigaki. 1987b. The spermatophore and its structural changes with time in the bursa copulatrix of the silkworm, *Bomby mori*. *Journal of Morphology* 193: 1–11.

Osanai, M., H. Kasuga, and T. Aigaki. 1990. Physiology of sperm maturation in the spermatophore of the silkworm, *Bombyx mori*. *Advances in Invertebrate Reproduction* 5: 531–536.

Osanai, M., and S. Nagaoka. 1992. Adenine compounds in the male reproductive tract and the spermatophore of the silkmoth, *Bombyx mori*. *Comparative Biochemistry and Physiology B—Biochemistry & Molecular Biology* 102: 49–55.

Otronen, M. 1990. Mating behavior and sperm competition in the fly, *Dryomyza anilis*. *Behavioral Ecology and Sociobiology* 26: 349–356.

Otronen, M. 1994a. Repeated copulations as a strategy to maximize fertilization in the fly, *Drymyza anilis* (Dryomyzidae). *Behavioral Ecology* 5: 51–56.

Otronen, M. 1994b. Fertilization success in the fly *Dryomyza anilis* (Dryomyzidae): Effects of male size and the mating situation. *Behavioral Ecology and Sociobiology* 35: 33–38.

Otronen, M. 1995. Energy reserves and mating success in males of the yellow dung fly, *Scatophaga stercoraria. Functional Ecology* 9: 683–688.

Otronen, M., and M. T. Sivajothy. 1991. The effect of postcopulatory male behavior on ejaculate distribution within the female sperm storage organs of the fly, *Dryomyza anilis* (Diptera: Dryomyzidae). *Behavioral Ecology and Sociobiology* 29: 33–37.

Otto, S. P., and D. B. Goldstein. 1992. Recombination and the evolution of dipoidy. *Genetics* 131: 745–751.

Owens, I.P.F., P. M. Bennett, and P. H. Harvey. 1999. Species richness among birds: Body size, life history, sexual selection or ecology? *Proceedings of the Royal Society of London Series B—Biological Sciences* 266: 933–939.

Owens, I.P.F., A. Dixon, T. Burke, and D. B. A. Thompson. 1995. Strategic paternity assurance in the sex-role reversed Eurasian dotterel (*Dharadrius morinellus*): Behavioral and genetic evidence. *Behavioral Ecology* 6: 14–21.

Owens, I.P.F., and I. R. Hartley. 1998. Sexual dimorphism in birds: Why are there so many different forms of dimorphism? *Proceedings of the Royal Society of London Series B—Biological Sciences* 265: 397–407.

Owens, I.P.F., and D.B.A. Thompson. 1994. Sex differences, sex ratios and sex roles. *Proceedings of the Royal Society of London Series B—Biological Sciences* 258: 93–99.

Packer, C., D. A. Collins, and L. E. Eberly. 2000. Problems with primate sex ratios. *Philosophical Transactions of the Royal Society of London Series B—Biological Sciences* 355: 1627–1635.

Paemen, L., A. Tips, L. Schoofs, P. Proost, J. Vandamme, and A. Deloof. 1991. Lomag-myotropin: A novel myotropic peptide from the male accessory glands of *Locusta migratoria. Peptides* 12: 7–10.

Palumbi, S. R. 1998. Species formation and the evolution of gamete recognition loci. In D. J. Howard and S. H. Berlocher (eds.), *Endless Forms: Species and Speciation*, Oxford University Press, New York, pp. 271–278.

Panhuis, T. M., R. Butlin, M. Zuk, and T. Tregenza. 2001. Sexual selection and speciation. *Trends in Ecology & Evolution* 16: 364–371.

Papaj, D. R. 1994. Oviposition site guarding by male walnut flies and its possible consequences for mating success. *Behavioral Ecology and Sociobiology* 34: 187–195.

Park, Y. I., S. B. Ramaswamy, and A. Srinivasan. 1998. Spermatophore formation and regulation of egg maturation and oviposition in female *Heliothis virescens* by the male. *Journal of Insect Physiology* 44: 903–908.

Parker, G. A. 1970a. Reproductive behaviour and nature of sexual selection in *Scatophaga stercoraria* L. (Diptera: Scatophagidae) II. Fertilization rate and spatial and temporal relationships of each sex around site of mating and oviposition. *Journal of Animal Ecology* 39: 205–208.

Parker, G. A. 1970b. Reproductive behaviour and nature of sexual selection in *Scatophaga stercoraria* L. (Diptera: Scatophagidae) VII. The origin and evolution of the passive phase. *Evolution* 24: 774–788.

Parker, G. A. 1970c. Sperm competition and its evolutionary consequences in insects. *Biological Reviews of the Cambridge Philosophical Society* 45: 525–567.

Parker, G. A. 1970d. Sperm competition and its evolutionary effect on copula duration in fly *Scatophaga stercoraria*. *Journal of Insect Physiology* 16: 1301–1328.

Parker, G. A. 1974. Courtship persistence and female guarding as male time investment strategies. *Behaviour* 48: 157–184.

Parker, G. A. 1978. Searching for mates. In J. R. Krebs and N. B. Davies (eds.), *Behavioural Ecology: An Evolutionary Approach*, Blackwell Scientific, Oxford, pp. 214–244.

Parker, G. A. 1979. Sexual selection and sexual conflict. In M. S. Blum and N. A. Blum (eds.), *Sexual Selection and Reproductive Competition in Insects*, Academic, New York, pp. 123–166.

Parker, G. A. 1983a. Arms races in evolution: An ESS to the opponent-independent costs game. *Journal of Theoretical Biology* 101: 619–648.

Parker, G. A. 1983b. Mate quality and mating decisions. In P.P.G. Bateson (ed.), *Mate Choice*, Cambridge University Press, Cambridge, pp. 141–166.

Parker, G. A. 1984. Sperm competition and the evolution of animal mating strategies. In R. L. Smith (ed.), *Sperm Competition and the Evolution of Animal Mating Systems*, Academic, Orlando, Fla., pp. 1–60.

Parker, G. A. 1985. Models of parent-offspring conflict. V. Effects of the behavior of the two parents. *Animal Behaviour* 33: 519–533.

Parker, G. A. 1998. Sperm competition and the evolution of ejaculates: Towards a theory base. In T. R. Birkhead and A. P. Møller (eds.), *Sperm Competition and Sexual Selection*, Academic, London, pp. 3–54.

Parker, G. A., and L. Partridge. 1998. Sexual conflict and speciation. *Philosophical Transactions of the Royal Society of London Series B—Biological Sciences* 353: 261–274.

Parker, G. A., N. J. Royle, and I. R. Hartley. 2002. Intrafamilial conflict and parental investment: A synthesis. *Philosophical Transactions of the Royal Society of London Series B—Biological Sciences* 357: 295–307.

Parker, G. A., and L. W. Simmons. 1989. Nuptial feeding in insects: Theoretical models of male and female interests. *Ethology* 82: 3–26.

Parker, G. A., and L. W. Simmons. 1994. Evolution of phenotypic optima and copula duration in dungflies. *Nature* 370: 53–56.

Parker, G. A., and L. W. Simmons. 1996. Parental investment and the control of sexual selection: Predicting the direction of sexual competition. *Proceedings of the Royal Society of London Series B—Biological Sciences* 263: 315–321.

Parker, G. A., V.G.F. Smith, and R. R. Baker. 1972. Origin and evolution of gamete dimorphism and male-female phenomenon. *Journal of Theoretical Biology* 36: 529–553.

Parmigiani, S., and F. S. Vom Saal. 1994. *Infanticide and Parental Care*. Harwood Academic, Chur, Switzerland.

Pärt, T., L. Gustafsson, and J. Moreno. 1992. "Terminal investment" and a sexual conflict in the collared flycatcher (*Ficedula albicollis*). *American Naturalist* 140: 868–882.

Partridge, L., and K. Fowler. 1990. Nonmating costs of exposure to males in female *Drosophila melanogaster*. *Journal of Insect Physiology* 36: 419–425.

Partridge, L., K. Fowler, S. Trevitt, and W. Sharp. 1986. An examination of the effects of males on the survival and egg-production rates of female *Drosophila melanogaster*. *Journal of Insect Physiology* 32: 925–929.

Partridge, L., A. Green, and K. Fowler. 1987. Effects of egg-production and of exposure to males on female survival in *Drosophila melanogaster*. *Journal of Insect Physiology* 33: 745–749.

Partridge, L., and L. D. Hurst. 1998. Sex and conflict. *Science* 281: 2003–2008.

Patel, C. S. 1993. Pineal, pineal indoles and photoperiodism in relation to seasonal reproductive functions and metabolism physiology of male feral blue rock pigeons, *Columba livia*. Ph.D. thesis, Maharaja Sayajirao University of Baroda, India.

Payne, R.J.H., and M. Pagel. 2001. Inferring the origins of state-dependent courtship traits. *American Naturalist* 157: 42–50.

Pedersen, H. C. 1989. Effects of exogenous prolactin on parental behavior in free-living female willow ptarmigan *Lagopus l. lagopus*. *Animal Behaviour* 38: 926–934.

Pen, I., and F. J. Weissing. 1999. Sperm competition and sex allocation in simultaneous hermaphrodites: A new look at Charnov's invariance principle. *Evolutionary Ecology Research* 1: 517–525.

Pennisi, E. 1998. A genomic battle of the sexes. *Science* 281: 1984–1985.

Persson, O., and P. Öhrström. 1989. A new avian mating system: Ambisexual polygamy in the penduline tit *Remiz pendulinus*. *Ornis Scandinavica* 20: 105–111.

Persson, O., and P. Öhrström. 1996. Female nest choice in the penduline tit: A comment. *Animal Behaviour* 51: 462–463.

Peter, M. G., P. D. Shirk, K. H. Dahm, and H. Roller. 1981. On the specificity of juvenile hormone biosynthesis in the male Hyalophora cecropia moth. *Zeitschrift Fur Naturforschung C—A Journal of Biosciences* 36: 579–585.

Peters, S. E., and D. A. Aulner. 2000. Sexual dimorphism in forelimb muscles of the bullfrog, *Rana catesbeiana*: A functional analysis of isometric contractile properties. *Journal of Experimental Biology* 203: 3639–3654.

Petkov, N., A. Yolov, G. Mladenov, and I. Nacheva. 1979. Influence of mating length of silkworm moths of some inbred silkworm (*Bombyx mori* L.) lines on silkworm seed quality and quantity. *Sericulture Experiment Station Vratza* 16: 116–122.

Petrie, M. 1992. Copulation frequency in birds: Why do females copulate more than once with the same male? *Animal Behaviour* 44: 790–792.

Petrie, M., and A. Williams. 1993. Peahens lay more eggs for peacocks with larger trains. *Proceedings of the Royal Society of London Series B—Biological Sciences* 251: 127–131.

Phillips, W. R., and S. J. Inwards. 1985. The annual activity and breeding cycles of Gould long-eared bat, *Nyctophilus gouldi* (Microchiroptera: Vespertilionidae). *Australian Journal of Zoology* 33: 111–126.

Piko, L. 1958. Etude de la polyspermie chez le rat. *Comptes Rendus des Seances de la Societe de Biologie et de ses Filiales* 152: 1356–1358.

Pilastro, A., S. Benetton, and A. Bisazza. 2003. Female aggregation and male competition reduce costs of sexual harassment in the mosquitofish *Gambusia holbrooki*. *Animal Behaviour* 65: 1161–1167.

Pitnick, S., and W. D. Brown. 2000. Criteria for demonstrating female sperm choice. *Evolution* 54: 1052–1056.

Pitnick, S., W. D. Brown, and G. T. Miller. 2001a. Evolution of female remating behaviour following experimental removal of sexual selection. *Proceedings of the Royal Society of London Series B—Biological Sciences* 268: 557–563.

Pitnick, S., and F. Garcia-Gonzalez. 2002. Harm to females increases with male body size in *Drosophila melanogaster*. *Proceedings of the Royal Society of London Series B—Biological Sciences* 269: 1821–1828.

Pitnick, S., T. A. Markow, and G. S. Spicer. 1995. Delayed male maturity is a cost of producing large sperm in *Drosophila*. *Proceedings of the National Academy of Sciences of the United States of America* 92: 10614–10618.

Pitnick, S., T. Markow, and G. S. Spicer. 1999. Evolution of multiple kinds of female sperm-storage organs in *Drosophila*. *Evolution* 53: 1804–1822.

Pitnick, S., G. T. Miller, J. Reagan, and B. Holland. 2001b. Males' evolutionary responses to experimental removal of sexual selection. *Proceedings of the Royal Society of London Series B—Biological Sciences* 268: 1071–1080.

Pitnick, S., G. S. Spicer, and T. Markow. 1997. Phylogenetic examination of female incorporation of ejaculate in *Drosophila*. *Evolution* 51: 833–845.

Pizzari, T., and R. R. Snook. 2003. Perspective: Sexual conflict and sexual selection: chasing away paradigm shifts. *Evolution* 57: 1223–1236.

Plath, M., J. Parzefall, and I. Schlupp. 2003. The role of sexual harassment in cave and surface dwelling populations of the Atlantic molly, *Poecilia mexicana* (Poeciliidae, Teleostei). *Behavioral Ecology and Sociobiology* 54: 303–309.

Polhemus, D. A. 1995. Two new species of Rhagovelia from the Philippines, with a discussion of zoogeographic relationships between the Philippines and New Guinea (Heteroptera: Veliidae). *Journal of the New York Entomological Society* 103: 55–68.

Polhemus, D. A., and J. T. Polhemus. 2000. Additional new genera and species of Microveliinae (Heteroptera: Veliidae) from New Guinea and adjacent regions. *Tijdschrift Voor Entomologie* 143: 91–123.

Polhemus, J. T., and D. A. Polhemus. 1996. The Trepobatinae (Heteroptera: Gerridae) of New Guinea and surrounding regions, with a review of the World fauna. 4. The marine tribe Stenobatini. *Entomologica Scandinavica* 27: 279–346.

Polis, G. A., and R. D. Farley. 1979. Behavior and ecology of mating in the cannibalistic scorpion, *Paruroctonus mesaensis* Stahnke (Scorpionida: Vaejovidae). *Journal of Arachnology* 7: 33–46.

Polis, G. A., and W. D. Sissom. 1990. Life history. In G. A. Polis (ed.), *The Biology of Scorpions*, Stanford University Press, Stanford, pp. 161–223.

Pomiankowski, A. 1987a. Sexual selection: the handicap principal does work sometimes. *Proceedings of the Royal Society of London Series B—Biological Sciences* 231: 123–145.

Pomiankowski, A. 1987b. The costs of choice in sexual selection. *Journal of Theoretical Biology* 128: 195–218.

Pomiankowski, A. 1988. The evolution of female mate preference for male genetic quality. *Oxford Survey of Evolutionary Biology* 5: 136–184.

Pomiankowski, A., and Y. Iwasa. 1993. Evolution of multiple sexual preferences by Fisher runaway process of sexual selection. *Proceedings of the Royal Society of London Series B—Biological Sciences* 253: 173–181.

Pomiankowski, A., Y. Iwasa, and S. Nee. 1991. The evolution of costly mate preferences. I. Fisher and biased mutation. *Evolution* 45: 1422–1430.

Pomiankowski, A., and A. P. Møller. 1995. A resolution of the lek paradox. *Proceedings of the Royal Society of London Series B—Biological Sciences* 260: 21–29.

Pomiankowski, A., and P. Reguera. 2001. The point of love. *Trends in Ecology & Evolution* 16: 533–534.

Pongratz, N., and N. K. Michiels. 2003. High multiple paternity and low last-male sperm precedence in a hermaphroditic planarian flatworm: Consequences for reciprocity patterns. *Molecular Ecology* 12: 1425–1433.

Pratt, H. L. 1979. Reproduction in the blue shark, *Prionace glauca. Fishery Bulletin* 77: 445–470.

Pratt, H. L., and J. C. Carrier. 2001. A review of elasmobranch reproductive behavior with a case study on the nurse shark, *Ginglymostoma cirratum. Environmental Biology of Fishes* 60: 157–188.

Prenter, J., R. W. Elwood, and W. I. Montgomery. 1994. Male exploitation of female predatory behavior reduces sexual cannibalism in male autumn spiders, *Metellina segmentata. Animal Behaviour* 47: 235–236.

Presgraves, D. C., R. H. Baker, and G. S. Wilkinson. 1999. Coevolution of sperm and female reproductive tract morphology in stalk-eyed flies. *Proceedings of the Royal Society of London Series B—Biological Sciences* 266: 1041–1047.

Preston-Mafham, K. G. 1999. Courtship and mating in *Empis (Xanthempis) trigramma* Meig., *E-tessellata* F. and *E. (Polyblepharis) opaca* F. (Diptera: Empididae) and the possible implications of "cheating" behaviour. *Journal of Zoology* 247: 239–246.

Pribil, S., and W. A. Searcy. 2001. Experimental confirmation of the polygyny threshold model for red-winged blackbirds. *Proceedings of the Royal Society of London Series B—Biological Sciences* 268: 1643–1646.

Price, C.S.C. 1997. Conspecific sperm precedence in *Drosophila. Nature* 388: 663–666.

Price, C.S.C., K. A. Dyer, and J. A. Coyne. 1999. Sperm competition between *Drosophila* males involves both displacement and incapacitation. *Nature* 400: 449–452.

Price, D. K., and N. T. Burley. 1994. Constratints on the evolution of attractive traits: selection in male and female zebra finches. *American Naturalist* 144: 908–934.

Price, T. 1998. Sexual selection and natural selection in bird speciation. *Philosophical Transactions of the Royal Society of London Series B—Biological Sciences* 353: 251–260.

Price, T., D. Schluter, and N. E. Heckman. 1993. Sexual selection when the female directly benefits. *Biological Journal of the Linnean Society* 48: 187–211.

Price, T. D. 2002. Domesticated birds as a model for the genetics of speciation by sexual selection. *Genetica* 116: 311–327.

Proctor H. C. 1991. Courtship in the water mit *Neumania papillator* males capitalize on female adaptations for predation. *Animal Behaviour* 42: 589–598.

Proctor, H. C. 1992. Sensory exploitation and the evolution of male matin behavior: a cladistic test using water mites (Acari: Parasitengona). *Animal Behaviour* 44: 745–752.

Promislow, D. 2003. Mate choice, sexual conflict, and evolution of senescence. *Behavior Genetics* 33: 191–201.

Promislow, D., R. Montgomerie, and T. E. Martin. 1994. Sexual selection and survival in North American waterfowl. *Evolution* 48: 2045–2050.

Promislow, D.E.L., R. Montgomerie, and T. E. Martin. 1992. Mortality costs of sexual dimorphism in birds. *Proceedings of the Royal Society of London Series B—Biological Sciences* 250: 143–150.

Promislow, D.E.L., E. A. Smith, and L. Pearse. 1998. Adult fitness consequences of sexual selection in *Drosophila melanogaster*. *Proceedings of the National Academy of Sciences of the United States of America* 95: 10687–10692.

Proulx, S. R. 1999. Matings systems and the evolution of niche breadth. *American Naturalist* 154: 89–98.

Proulx, S. R., T. Day, and L. Rowe. 2002. Older males signal more reliably. *Proceedings of the Royal Society of London Series B—Biological Sciences* 269: 2291–2299.

Prout, T., and A. G. Clark. 1996. Polymorphism in genes that influence sperm displacement. *Genetics* 144: 401–408.

Prout, T., and A. G. Clark. 2000. Seminal fluid causes temporarily reduced egg hatch in previously mated females. *Proceedings of the Royal Society of London Series B—Biological Sciences* 267: 201–203.

Pryke, S. R., S. Andersson, and M. J. Lawes. 2001. Sexual selection of multiple handicaps in the red-collared widowbird: Female choice of tail length but not carotenoid display. *Evolution* 55: 1452–1463.

Ptacek, M. B. 1998. Interspecific mate choice in sailfin and shortfin species of mollies. *Animal Behaviour* 56: 1145–1154.

Punitham, M. T., M. A. Haniffa, and S. Arunachalam. 1987. Effect of mating duration on fecundity and fertility of egges in *Bombyx mori* L. (Lepidoptera: Bombycidae). *Entomon* 12: 55–58.

Queller, D. C. 1987. Sexual selection in flowering plants. In J. W. Bradbury and M. B. Andersson (eds.), *Sexual Selection: Testing the Alternatives*, Wiley, Chichester, pp. 165–179.

Queller, D. C. 1994. Male-female conflict and parent-offspring conflict. *American Naturalist* 144: S84–S99.

Queller, D. C. 1997. Why do females care more than males? *Proceedings of the Royal Society of London Series B—Biological Sciences* 264: 1555–1557.

Questiau, S. 1999. How can sexual selection promote population divergence? *Ethology Ecology & Evolution* 11: 313–324.

Raabe, M. 1986. Insect reproduction: Regulation of successive steps. *Advances in Insect Physiology* 19: 29–154.

Radwan, J. 1996. Intraspecific variation in sperm competition success in the bulb mite: A role for sperm size. *Proceedings of the Royal Society of London Series B—Biological Sciences* 263: 855–859.

Radwan, J. 1998. Heritability of sperm competition success in the bulb mite, *Rhizoglyphus robini*. *Journal of Evolutionary Biology* 11: 321–327.

Raina, A. K., T. G. Kingan, and J. M. Giebultowicz. 1994. Mating induced loss of sex-pheromone and sexual receptivity in insects with emphasis on *Helicoverpa zea* and *Lymantria dispar*. *Archives of Insect Biochemistry and Physiology* 25: 317–327.

Ramaswamy, S. B., S. Q. Shu, Y. I. Park, and F. R. Zeng. 1997. Dynamics of juvenile hormone-mediated gonadotropism in the lepidoptera. *Archives of Insect Biochemistry and Physiology* 35: 539–558.

Ratti, O., and R. V. Alatalo. 1993. Determinants of the mating success of polyterritorial pied flycatcher males. *Ethology* 94: 137–146.

Raz, T., and R. Shalgi. 1998. Early events in mammalian egg activation. *Human Repro-duction* 13: 133–145.

Redline, R. W., T. Hassold, and M. V. Zaragoza. 1998. Prevalence of the partial molar phenotype in triploidy of maternal and paternal origin. *Human Pathology* 29: 505–511.

Reed, K. J., and A. H. Sinclair. 2002. FET-1: A novel W-linked, female specific gene up-regulated in the embryonic chicken ovary. *Gene Expression Patterns* 2: 83–86.

Reeve, H. K., and D. W. Pfennig. 2003. Genetic biases for showy males: Are some genetic systems especially conducive to sexual selection? *Proceedings of the National Academy of Sciences of the United States of America* 100: 1089–1094.

Reeve, H. K., and P. W. Sherman. 1993. Adaptation and the goals of evolutionary research. *Quarterly Review of Biology* 68: 1–32.

Reik, W., and A. Surani. 1997. *Genomic imprinting*. IRL Press at Oxford University Press, Oxford.

Reinhardt, K., R. Naylor, and M. T. Siva-Jothy. 2003. Reducing a cost of traumatic insemination: Female bedbugs evolve a unique organ. *Proceedings of the Royal Society of London Series B—Biological Sciences* 270: 2371–2375.

Reinhold, K. 1998. Sex linkage among genes controlling sexually selected traits. *Behavioral Ecology and Sociobiology* 44: 1–7.

Reinhold, K. 1999. Evolutionary genetics of sex limited traits under fluctuating selection. *Journal of Evolutionary Biology* 12: 897–902.

Reise, H., and J.M.C. Hutchinson. 2002. Penis-biting slugs: Wild claims and confusions. *Trends in Ecology & Evolution* 17: 163.

Reynolds, J. D., and M. R. Gross. 1990. Costs and benefits of female mate choice: Is there a lek paradox? *American Naturalist* 136: 230–243.

Reznick, D., and J. A. Endler. 1982. The impact of predation on life history evolution of Trinidadian guppies (*Poecilia reticulata*). *Evolution* 36: 160–177.

Reznick, D. A., H. Bryga, and J. A. Endler. 1990. Experimentally induced life history evolution in a natural population. *Nature* 346: 357–359.

Rhen, T. 2000. Sex-limited mutations and the evolution of sexual dimorphism. *Evolution* 54: 37–43.

Ribeiro, J.M.C., and A. Spielman. 1986. The satyr effect: A model predicting parapatry and species extinction. *American Naturalist* 128: 513–528.

Rice, W. R. 1984. Sex-chromosomes and the evolution of sexual dimorphism. *Evolution* 38: 735–742.

Rice, W. R. 1992. Sexually antagonistic genes: Experimental evidence. *Science* 256: 1436–1439.

Rice, W. R. 1996a. Evolution of the Y sex chromosome in animals. *Bioscience* 46: 331–343.

Rice, W. R. 1996b. Sexually antagonistic male adaptation triggered by experimental arrest of female evolution. *Nature* 381: 232–234.

Rice, W. R. 1998a. Intergenomic conflict, interlocus antagonistic coevolution, and the evolution of reproductive isolation. In D. J. Howard and S. H. Berlocher (eds.), *Endless Forms: Species and Speciation*, Oxford University Press, New York, pp. 261–270.

Rice, W. R. 1998b. Male fitness increases when females are eliminated from gene pool: Implications for the Y chromosome. *Proceedings of the National Academy of Sciences of the United States of America* 95: 6217–6221.

Rice, W. R. 2000. Dangerous liaisons. *Proceedings of the National Academy of Sciences of the United States of America* 97: 12953–12955.

Rice, W. R., and A. K. Chippindale. 2001. Intersexual ontogenetic conflict. *Journal of Evolutionary Biology* 14: 685–693.

Rice, W. R., and B. Holland. 1997. The enemies within: intergenomic conflict, interlocus contest evolution (ICE), and the intraspecific Red Queen. *Behavioral Ecology and Sociobiology* 41: 1–10.

Rice, W. R., and B. Holland. 1999. Reply to comments on the chase-away model of sexual selection. *Evolution* 53: 302–306.

Richards, O. W. 1927. Sexual selection and allied problems in the insects. *Biological Reviews* 2: 298–364.

Richter, H. J. 1986. The swordtail characin, *Coynopom riisei*. *Tropical Fish Hobbyist* 35: 46–49.

Ridley, M. 1988. Mating frequency and fecundity in insects. *Biological Reviews of the Cambridge Philosophical Society* 63: 509–549.

Ridley, M. 1990. The control and frequency of mating in insects. *Functional Ecology* 4: 75–84.

Ridley, M., and D. J. Thompson. 1979. Size and mating in *Asellus aquaticus* (Crustacea: Isopoda). *Zeitschrift Fur Tierpsychologie—Journal of Comparative Ethology* 51: 380–397.

Riemann, J. G., D. J. Moen, and B. J. Thorson. 1967. Female monogamy and its control in houseflies. *Journal of Insect Physiology* 13: 407–418.

Riemann, J. G., and B. J. Thorson. 1969. Effect of male accessory material on oviposition and mating by female house flies. *Annals of the Entomological Society of America* 62: 828–834.

Ringo, J. M. 1977. Why 300 species of Hawaiian *Drosophila*? Sexual selection hypothesis. *Evolution* 31: 694–696.

Rink, L., and H. Kirchner. 2000. Zinc-altered immune function and cytokine production. *Journal of Nutrition* 130: 1407S–1411S.

Rintamäki, P. T., A. Lundberg, R. V. Alatalo, and J. Hoglund. 1998. Assortative mating and female clutch investment in black grouse. *Animal Behaviour* 56: 1399–1403.

Ritchie, M. G., and S. D. F. Phillips. 1998. The genetics of sexual isolation. In D. J. Howard and S. H. Berlocher (eds.), *Endless Forms: Species and Speciation*, Oxford University Press, New York, pp. 291–308.

Robertson, H. M. 1985. Female dimorphism and mating behavior in a damselfly, *Ischnura ramburi*: Females mimicking males. *Animal Behaviour* 33: 805–809.

Robinson, M. H. 1982. Courtship and mating behavior in spiders. *Annual Review of Entomology* 27: 1–20.

Robinson, M. H., and B. Robinson. 1980. Comparative studies of the courtship and mating behaviour of tropical Araneid spiders. *Pacific Insects Monographs* 36.

Rodd, F. H., K. A. Hughes, G. F. Grether, and C. T. Baril. 2002. A possible non-sexual origin of mate preference: Are male guppies mimicking fruit? *Proceedings of the Royal Society of London Series B—Biological Sciences* 269: 475–481.

Rodd, F. H., and M. B. Sokolowski. 1995. Complex origins of variation in the sexual behavior of male Trinidad Guppies, *Poecilia reticulata*: Interactions between social environment, heredity, body size and age. *Animal Behaviour* 49: 1139–1159.

Roff, D. A. 1992. *The Evolution of Life Histories: Theory and Analysis.* Chapman & Hall, New York.

Rogers, D. W., and R. Chase. 2001. Dart receipt promotes sperm storage in the garden snail *Helix aspersa. Behavioral Ecology and Sociobiology* 50: 122–127.

Roldan, E.R.S., and M. Gomendio. 1999. The Y chromosome as a battle ground for sexual selection. *Trends in Ecology & Evolution* 14: 58–62.

Roldan, E.R.S., M. Gomendio, and A. D. Vitullo. 1992. The evolution of eutherian spermatozoa and underlying selective forces: Female selection and sperm competition. *Biological Reviews of the Cambridge Philosophical Society* 67: 551–593.

Rolff, J., and M. T. Siva-Jothy. 2002. Copulation corrupts immunity: A mechanism for a cost of mating in insects. *Proceedings of the National Academy of Sciences of the United States of America* 99: 9916–9918.

Rooney, J., and S. M. Lewis. 1999. Differential allocation of male-derived nutrients in two lampyrid beetles with contrasting life-history characteristics. *Behavioral Ecology* 10: 97–104.

Rosenblatt, J. S., and C. T. Snowdon (eds.). 1996. *Parental Care, Evolution Mechanisma, and Adaptive Significance.* Advances in the study of behavior, vol. 25. Academic, London.

Rosenthal, G. G., and M. R. Servedio. 1999. Chase-away sexual selection: Resistance to "resistance." *Evolution* 53: 296–299.

Rovio, A. T., D. R. Marchington, S. Donat, H. C. Schuppe, J. Abe, E. Fritsche, D. J. Elliott, P. Laippala, A. L. Ahola, D. McNay, R. F. Harrison, B. Hughes, T. Barrett, D.M.D. Bailey, D. Mehmet, A. M. Jequier, T. B. Hargreave, S. H. Kao, J. M. Cummins, D. E. Barton, H. J. Cooke, Y. H. Wei, L. Wichmann, J. Poulton, and H. T. Jacobs. 2001. Mutations at the mitochondrial DNA polymerase (POLG) locus associated with male infertility. *Nature Genetics* 29: 261–262.

Rowe, C. 1999. Receiver psychology and the evolution of multicomponent signals. *Animal Behaviour* 58: 921–931.

Rowe, L. 1992. Convenience polyandry in a water strider: foraging conflicts and female control of copulation frequency and guarding duration. *Animal Behaviour* 44: 189–202.

Rowe, L. 1994. The costs of mating and mate choice in water striders. *Animal Behaviour* 48: 1049–1056.

Rowe, L., and G. Arnqvist. 2002. Sexually antagonistic coevolution in a mating system: Combining experimental and comparative approaches to address evolutionary processes. *Evolution* 56: 754–767.

Rowe, L., G. Arnqvist, A. Sih, and J. Krupa. 1994. Sexual conflict and the evolutionary ecology of mating patterns: Water striders as a model system. *Trends in Ecology & Evolution* 9: 289–293.

Rowe, L., E. Cameron, and T. Day. 2003. Detecting sexually antagonistic coevolution with population crosses. *Proceedings of the Royal Society of London Series B— Biological Sciences* 270: 2009–2016.

Rowe, L., and D. Houle. 1996. The lek paradox and the capture of genetic variance by condition dependent traits. *Proceedings of the Royal Society of London Series B—Biological Sciences* 263: 1415–1421.

Rowe, L., J. J. Krupa, and A. Sih. 1996. An experimental test of condition-dependent mating behavior and habitat choice by water striders in the wild. *Behavioral Ecology* 7: 474–479.

Royle, N. J., I. R. Hartley, and G. A. Parker. 2002. Sexual conflict reduces offspring fitness in zebra finches. *Nature* 416: 733–736.

Rubenstein, D. I. 1987. Alternative reproductive tactics in the spider *Meta segmentata*. *Behavioral Ecology and Sociobiology* 20: 229–237.

Ruiz-Pesini, E., A. C. Lapena, C. Diez-Sanchez, A. Perez-Martos, J. Montoya, E. Alvarez, M. Diaz, A. Urrieis, L. Montoro, M. J. Lopez-Perez, and J. A. Enriquez. 2000. Human mtDNA haplogroups associated with high or reduced spermatozoa motility. *American Journal of Human Genetics* 67: 682–696.

Rutowski, R. L. 1979. Butterfly as an honest salesman. *Animal Behaviour* 27: 1269–1270.

Rutowski, R. L. 1980. Courtship solicitation by females of the checkered white butterfly, *Pieris protodice*. *Behavioral Ecology and Sociobiology* 7: 113–117.

Rutowski, R. L. 1982. Epigamic seletion by males as evidenced by courtship partner preferences in the checkered white butterfly (*Pieris protodice*). *Animal Behaviour* 30: 108–112.

Rutowski, R. L., G. W. Gilchrist, and B. Terkanian. 1987. Female butterflies mated with recently mated males show reduced reproductive output. *Behavioral Ecology and Sociobiology* 20: 319–322.

Ryan, M. J. 1990. Sexual selection, sensory systems, and sensory exploitation. *Oxford Surveys in Evolutionary Biology* 7: 157–195.

Ryan, M. J. 1998. Sexual selection, receiver biases, and the evolution of sex differences. *Science* 281: 1999–2003.

Ryan M. J, J. H. Fox, W. Wilczynski, and A. S. Rand. 1990. Sexual selection for sensory exploitation in the frog *Physalaemus pustulosus*. *Nature* 343: 66–67.

Rybak, F., G. Sureau, and T. Aubin. 2002. Functional coupling of acoustic and chemical signals in the courtship behaviour of the male *Drosophila melanogaster*. *Proceedings of the Royal Society of London Series B—Biological Sciences* 269: 695–701.

Ryckman, R. E., and N. Ueshima. 1964. Biosystematics of the *Hesperocimex* complex (Hemiptera: Cimicidae) and avian hosts. *Annals of the Entomological Society of America* 57: 624–638.

Sadowski, J. A., A. J. Moore, and E. D. Brodie. 1999. The evolution of empty nuptial gifts in a dance fly, *Empis snoddyi* (Diptera: Empididae): Bigger isn't always better. *Behavioral Ecology and Sociobiology* 45: 161–166.

Sahara, K., and Y. Takemura. 2003. Application of artificial insemination technique to eupyrene and/or apyrene sperm in *Bombyx mori*. *Journal of Experimental Zoology Part A—Comparative Experimental Biology* 297A: 196–200.

Saifi, G. M., and H. S. Chandra. 1999. An apparent excess of sex- and reproduction-related genes on the human X chromosome. *Proceedings of the Royal Society of London Series B—Biological Sciences* 266: 203–209.

Sakaluk, S. K. 2000. Sensory exploitation as an evolutionary origin to nuptial food gifts in insects. *Proceedings of the Royal Society of London Series B—Biological Sciences* 267: 339–343.

Sakaluk, S. K., P. J. Bangert, A. K. Eggert, C. Gack, and L. V. Swanson. 1995. The gin trap as a device facilitating coercive mating in sagebrush crickets. *Proceedings of the Royal Society of London Series B—Biological Sciences* 261: 65–71.

Sakaluk, S. K., and A. K. Eggert. 1996. Female control of sperm transfer and intraspecific variation in sperm precedence: Antecedents to the evolution of a courtship food gift. *Evolution* 50: 694–703.

Saleuddin, A.S.M., M. L. Ashton, and H. R. Khan. 1989. Mating-induced release of granules by the endocrine dorsal body cells of the snail *Helisoma duryl* (Mollusca). *Journal of Experimental Zoology* 250: 206–213.

Sandell, M. I. 1998. Female aggression and the maintenance of monogamy: female behaviour predicts male mating status in European starlings. *Proceedings of the Royal Society of London Series B—Biological Sciences* 265: 1307–1311.

Sandell, M. I., and H. G. Smith. 1996. Already mated females constrain male mating success in the European starling. *Proceedings of the Royal Society of London Series B—Biological Sciences* 263: 743–747.

Sandell, M. I., and H. G. Smith. 1997. Female aggression in the European starling during the breeding season. *Animal Behaviour* 53: 13–23.

Sankai, T., F. Cho, and Y. Yoshikawa. 1997. In vitro fertilization and preimplantation embryo development of African green monkeys (*Cercopithecus aethiops*). *American Journal of Primatology* 43: 43–50.

Sargent, R. C. 1996. Parental care. In J.G.J. Godin (ed.), *Behavioural Ecology of Teleost Fishes*, Oxford University Press, Oxford, pp. 292–315.

Savalli, U. M., and C. W. Fox. 1998a. Sexual selection and the fitness consequences of male body size in the seed beetle *Stator limbatus*. *Animal Behaviour* 55: 473–483.

Savalli, U. M., and C. W. Fox. 1998b. Genetic variation in paternal investment in a seed beetle. *Animal Behaviour* 56: 953–961.

Schleicher, B., H. Hoi, F. Valera, and M. Hoileitner. 1997. The importance of different paternity guards in the polygynandrous penduline tit (*Remiz pendulinus*). *Behaviour* 134: 941–959.

Schleicher, B., F. Valera, and H. Hoi. 1993. The conflict between nest guarding and mate guarding in penduline tits (*Remiz pendulinus*). *Ethology* 95: 157–165.

Schlupp, I., R. Mcknab, and M. J. Ryan. 2001. Sexual harassment as a cost for molly females: Bigger males cost less. *Behaviour* 138: 277–286.

Schluter, D. 2000. *The Ecology of Adaptive Radiation*. Oxford University Press, Oxford.

Schluter, D., and T. Price. 1993. Honesty, perception and population divergence in sexually selected traits. *Proceedings of the Royal Society of London Series B—Biological Sciences* 253: 117–122.

Schmid-Hempel, R., and P. Schmid-Hempel. 2000. Female mating frequencies in *Bombus* spp. from Central Europe. *Insectes Sociaux* 47: 36–41.

Schmidt, G. 1953. Eine deutsche Spinne, die Wirbeltiere frisst. *Orion* 8: 7–8.

Schmidt, G. 1957. Einige Notize über *Dolomedes fimbriatus* (Cl.). *Zoologische Anzeiger* 158: 83–97.

Schneider, J. M. 1999. Delayed oviposition: A female strategy to counter infanticide by males? *Behavioral Ecology* 10: 567–571.

Schneider, J. M., and M. A. Elgar. 2001. Sexual cannibalism and sperm competition in the golden orb-web spider *Nephila plumipes* (Araneoidea): Female and male perspectives. *Behavioral Ecology* 12: 547–552.

Schneider, J. M., and M. A. Elgar. 2002. Sexual cannibalism in *Nephila plumipes* as a consequence of female life history strategies. *Journal of Evolutionary Biology* 15: 84–91.

Schneider, J. M., and Y. Lubin. 1996. Infanticidal male eresid spiders. *Nature* 381: 655–656.

Schneider, J. M., and Y. Lubin. 1997. Infanticide by males in a spider with suicidal maternal care, *Stegodyphus lineatus* (Eresidae). *Animal Behaviour* 54: 305–312.

Schneider, J. M., and Y. Lubin. 1998. Intersexual conflict in spiders. *Oikos* 83: 496–506.

Schneider, J. M., M. L. Thomas, and M. A. Elgar. 2001. Ectomised conductors in the golden orb-web spider, *Nephila plumipes* (Araneoidea): A male adaptation to sexual conflict? *Behavioral Ecology and Sociobiology* 49: 410–415.

Schoech, S. J., R. L. Mumme, and J. C. Wingfield. 1996. Delayed breeding in the cooperatively breeding Florida scrub-jay (*Aphelocoma coerulescens*): Inhibition or the absence of stimulation? *Behavioral Ecology and Sociobiology* 39: 77–90.

Schoofs, L., D. Veelaert, J. Vandenbroeck, and A. Deloof. 1997. Peptides in the locusts, *Locusta migratoria* and *Schistocerca gregaria*. *Peptides* 18: 145–156.

Schrader, F. 1923. The sex ratio and oogenesis of *Pseudococcus citri*. *Zeitschrift für Induktive Abstammungs und Vererbungslehre* 30: 163–182.

Schradin, C., and J. Lamprecht. 2000. Female-biased immigration and male peacekeeping in groups of the shell-dwelling cichlid fish *Neolamprologus multifasciatus*. *Behavioral Ecology and Sociobiology* 48: 236–242.

Schroder, I. 1993. Concealed ovulation and clandestine copulation: A female contribution to human evolution. *Ethology and Sociobiology* 14: 381–389.

Schuh, R. T., and P. Stys. 1991. Phylogenetic analysis of cimicomorphan family relationships (Heteroptera). *Journal of the New York Entomological Society* 99: 298–350.

Searcy, W. A., D. Eriksson, and A. Lundberg. 1991. Deceptive behavior in pied flycatchers. *Behavioral Ecology and Sociobiology* 29: 167–175.

Seehausen, O. 2000. Explosive speciation rates and unusual species richness in haplochromine cichlid fishes: Effects of sexual selection. *Advances in Ecological Research* 31: 237–274.

Segerstråle, U.C.O. 2000. *Defenders of the Truth: The Battle for Science in the Sociobiology Debate and Beyond*. Oxford University Press, Oxford.

Sella, G., and M. C. Lorenzi. 2000. Partner fidelity and egg reciprocation in the simultaneously hermaphroditic polychaete worm *Ophryotrocha diadema*. *Behavioral Ecology* 11: 260–264.

Sella, G., M. C. Premoli, and F. Turri. 1997. Egg trading in the simultaneously hermaphroditic polychaete worm *Ophryotrocha gracilis* (Huth). *Behavioral Ecology* 8: 83–86.

Service, P. M. 1993. Laboratory evolution of longevity and reproductive fitness components in male fruit flies: Mating ability. *Evolution* 47: 387–399.

Sgro, C. M., T. Chapman, and L. Partridge. 1998. Sex-specific selection on time to remate in *Drosophila melanogaster*. *Animal Behaviour* 56: 1267–1278.

Shaanker, R. U., and K. N. Ganeshaiah. 1997. Conflict between parent and offspring in plants: Predictions, processes and evolutionary consequences. *Current Science* 72: 932–939.

Shalgi, R., and T. Raz. 1997. The role of carbohydrate residues in mammalian fertilization. *Histology and Histopathology* 12: 813–822.

Shaw, K. 1995. Phylogenetic tests of the sensory exploitation model of sexual selection. *Trends in Ecology & Evolution* 10: 117–120.

Sheldon, B. C. 2000. Differential allocation: Tests, mechanisms and implications. *Trends in Ecology & Evolution* 15: 397–402.

Sherman, P. W., and L. L. Wolfenbarger. 1995. Sensory biases and the evolution of sensory systems: Reply. *Trends in Ecology & Evolution* 10: 489.

Sherratt, T. N. 2001. The evolution of female-limited polymorphisms in damselflies: A signal detection model. *Ecology Letters* 4: 22–29.

Shields, K. M., J. T. Yamamoto, and J. R. Millam. 1989. Reproductive behavior and LH levels of cockatiels (*Nymphicus hollandicus*) associated with photostimulation, nest-box presentation, and degree of mate access. *Hormones and Behavior* 23: 68–82.

Shimizu, K. K., and K. Okada. 2000. Attractive and repulsive interactions between female and male gametophytes in Arabidopsis pollen tube guidance. *Development* 127: 4511–4518.

Shine, R. 1979. Sexual selection and sexual dimorphism in the Amphibia. *Copeia* 297–306.

Shine, R., T. Langkilde, and R. T. Mason. 2003. Cryptic forcible insemination: male snakes exploit female physiology, anatomy, and behavior to obtain coercive matings. *American Naturalist* 162: 653–667.

Shine, R., D. O'Connor, and R. T. Mason. 2000. Sexual conflict in the snake den. *Behavioral Ecology and Sociobiology* 48: 392–401.

Shirk, P. D., G. Bhaskaran, and H. Roller. 1980. The transfer of juvenile-hormone from male to female during mating in the cecropia silkmoth. *Experientia* 36: 682–683.

Shirk, P. D., G. Bhaskaran, and H. Roller. 1983. Developmental physiology of corpora allata and accessory sex glands in the cecropia silkmoth. *Journal of Experimental Zoology* 227: 69–79.

Shuker, D. M., and T. H. Day. 2001. The repeatability of a sexual conflict over mating. *Animal Behaviour* 61: 755–762.

Shuker, D. M., and T. H. Day. 2002. Mate sampling and the sexual conflict over mating in seaweed flies. *Behavioral Ecology* 13: 83–86.

Shuster, S. M., and M. J. Wade. 2003. *Mating Systems and Strategies*. Princeton University Press, Princeton, N.J.

Shykoff, J. A., and P. Schmid-Hempel. 1991. Parasites and the advantage of genetic-variability within social insect colonies. *Proceedings of the Royal Society of London Series B—Biological Sciences* 243: 55–58.

Sih, A., J. Krupa, and S. Travers. 1990. An experimental study on the effects of predation risk and feeding regime on the mating behavior of the water strider. *American Naturalist* 135: 284–290.

Sillén-Tullberg, B. 1981. Prolonged copulation—a male "post-copulatory" strategy in a promiscuous species, *Lygaeus equestris* (Heteroptera: Lygaeidae). *Behavioral Ecology and Sociobiology* 9: 283–289.

Siller, S. 2001. Sexual selection and the maintenance of sex. *Nature* 411: 689–692.

Simmons, L. W. 1990. Nuptial feeding in tettigoniids: Male costs and the rates of fecundity increase. *Behavioral Ecology and Sociobiology* 27: 43–47.

Simmons, L. W. 2001a. *Sperm Competition and Its Evolutionary Consequences in the Insects*. Princeton University Press, Princeton, N.J.

Simmons, L. W. 2001b. The evolution of polyandry: an examination of the genetic incompatibility and good-sperm hypotheses. *Journal of Evolutionary Biology* 14: 585–594.

Simmons, L. W. 2003. The evolution of polyandry: Patterns of genotypic variation in female mating frequency, male fertilization success and a test of the sexy-sperm hypothesis. *Journal of Evolutionary Biology* 16: 624–634.

Simmons, L. W., and D. T. Gwynne. 1991. The refractory period of female katydids (Orthoptera: Tettigoniidae): Sexual conflict over the remating interval. *Behavioral Ecology* 2: 276–282.

Simmons, L. W., and J. S. Kotiaho. 2002. Evolution of ejaculates: Patterns of phenotypic and genotypic variation and condition dependence in sperm competition traits. *Evolution* 56: 1622–1631.

Simmons, L. W., and G. A. Parker. 1989. Nuptial feeding in insects: Mating effort versus paternal investment. *Ethology* 81: 332–343.

Simmons, L. W., and G. A. Parker. 1992. Individual variation in sperm competition success of yellow dung flies, *Scatophaga stercoraria*. *Evolution* 46: 366–375.

Simmons, L. W., and M. T. Siva-Jothy. 1998. Sperma competition in insects: Mechanisms and the potential for selection. In T. R. Birkhead and A. P. Møller (eds.), *Sperm Competition and Sexual Selection*, Academic, London, pp. 341–434.

Simmons, L. W., P. Stockley, R. L. Jackson, and G. A. Parker. 1996. Sperm competition or sperm selection: No evidence for female influence over paternity in yellow dung flies *Scatophaga stercoraria*. *Behavioral Ecology and Sociobiology* 38: 199–206.

Singer, F., S. E. Riechert, H. F. Xu, A. W. Morris, E. Becker, J. A. Hale, and M. A. Noureddine. 2000. Analysis of courtship success in the funnel-web spider *Agelenopsis aperta*. *Behaviour* 137: 93–117.

Singh, T. 1998. Behavioural aspects of oviposition in the silkworm, *Bombyx mori*: A review. *Indian Journal of Sericulture* 37: 101–108.

Sirot, L. K., and H. J. Brockmann. 2001. Costs of sexual interactions to females in Rambur's forktail damselfly, *Ischnura ramburi* (Zygoptera: Coenagrionidae). *Animal Behaviour* 61: 415–424.

Sirot, L. K., H. J. Brockmann, C. Marinis, and G. Muschett. 2003. Maintenance of a female-limited polymorphism in *Ischnura ramburi* (Zygoptera: Coenagrionidae). *Animal Behaviour* 66: 763–775.

Siva-Jothy, M. T. 1987. Variation in copulation duration and the resultant degree of sperm removal in *Orthetrum cancellatum* (L) (Libellulidae, Odonata). *Behavioral Ecology and Sociobiology* 20: 147–151.

Siva-Jothy, M. T., and A. D. Stutt. 2003. A matter of taste: Direct detection of female mating status in the bedbug. *Proceedings of the Royal Society of London Series B— Biological Sciences* 270: 649–652.

Siva-Jothy, M. T., and Y. Tsubaki. 1989. Variation in copulation duration in *Mnais pruinosa pruinosa* Selys (Odonata: Caolpterygidae) I. Altenative mate securing tactics and sperm precedence. *Behavioral Ecology and Sociobiology* 24: 39–45.

Slabbekoorn, H., and T. B. Smith. 2002. Bird song, ecology and speciation. *Philosophical Transactions of the Royal Society of London Series B—Biological Sciences* 357: 493–503.

Slagsvold, T. 1993. Female-female aggression and monogamy in great tits *Parus major. Ornis Scandinavica* 24: 155–158.

Slagsvold, T., T. Amundsen, and S. Dale. 1994. Selection by sexual conflict for evenly spaced offspring in blue tits. *Nature* 370: 136–138.

Slagsvold, T., T. Amundsen, and S. Dale. 1995. Costs and benefits of hatching asynchrony in blue tits *Parus caeruleus. Journal of Animal Ecology* 64: 563–578.

Slagsvold, T., S. Dale, and H. M. Lampe. 1999. Does female aggression prevent polygyny? An experiment with pied flycatchers (Ficedula hypoleuca). *Behavioral Ecology and Sociobiology* 45: 403–410.

Slagsvold, T., and T. Drevon. 1999. Female pied flycatchers trade between male quality and mating status in mate choice. *Proceedings of the Royal Society of London Series B—Biological Sciences* 266: 917–921.

Slagsvold, T., and J. T. Lifjeld. 1989. Hatching asynchrony in birds: the hypothesis of sexual conflict over parental investment. *American Naturalist* 134: 239–253.

Slagsvold, T., and J. T. Lifjeld. 1994. Polygyny in birds: The role of competition between females for male parental care. *American Naturalist* 143: 59–94.

Slater, P.J.B., and T. R. Halliday (eds.). 1994. *Behaviour and Evolution.* Cambridge University Press, Cambridge.

Slooten, E., and D. M. Lambert. 1983. Evolutionary studies of the New Zealand coastal mosquito *Opifex fuscus* (Hutton). I. Mating behavior. *Behaviour* 84: 157–172.

Smid, H. M. 1997. Chemical mate guarding and oviposition stimulation in insects: A model mechanism alternative to the *Drosophila* sex-peptide paradigm. *Proceedings of the Koninklijke Nederlandse Akademie Van Wetenschappen* 100: 269–278.

Smid, H. M. 1998a. A peptide from the male accessory glands of the Colorado potato beetle. Ph.D. thesis, Agricultural University, Wageningen, Germany.

Smid, H. M. 1998b. Transfer of a male accessory gland peptide to the female during mating in *Leptinotarsa decemlineata. Invertebrate Reproduction & Development* 34: 47–53.

Smid, H. M., A. B. Koopmanschap, C.A.D. Dekort, and H. Schooneveld. 1997. A peptide from the male accessory gland in Leptinotarsa decemlineata: Purification, characterization and molecular cloning. *Journal of Insect Physiology* 43: 355–362.

Smid, H. M., and H. Schooneveld. 1992. Male accessory sex glands contain a new class of exocrine peptidergic cells in *Leptinotarsa decemlineata* (Say), identified with the neuropeptide-specific monoclonal antibody MAC-18. *Invertebrate Reproduction & Development* 21: 141–148.

Smith, C. C., and S. D. Fretwell. 1974. Optimal balance between size and number of offspring. *American Naturalist* 108: 499–506.

Smith, N.G.C. 1998. The dynamics of maternal-effect selfish genetic elements. *Journal of Theoretical Biology* 191: 173–180.

Smith, P. H., L. B. Brown, and A.C.M. Vangerwen. 1989. Causes and correlates of loss and recovery of sexual receptivity in *Lucilia cuprina* females after their first mating. *Journal of Insect Behavior* 2: 325–337.

Smith, P. H., C. Gillott, L. B. Browne, and A.C.M. Vangerwen. 1990. The mating-induced refractoriness of *Lucilia cuprina* females: Manipulating the male contribution. *Physiological Entomology* 15: 469–481.

Smuts, B. B., and R. W. Smuts. 1993. Male-aggression and sexual coercion of females in nonhuman-primates and other mammals—evidence and theoretical implications. *Advances in the Study of Behavior* 22: 1–63.

Snow, A. A. 1994. Postpollination selection and male fitness in plants. *American Naturalist* 144: S69–S83.

Soltis, J., R. Thomsen, K. Matsubayashi, and O. Takenaka. 2000. Infanticide by resident males and female counter-strategies in wild Japanese macaques (*Macaca fuscata*). *Behavioral Ecology and Sociobiology* 48: 195–202.

Sorci, G., A. P. Møller, and J. Clobert. 1998. Plumage dichromatism of birds predicts introduction success in New Zealand. *Journal of Animal Ecology* 67: 263–269.

Sparkes, T. C., D. P. Keogh, and T. H. Orsburn. 2002. Female resistance and mating outcomes in a stream-dwelling isopod: Effects of male energy reserves and mating history. *Behaviour* 139: 875–895.

Sparkes, T. C., D. P. Keogh, and R. A. Pary. 1996. Energetic costs of mate guarding behavior in male stream-dwelling isopods. *Oecologia* 106: 166–171.

Speicher, B. R. 1936. Oogenesis, fertilization and early cleavage in *Habrobracon*. *Journal of Morphology* 59: 401–421.

Spencer, J. L., M. P. Candolfi, J. E. Keller, and J. R. Miller. 1995. Onion fly, *Delia antiqua*, oviposition and mating as influenced by insect age and dosage of male reproductive tract extract (Diptera: anthomyiidae). *Journal of Insect Behavior* 8: 617–635.

Sperling, F.A.H. 1994. Sex-linked genes and species differences in Lepidoptera. *Canadian Entomologist* 126: 807–818.

Springer, S. 1960. Natural history of the sandbar shark, *Eulamia milberti*. *Fishery Bulletin of the Fish and Wildlife Service of the United States* 61: 1–38.

Stanley-Samuelson, D. W., R. A. Jurenka, G. J. Blomquist, and W. Loher. 1987. Sexual transfer of prostaglandin precursor in the field cricket, *Teleogryllus commodus*. *Physiological Entomology* 12: 347–354.

Stanley-Samuelson, D. W., and W. Loher. 1983. Arachidonic and other long-chain polyunsaturated fatty-acids in spermatophores and spermathecae of *Teleogryllus commodus*: Significance in prostaglandin-mediated reproductive behavior. *Journal of Insect Physiology* 29: 41–45.

Stanley-Samuelson, D. W., and W. Loher. 1985. The disappearance of injected prostaglandins from the circulation of adult female Australian field crickets, *Teleogryllus commodus*. *Archives of Insect Biochemistry and Physiology* 2: 367–374.

Stanley-Samuelson, D. W., and J. J. Peloquin. 1986. Egg-laying in response to prostaglandin injections in the Australian field cricket, *Teleogryllus commodus*. *Physiological Entomology* 11: 213–219.

Stanton, M. L. 1994. Male-male competition during pollination in plant populations. *American Naturalist* 144: S40–S68.

Stearns, S. C. 1992. *The Evolution of Life Histories*. Oxford University Press, Oxford.

Stephenson, A. G. 1981. Flower and fruit abortion: Proximate causes and ultimate functions. *Annual Review of Ecology and Systematics* 12: 253–279.

Sterck, E.H.M., D. P. Watts, and C. P. Vanschaik. 1997. The evolution of female social relationships in nonhuman primates. *Behavioral Ecology and Sociobiology* 41: 291–309.

Stjernholm, F., and B. Karlsson. 2000. Nuptial gifts and the use of body resources for reproduction in the green-veined white butterfly *Pieris napi*. *Proceedings of the Royal Society of London Series B—Biological Sciences* 267: 807–811.

Stockley, P. 1997. Sexual conflict resulting from adaptations to sperm competition. *Trends in Ecology & Evolution* 12: 154–159.

Stockley, P., M.J.G. Gage, G. A. Parker, and A. P. Moller. 1997. Sperm competition in fishes: The evolution of testis size and ejaculate characteristics. *American Naturalist* 149: 933–954.

Stockley, P., and L. W. Simmons. 1998. Consequences of sperm displacement for female dung flies, Scatophaga stercoraria. *Proceedings of the Royal Society of London Series B—Biological Sciences* 265: 1755–1760.

Stone, G. N. 1995. Female foraging responses to sexual harassment in the solitary bee *Anthophora plumipes*. *Animal Behaviour* 50: 405–412.

Stricker, S. A. 1999. Comparative biology of calcium signaling during fertilization and egg activation in animals. *Developmental Biology* 211: 157–176.

Strong, D. R. 1973. Amphipod amplexus, significance of ecotypic variation. *Ecology* 54: 1383–1388.

Stutt, A. D. 1999. Reproductive strategies and sexual conflict in the bed bug. Ph.D. thesis, University of Sheffield, England.

Stutt, A. D., and M. T. Siva-Jothy. 2001. Traumatic insemination and sexual conflict in the bed bug *Cimex lectularius*. *Proceedings of the National Academy of Sciences of the United States of America* 98: 5683–5687.

Styan, C. A. 1998. Polyspermy, egg size, and the fertilization kinetics of free-spawning marine invertebrates. *American Naturalist* 152: 290–297.

Sugawara, T. 1987. Cuticular lining in the genital chamber of the cricket: an obstacle to prostaglandin diffusing? *International Journal of Invertebrate Reproduction and Development* 12: 213–216.

Sundström, L., and J. J. Boomsma. 2001. Conflicts and alliances in insect families. *Heredity* 86: 515–521.

Surai, P., I. Kostjuk, G. Wishart, A. Macpherson, B. Speake, R. Noble, I. Ionov, and E. Kutz. 1998. Effect of vitamin E and selenium supplementation of cockerel diets on glutathione peroxidase activity and lipid peroxidation susceptibility in sperm, testes, and liver. *Biological Trace Element Research* 64: 119–132.

Sutherland, S. 1986. Patterns of fruit set: What controls fruit-flower ratios in plants. *Evolution* 40: 117–128.

Suzuki, H., M. Moriguch, R. Kida, and Y. Moro. 1996a. Delay in ovulation and fertilization and asynchronous pronuclear development in aged hamsters. *Journal of Reproduction and Development* 42: 15–22.

Suzuki, N., T. Okuda, and H. Shinbo. 1996b. Sperm precedence and sperm movement under different copulation intervals in the silkworm, *Bombyx mori*. *Journal of Insect Physiology* 42: 199–204.

Svärd, L., and J. N. Mcneil. 1994. Female benefit, male risk: Polyandry in the true armyworm *Pseudaletia unipuncta. Behavioral Ecology and Sociobiology* 35: 319–326.

Svärd, L., and C. Wiklund. 1988. Fecundity, egg weight and longevity in relation to multipel matings in females of the monarch butterfly. *Behavioral Ecology and Sociobiology* 23: 39–43.

Svärd, L., and C. Wiklund. 1989. Mass and production rate of ejaculates in relation to monandry polyandry in butterflies. *Behavioral Ecology and Sociobiology* 24: 395–402.

Svensson, B. G. 1997. Swarming behavior, sexual dimorphism, and female reproductive status in the sex role-reversed dance fly species *Rhamphomyia marginata. Journal of Insect Behavior* 10: 783–804.

Svensson, B. G., and E. Petersson. 1987. Sex role reversed courtship behavior, sexual dimorphism and nuptial gifts in the dance fly, *Empis borealis* (L). *Annales Zoologici Fennici* 24: 323–334.

Swaddle, J. P. 1996. Reproductive success and symmetry in zebra finches. *Animal Behaviour* 51: 203–210.

Swanson, W. J. 2003. Sex peptide and the sperm effect in *Drosophila melanogaster. Proceedings of the National Academy of Sciences of the United States of America* 100: 9643–9644.

Swanson, W. J., A. G. Clark, H. M. Waldrip-Dail, M. F. Wolfner, and C. F. Aquadro. 2001a. Evolutionary EST analysis identifies rapidly evolving male reproductive proteins in *Drosophila. Proceedings of the National Academy of Sciences of the United States of America* 98: 7375–7379.

Swanson, W. J., R. Nielsen, and Q. F. Yang. 2003. Pervasive adaptive evolution in mammalian fertilization proteins. *Molecular Biology and Evolution* 20: 18–20.

Swanson, W. J., and V. D. Vacquier. 2002. The rapid evolution of reproductive proteins. *Nature Reviews Genetics* 3: 137–144.

Swanson, W. J., Z. H. Zhang, M. F. Wolfner, and C. F. Aquadro. 2001b. Positive Darwinian selection drives the evolution of several female reproductive proteins in mammals. *Proceedings of the National Academy of Sciences of the United States of America* 98: 2509–2514.

Szecsi, P. B., and H. Lilja. 1993. Gastricsin-mediated proteolytic degradation of human seminal fluid proteins at PH levels found in the human vagina. *Journal of Andrology* 14: 351–358.

Székely, T., J. N. Webb, and I. C. Cuthill. 2000. Mating patterns, sexual selection and parental care: an integral approach. In M. Appolonio, M. Festa-Bianchet, and M. Mainardi (eds.), *Vertebrate Mating Systems*, World Scientific, Singapore, pp. 159–185.

Székely, T., J. N. Webb, A. I. Houston, and J. M. McNamara. 1996. An evolutionary aproach to offspring desertion in birds. In V. Nolan and E. Ketterson (eds.), *Current Ornithology*, vol. 13, Plenum, New York, pp. 271–330.

Székely, T., and T. D. Williams. 1994. Factors affecting timing of brood desertion by female Kentish plovers *Charadrius alexandrinus. Behaviour* 130: 17–28.

Tadler, A. 1993. Genitalia fitting, mating behavior and possible hybridization in millipedes of the genus *Craspedosoma* (Diplopoda, Chordeumatida, Craspedosomatidae). *Acta Zoologica* 74: 215–225.

Tallarovic, S. K., J. M. Melville, and P. H. Brownell. 2000. Courtship and mating in the giant hairy desert scorpion, *Hadrurus arizonensis* (Scorpionida, Iuridae). *Journal of Insect Behavior* 13: 827–838.

Tanaka, Y. 1964. *Sericology*, Central Silk Board, India.

Tanaka, Y. 1996. Sexual selection enhances population extinction in a changing environment. *Journal of Theoretical Biology* 180: 197–206.

Tatar, M., and D.E.L. Promislow. 1997. Fitness costs of female reproduction. *Evolution* 51: 1323–1326.

Taylor, E. H. 1949. Costa Rican frogs of the genera Centrolene and Centrolenalla. *University of Kansas Scientific Bulletin* 33: 257–270.

Taylor, P. D., and G. C. Williams. 1982. The lek paradox is not resolved. *Theoretical Population Biology* 22: 392–409.

Thibault, C. 1959. Analyse de la fécondation de l'oeuf de la Truie aprés accouplement ou insémination artficielle. *Annales de Zootechnie Supplement* 165–177.

Thiel, M., and I. A. Hinojosa. 2003. Mating behavior of female rock shrimp *Rhynchocinetes typus* (Decapoda: Caridea)—indication for convenience polyandry and cryptic female choice. *Behavioral Ecology and Sociobiology* 55: 113–121.

Thomas, D. W., and M. B. Fenton. 1979. Social behavior of the little brown bat, *Myotis lucifugus*. I. Mating behavior. *Behavioral Ecology and Sociobiology* 6: 129–136.

Thorne, M. H., and B. L. Sheldon. 1991. Cytological evidence of maternal meiotic errors in a line of chickens with a high incidence of triploidy. *Cytogenetics and Cell Genetics* 57: 206–210.

Thornhill, R. 1976. Sexual selection and paternal investment in insects. *American Naturalist* 110: 153–163.

Thornhill, R. 1980. Rape in *panorpa* scorpionflies and a general rape hypothesis. *Animal Behaviour* 28: 52–59.

Thornhill, R. 1983. Cryptic female choice and its implications in the scorpionfly *Harpobittacus nigriceps*. *American Naturalist* 122: 765–788.

Thornhill, R. 1984. Alternative hypotheses for traits believed to have evolved by sperm competition. In R. L. Smith (ed.), *Sperm Competition and the Evolution of Animal Mating Systems*, Academic, Orlando, Fla.

Thornhill, R. 1990. The study of adaptation. In M. Bekoff and D. Jamieson (eds.), *Interpretation and Explanation in the Study of Behavior*, vol. II, Westview, Boulder, Colo., pp. 31–62.

Thornhill, R. 1991. The notal organ of the scorpionfly (*Panorpa vulgaris*): An adaptation to coerce mating duration. *Behavioral Ecology* 2: 156–164.

Thornhill, R., and J. Alcock. 1983. *The Evolution of Insect Mating Systems*, Harvard University Press, Cambridge, Mass.

Thornhill, R., and K. P. Sauer. 1991. The notal organ of the scorpionfly (*Panorpa vulgaris*): An adaptation to coerce mating duration. *Behavioral Ecology* 2: 156–164.

Tidemann, C. R. 1982. Sex differences in seasonal changes of brown adipose tissue and activity of the Australian vespertilionid bat *Eptesicus vulturnus*. *Australian Journal of Zoology* 30: 15–22.

Tilghman, S. M. 1999. The sins of the fathers and mothers: Genomic imprinting in mammalian development. *Cell* 96: 185–193.

Tobe, S. S., and W. Loher. 1983. Propeties of the prostaglandin synthetase complex in the cricket *Teleogryllus commodus*. *Insect Biochemistry* 13: 137–141.

Todte, I. 1994. Population dynamics, breeding biology and mating system of penduline tits *Remiz pendulinus* in Sachsen-Anhalt, eastern Germany. *Vogelwelt* 115: 299–308.

Tompa, A. 1982. X-ray radiographic examination of dart formation in *Helix aspersa*. *Netherlands Journal of Zoology* 32: 63–71.

Torgerson, D. G., R. J. Kulathinal, and R. S. Singh. 2002. Mammalian sperm proteins are rapidly evolving: Evidence of positive selection in functionally diverse genes. *Molecular Biology and Evolution* 19: 1973–1980.

Torres-Vila, L. M., J. Stockel, and M. C. Rodriguez-Molina. 1997. Physiological factors regulating polyandry in *Lobesia botrana* (Lepidoptera: Tortricidae). *Physiological Entomology* 22: 387–393.

Tregenza, T., R. K. Butlin, and N. Wedell. 2000. Evolutionary biology—sexual conflict and speciation. *Nature* 407: 149–150.

Tregenza, T., and N. Wedell. 1998. Benefits of multiple mates in the cricket *Gryllus bimaculatus*. *Evolution* 52: 1726–1730.

Tregenza, T., and N. Wedell. 2000. Genetic compatibility, mate choice and patterns of parentage: Invited review. *Molecular Ecology* 9: 1013–1027.

Tregenza, T., and N. Wedell. 2002. Polyandrous females avoid costs of inbreeding. *Nature* 415: 71–73.

Tricas, T. C., and E. M. Lefeuvre. 1985. Mating in the reef white-tip shark *Triaenodon obesus*. *Marine Biology* 84: 233–237.

Trivers, R. L. 1972. Parental investment and sexual selection. In B. Campbell (ed.), *Sexual Selection and the Descent of Man*, Aladine, Chicago, pp. 136–179.

Trivers, R. L. 1974. Parent-offspring conflict. *American Zoologist* 14: 249–264.

Trivers, R. L., and H. Hare. 1976. Haplodiploidy and evolution of social insects. *Science* 191: 249–263.

Trowbridge, C. D. 1995. Hypodermic insemination, oviposition, and embyonic development of a pool-dwelling ascoglossan (*Equals sacoglossan*) opisthobranch—*Ercolania felina* (Hutton, 1882) on New Zealand shores. *Veliger* 38: 203–211.

Truckenbrodt, W. 1964. Zytologische und entwicklungsphysiologische Untersuchungen am besamten und am partenogenetischen Ei von *Kalotermes flavicollis*. *Zoologische Jahrbücher. Abteilung für Anatomie und Ontogenie der Tiere* 81: 359–434.

Trumbo, S. T. 1992. Monogamy to communal breeding: Exploitation of a broad resource base by burying beetles (*Nicrophorus*). *Ecological Entomology* 17: 289–298.

Trumbo, S. T., and A. K. Eggert. 1994. Beyond monogamy: Territory quality influences sexual advertisement in male burying beetles. *Animal Behaviour* 48: 1043–1047.

Tsaur, S. C., C. T. Ting, and C. I. Wu. 1998. Positive selection driving the evolution of a gene of male reproduction, Acp26Aa, of *Drosophila*: II. Divergence versus polymorphism. *Molecular Biology and Evolution* 15: 1040–1046.

Tsaur, S. C., C. T. Ting, and C. I. Wu. 2001. Sex in *Drosophila mauritiana*: A very high level of amino acid polymorphism in a male reproductive protein gene, Acp26Aa. *Molecular Biology and Evolution* 18: 22–26.

Tsaur, S. C., and C. I. Wu. 1997. Positive selection and the molecular evolution of a gene of male reproduction, Acp26Aa of *Drosophila*. *Molecular Biology and Evolution* 14: 544–549.

Tsubaki, Y., M. T. Siva-Jothy, and T. Ono. 1994. Re-copulation and postcopulatory mate guarding increase immediate female reproductive output in the dragonfly *Nannophya pygmaea* Rambur. *Behavioral Ecology and Sociobiology* 35: 219–225.

Turelli, M., N. H. Barton, and J. A. Coyne. 2001. Theory and speciation. *Trends in Ecology & Evolution* 16: 330–343.

Turner, G. F. 1994. Speciation mechanisma in Lake Malawi cichlids: A critical review. In K. Martens, B. Godderis, and G. Coulter (eds.), *Speciation in Ancient Lakes*, Schweitzerbart, Stuttgart, pp. 139–160.

Turner, J.R.G. 1987. The evolutionary dynamics of Batesian and Müllerian mimicry: Similarities and differences. *Ecological Entomology* 12: 81–95.

Usinger, R. L. 1966. *Monograph of Cimicidae (Hemiptera, Heteroptera)*. Entomological Society of America, College Park, Md.

Uyenoyama, M. K., K. E. Holsinger, and D. M. Waller. 1993. Ecological and genetic factors directing the evolution of self-fertilization. *Oxford Surveys in Evolutionary Biology* 9: 327–381.

Vacquier, V. D. 1998. Evolution of gamete recognition proteins. *Science* 281: 1995–1998.

Vacquier, V. D., K. R. Carner, and C. D. Stout. 1990. Species-specific sequences of abalone lysin, the sperm protein that creates a hole in the egg envelope. *Proceedings of the National Academy of Sciences of the United States of America* 87: 5792–5796.

Vahed, K. 1998. The function of nuptial feeding in insects: Review of empirical studies. *Biological Reviews of the Cambridge Philosophical Society* 73: 43–78.

Vahed, K. 2003. Increases in egg production in multiply mated female bushcrickets *Leptophyes punctatissima* are not due to substances in the nuptial gift. *Ecological Entomology* 28: 124–128.

Valera, F., H. Hoi, and B. Schleicher. 1997. Egg burial in penduline tits, *Remiz pendulinus*: Its role in mate desertion and female polyandry. *Behavioral Ecology* 8: 20–27.

Vamosi, J. C., and S. P. Otto. 2002. When looks can kill: The evolution of sexually dimorphic floral display and the extinction of dioecious plants. *Proceedings of the Royal Society of London Series B—Biological Sciences* 269: 1187–1194.

Van Den Berg, M. J., G. Thomas, H. Hendriks, and W. Van Delden. 1984. A reexamination of the negative assortative mating phnomenon and its underlying mechanism in *Drosophila melanogaster*. *Behavior Genetics* 14: 45–61.

Van Duivenboden, Y. A., A. W. Pieneman, and A. Termaat. 1985. Multiple mating suppresses fecundity in the hermaphrodite fresh-water snail *Lymnaea stagnalis*: A laboratory study. *Animal Behaviour* 33: 1184–1191.

Van Gossum, H., R. Stoks, and L. De Bruyn. 2001. Reversible frequency-dependent switches in male mate choice. *Proceedings of the Royal Society of London Series B—Biological Sciences* 268: 83–85.

Van Gossum, H., R. Stoks, E. Matthysen, F. Valck, and L. De Bruyn. 1999. Male choice for female colour morphs in *Ischnura elegans* (Odonata, Coenagrionidae): Testing the hypotheses. *Animal Behaviour* 57: 1229–1232.

van Schaik, C., and C. H. Janson. 2000. *Infanticide by Males and Its Implications*. Cambridge University Press, Cambridge.

van Valen, L. 1973. A new evolutionary law. *Evolutionary Theory* 1: 1–30.

Veiga, J. P. 1990. Sexual conflict in the house sparrow: Interference between polysynously mated females versus asymmetric male investement. *Behavioral Ecology and Sociobiology* 27: 345–350.

Veiga, J. P. 1992. Why are house sparrows predominantly monogamous? A test of hypotheses. *Animal Behaviour* 43: 361–370.

Vepsalainen, K., and R. Savolainen. 1995. Operational sex ratios and mating conflict between the sexes in the water strider *Gerris lacustris. American Naturalist* 146: 869–880.

Verner, J. 1964. Evolution of polygamy in long-billed marsh wren. *Evolution* 18: 252–261.

Verner, J., and M. F. Willson. 1966. The influence of habitats on mating systems of North American passerine birds. *Ecology* 47: 143–147.

Vick, K. W. 1973. Effects of interspecific matings of *Trogoderma glabrum* and *T. inclusum* on oviposition and remating. *Annals of the Entomological Society of America* 66: 237–239.

Vieira, C., E. G. Pasyukova, Z. B. Zeng, J. B. Hackett, R. F. Lyman, and T.F.C. Mackay. 2000. Genotype-environment interaction for quantitative trait loci affecting life span in *Drosophila melanogaster. Genetics* 154: 213–227.

Viriyapanich, P., and J. M. Bedford. 1981. Sperm capacitation in the fallopian tube of the hamster and its suppression by endocrine factors. *Journal of Experimental Zoology* 217: 403–407.

Von Helversen, D., and O. Von Helversen. 1991. Pre-mating sperm removal in the bush-cricket *Metaplastes ornatus* Ramme 1931 (Orthoptera: Tettigonoidea: Phaneropteridae). *Behavioral Ecology and Sociobiology* 28: 391–396.

Vreys, C., and N. K. Michiels. 1998. Sperm trading by volume in a hermaphroditic flatworm with mutual penis intromission. *Animal Behaviour* 56: 777–785.

Vueille, M. 1980. Sexual behaviour and evolution of sexual dimorphism in body size of Jaera (*Isopoda asellota*). *Biological Journal of the Linnean Society* 13: 89–100.

Wachtmeister, C. A. 2001. Display in monogamous pairs: A review of empirical data and evolutionary explanations. *Animal Behaviour* 61: 861–868.

Wachtmeister, C. A., and M. Enquist. 2000. The evolution of courtship rituals in monogamous species. *Behavioral Ecology* 11: 405–410.

Wade, M. J., and S. J. Arnold. 1980. The intensity of sexual selection in relation to male sexual behavior female choice and sperm precedence. *Animal Behaviour* 28: 446–461.

Wade, M. J., and R. W. Beeman. 1994. The population dynamics of maternal effect selfish genes. *Genetics* 138: 1309–1314.

Wade, M. J., and N. W. Chang. 1995. Increased male fertility in *Tribolium confusm* beetles after infection with the intracellular parasite *Wolbachia. Nature* 373: 72–74.

Wade, M. J., and S. M. Shuster. 2002. The evolution of parental care in the context of sexual selection: A critical reassessment of parental investment theory. *American Naturalist* 160: 285–292.

Wagner, W. E. 1998. Measuring female mating preferences. *Animal Behaviour* 55: 1029–1042.

Walker, W. F. 1980. Sperm utilization strategies in nonsocial insects. *American Naturalist* 115: 780–799.

Walpole, D. E. 1988. Cross-mating studies between two species of bedbugs (Hemiptera: Cimicidae) with a description of a marker of inter-specific mating. *South African Journal of Science* 84: 215–216.

Walsh, N. E., and D. Charlesworth. 1992. Evolutionary interpretations of differences in pollen-tube growth-rates. *Quarterly Review of Biology* 67: 19–37.

Walter, B., and F. Trillmich. 1994. Female aggression and male peace-keeping in a cichlid fish harem: Conflict between and within the sexes in *Lamprologus ocellatus*. *Behavioral Ecology and Sociobiology* 34: 105–112.

Wang, P. J., J. R. McCarrey, F. Yang, and D. C. Page. 2001. An abundance of X-linked genes expressed in spermatogonia. *Nature Genetics* 27: 422–426.

Wang, Q., and J. G. Millar. 1997. Reproductive behavior of *Thyanta pallidovirens* (Heteroptera: Pentatomidae). *Annals of the Entomological Society of America* 90: 380–388.

Wang, Z., and T. R. Insel. 1996. Parental behaviour in voles. In J. S. Rosenblatt and C. T. Snowdon (eds.), *Parental Care, Evolution Mechanisma, and Adaptive Significance*. Advances in the Study of Behavior, vol. 25, Academic, London, pp. 361–384.

Ward, P. I. 1984. The effects of size on the mating decisions of *Gammarus pulex* (Crustacea: Amphipoda). *Zeitschrift Fur Tierpsychologie—Journal of Comparative Ethology* 64: 174–184.

Ward, P. I. 1986. A comparative field sudy of the breeding behavior of a stream and a pond population of *Gammarus pulex* (Amphipoda). *Oikos* 46: 29–36.

Ward, P. I. 1993. Females influence sperm storage and use in the yellow dung fly *Scathophaga stercoraria* (L). *Behavioral Ecology and Sociobiology* 32: 313–319.

Ward, P. I. 1998. Intraspecific variation in sperm size characters. *Heredity* 80: 655–659.

Ward, P. I., J. Hemmi, and T. Roosli. 1992. Sexual conflict in the dung fly *Sepsis cynipsea*. *Functional Ecology* 6: 649–653.

Ward, S., and J. S. Carrel. 1979. Fertilization and sperm competition in the nematode *Coenorhabditis elegans*. *Developmental Biology* 73: 304–321.

Warner, R. R., D. Y. Shapiro, A. Marcanato, and C. W. Petersen. 1995. Sexual conflict: Males with honest mating success convey the lowest fertilization benefits to females. *Proceedings of the Royal Society of London Series B—Biological Sciences* 262: 135–139.

Watanabe, M. 1988. Multiple matings increase the fecundity of the yellow swallowtail butterfly, *Papilio xuthus* L., in summer generations. *Journal of Insect Behavior* 1: 17–27.

Watson, P. J. 1986. Transmission of a female sex-pheromone thwarted by males in the spider *Linyphia litigiosa* (Linyphiidae). *Science* 233: 219–221.

Watson, P. J. 1991. Multiple paternity as genetic bet-hedging in female Sierra dome spiders, *Lynyphia litigiosa* (Linyphiidae). *Animal Behaviour* 41: 343–360.

Watson, P. J. 1993. Foraging advantage of polyandry for female sierra dome spiders (*Linyphia litigiosa* Linyphiidae) and assessment of alternative direct benefit hypotheses. *American Naturalist* 141: 440–465.

Watson, P. J. 1998. Multi-male mating and female choice increase offspring growth in the spider *Neriene litigiosa* (Linyphiidae). *Animal Behaviour* 55: 387–403.

Watson, P. J., G. Arnqvist, and R. R. Stallmann. 1998. Sexual conflict and the energetic costs of mating and mate choice in water striders. *American Naturalist* 151: 46–58.

Weatherhead, P. J., and R. J. Robertson. 1979a. Offspring quality and the polygyny threshold: "Sexy son" hypothesis. *American Naturalist* 113: 201–208.

Weatherhead, P. J., and R. J. Robertson. 1979b. Offspring quality and the polygyny threshold: The "sexy son" hypothesis. *American Naturalist* 113: 201–208.

Webb, J. N., A. I. Houston, J. M. McNamara, and T. Székely. 1999. Multiple patterns of parental care. *Animal Behaviour* 58: 983–993.

Wedell, N. 1991. Sperm competition selects for nuptial feeding in a bush-cricket. *Evolution* 45: 1975–1978.

Wedell, N. 1993. Spermatophore size in bush-crickets: Comparative evidence for nuptial gifts as a sperm protection device. *Evolution* 47: 1203–1212.

Wedell, N. 1996. Mate quality affects reproductive effort in a paternally investing species. *American Naturalist* 148: 1075–1088.

Wedell, N. 1998. Sperm protection and mate assessment in the bushcricket *Coptaspis* sp. 2. *Animal Behaviour* 56: 357–363.

Wedell, N. 2001. Female remating in butterflies: interaction between female genotype and nonfertile sperm. *Journal of Evolutionary Biology* 14: 746–754.

Wedell, N., and B. Karlsson. 2003. Paternal investment directly affects female reproductive effort in an insect. *Proceedings of the Royal Society of London Series B—Biological Sciences* 270: 2065–2071.

Wedell, N., and T. Tregenza. 1999. Successful fathers sire successful sons. *Evolution* 53: 620–625.

Weigensberg, I., and D. J. Fairbairn. 1994. Conflicts of interest between the sexes: A study of mating interactions in a semiaquatic bug. *Animal Behaviour* 48: 893–901.

Weigensberg, I., and D. J. Fairbairn. 1996. The sexual arms race and phenotypic correlates of mating success in the waterstrider, *Aquarius remigis* (Hemiptera: Gerridae). *Journal of Insect Behavior* 9: 307–319.

Weitzman, S. H. 1975. Der Flügelschuppensalmler, *Pternobrycon myrnae*, ein beqaubernder Aqurienfisch der Zukunft aus Costa Rica. *Die Aquarien- un Terrarien-Zeitschrift* 28: 407–411.

Weitzman, S. H., and S. V. Fink. 1985. Xenurobryconin phylogeny and putative pheromone pumps in glandulocaudin fishes (Teleosteti: Characidae). *Smithsonian Contributions to Zoology* 421: 1–121.

Weitzman, S. H., and N. A. Menezes. 1998. Relationships of the tribes and genera of the Glandulocaudiae (Ostariophysi: Characidae) with a description of a new genus, *Chrysobrycon*. In L. R. Malabarba, R. E. Reis, R. P. Vari, Z. M. Lucena, and C. A. S. Lucena (eds.), *Phylogeny and Classification of Neotropical Fishes Part 2—Characiformes*, EDIPUCRS, Porto Alegre, Brazil, pp. 171–192.

Wells, M. M., and C. S. Henry. 1998. Songs, reproductive isolation, and speciation in cryptic species of insects: A case study using green lacewings. In D. J. Howard and S. H. Berlocher (eds.), *Endless Forms: Species and Speciation*, Oxford University Press, New York, pp. 217–233.

Werren, J. H., and L. W. Beukeboom. 1998. Sex determination, sex ratios, and genetic conflict. *Annual Review of Ecology and Systematics* 29: 233–261.

Westcott, D. A. 1994. Leks of leks: A role for hotspots in lek evolution. *Proceedings of the Royal Society of London Series B—Biological Sciences* 258: 281–286.

West-Eberhard, M. J. 1979. Sexual selection, social competition, and evolution. *Proceedings of the American Philosophical Society* 123: 222–234.

West-Eberhard, M. J. 1983. Sexual selection, social competition, and speciation. *Quarterly Review of Biology* 58: 155–183.

West-Eberhard, M. J. 1984. Sexual selection, competitive communication and species-specific signals in insects. In T. Lewis (ed.), *Insect Communication*, Academic, New York, pp. 283–342.

West-Eberhard, M. J., J. W. Bradbury, N. B. Davies, P. H. Gouyon, P. Hamerstein, B. König, G. A. Parker, D. C. Queller, N. Sachser, T. Slagsvold, F. Trillmich, and C. Vogel. 1987. Conflicts betweeen and within the sexes in sexual selection. In J. W. Bradbury and M. B. Andersson (eds.), *Sexual Selection: Testing the Alternatives*, Wiley, Chichester, pp. 181–195.

Westlake, K. P., and L. Rowe. 1999. Developmental costs of male sexual traits in the water strider *Rheumatobates rileyi*. *Canadian Journal of Zoology* 77: 917–922.

Westlake, K. P., L. Rowe, and D. C. Currie. 2000. Phylogeny of the water strider genus *Rheumatobates* (Heteroptera: Gerridae). *Systematic Entomology* 25: 125–145.

Westneat, D. F., and R. C. Sargent. 1996. Sex and parenting: The effects of sexual conflict and parentage on parental strategies. *Trends in Ecology & Evolution* 11: A87–A91.

Westneat, D. F., and I. R. K. Stewart. 2003. Extra-pair paternity in birds: Causes, correlates, and conflict. *Annual Review of Ecology Evolution and Systematics* 34: 365–396.

Wethington, A. R., and R. T. Dillon. 1996. Gender choice and gender conflict in a non-reciprocally mating simultaneous hermaphrodite, the freshwater snail, *Physa*. *Animal Behaviour* 51: 1107–1118.

Wheeler, M. J., V. E. Franklin-Tong, and F. C. H. Franklin. 2001. The molecular and genetic basis of pollen-pistil interactions. *New Phytologist* 151: 565–584.

Whitfield, P. 1989. Sexual conflict in the red-necked phalarope. *Bliki* 7: 59–62.

Whitlock, M. C. 2000. Fixation of new alleles and the extinction of small populations: Drift load, beneficial alleles, and sexual selection. *Evolution* 54: 1855–1861.

Wickler, W. 1968. *Mimicry in Plants and Animals*. McGraw-Hill, New York.

Wickler, W. 1985. Stepfathers in insects and their pseudo-parental investment. *Zeitschrift Fur Tierpsychologie—Journal of Comparative Ethology* 69: 72–78.

Wickler, W. 1994. On nuptial gifts and paternity. *Ethology* 98: 165–170.

Wickman, P. O. 1986. Courtship solicitation by females of the small heath butterfly, *Coenonympha pamphilus* (L) (Lepidoptera: Satyridae) and their behavior in relation to male territories before and after copulation. *Animal Behaviour* 34: 153–157.

Wiens, B. L., and P. H. Brownell. 1994. Neuroendocrine control of egg-laying behavior in the nudibranch, *Archidoris montereyensis*. *Journal of Comparative Neurology* 344: 619–625.

Wiklund, C., and J. Forsberg. 1991. Sexual size dimorphism in relation to female polygay and protandry in butterflies: A comparative study of Swedish Pieridae and Satyridae. *Oikos* 60: 373–381.

Wiklund, C., and A. Kaitala. 1995. Sexual selection for large male size in a polyandrous butterfly: The effect of body size on male versus female reproductive success in *Pieris napi*. *Behavioral Ecology* 6: 6–13.

Wiklund, C., A. Kaitala, V. Lindfors, and J. Abenius. 1993. Polyandry and its effect on female reproduction in the green-veined white butterfly (*Pieris napi* L.). *Behavioral Ecology and Sociobiology* 33: 25–33.

Wiklund, C., B. Karlsson, and O. Leimar. 2001. Sexual conflict and cooperation in butterfly reproduction: A comparative study of polyandry and female fitness. *Proceedings of the Royal Society of London Series B—Biological Sciences* 268: 1661–1667.

Wilcock, C., and R. Neiland. 2002. Pollination failure in plants: Why it happens and when it matters. *Trends in Plant Science* 7: 270–277.

Wilcox, R. S. 1984. Male copulatory guarding enhances female foraging in a water strider. *Behavioral Ecology and Sociobiology* 15: 171–174.

Wiley, R. H., and J. Poston. 1996. Perspective: Indirect mate choice, competition for mates, and coevolution of the sexes. *Evolution* 50: 1371–1381.

Wilkins, J. F., and D. Haig. 2001. Genomic imprinting of two antagonistic loci. *Proceedings of the Royal Society of London Series B—Biological Sciences* 268: 1861–1867.

Willard, H. F. 2003. Genome biology—tales of the Y chromosome. *Nature* 423: 810–813.

Willey, M. B., and F. A. Coyle. 1992. Female spiders (Araneae, Dipluridae, Desidae, Linyphiidae) eat their own eggs. *Journal of Arachnology* 20: 151–152.

Williams, G. C. 1966. *Adaptation and Natural Selection; A Critique of Some Current Evolutionary Thought.* Princeton University Press, Princeton, N.J.

Willson, M. F. 1994. Sexual selection in plants: Perspective and overview. *American Naturalist* 144: S13–S39.

Willson, M. F., and N. Burley. 1983. *Mate Choice in Plants: Tactics, Mechanisms, and Consequences.* Princeton University Press, Princeton, N.J.

Wilson, N., S. C. Tubman, P. E. Eady, and G. W. Robertson. 1997. Female genotype affects male success in sperm competition. *Proceedings of the Royal Society of London Series B—Biological Sciences* 264: 1491–1495.

Wilson, N., T. J. Tufton, and P. E. Eady. 1999. The effect of single, double, and triple matings on the lifetime fecundity of *Callosobruchus analis* and *Callosobruchus maculatus* (Coleoptera: Bruchidae). *Journal of Insect Behavior* 12: 295–306.

Wing, S., J. E. Lloyd, and T. Hongtrakul. 1983. Male competition in *Pteroptyx* fireflies: Wing-cover clamps, female anatomy, and mating plugs. *Florida Entomologist* 66: 86–91.

Wing, S. R. 1988. Cost of mating for female insects: Risk of predation in *Photinus collustrans* (Coeoptera: Lampyridae). *American Naturalist* 131: 139–142.

Wingfeld, J. C. 1993. Endocrinology of reproduction in wild species. In D. S. Farner, J. R. King and K. C. Parks (eds.), *Avian Biology,* vol. IX, Academic, New York, pp. 163–327.

Wingfeld, J. C. 1994. Communication in vertebrate aggression and reproduction: the role of hormones. In E. Knobil and J. D. Neill (eds.), *The Physiology of Reproduction,* 2nd Ed., Raven, New York, pp. 303–342.

Wingfield, J. C., E. Ronchi, A. R. Goldsmith, and C. Marler. 1989. Interactions of sex sterioid hormone and prolactin in male and female song sparrows, *Melospiz mododia. Physiological Zoology* 62: 11–24.

Wisenden, B. D. 1999. Alloparental care in fishes. *Reviews in Fish Biology and Fisheries* 9: 45–70.

Witte, K., and E. Curio. 1999. Sexes of a monomorphic species differ in preference for mates with a novel trait. *Behavioral Ecology* 10: 15–21.

Wolf, J. B., A. J. Moore, and E. D. Brodie. 1997. The evolution of indicator traits for parental quality: The role of maternal and paternal effects. *American Naturalist* 150: 639–649.

Wolf, L. L., E. C. Waltz, K. Wakeley, and D. Klockowski. 1989. Copulation duration and sperm competition in white-faced dragonflies (*Leucorrhinia intacta* Odonoata: Libellulidae). *Behavioral Ecology and Sociobiology* 24: 63–68.

Wolff, J. O., and D. W. Macdonald. 2004. Promiscuous females protect their offspring. *Trends in Ecology & Evolution* 19: 127–134.

Wolfner, M. F. 1997. Tokens of love: Functions and regulation of *Drosophila* male accessory gland products. *Insect Biochemistry and Molecular Biology* 27: 179–192.

Wolfner, M. F. 2002. The gifts that keep on giving: physiological functions and evolutionary dynamics of male seminal proteins in *Drosophila*. *Heredity* 88: 85–93.

Wu, C. I. 1985. A stochastic simulation study on speciation by sexual selection. *Evolution* 39: 66–82.

Xu, X., J. Ding, P. C. Seth, D. S. Harbison, and G. R. Foxcroft. 1996a. In vitro fertilization of in vitro matured pig oocytes: Effects of boar and ejaculate fraction. *Theriogenology* 45: 745–755.

Xu, X., P. C. Seth, D. S. Harbison, A. P. Cheung, and G. R. Foxcroft. 1996b. Semen dilution for assessment of boar ejaculate quality in pig RVM and IVF systems. *Theriogenology* 46: 1325–1337.

Yamaja-Setty, B. N., and T. R. Ramaiah. 1980. Effect of prostaglandins and inhibitors of prostaglandin biosynthesis on oviposition in the silkworm moth, *Bombyx mori*. *Indian Journal of Experimental Biology* 18: 539–541.

Yamamura, N. 1986. An evolutionary stable strategy (ESS) model of postcopulatory guarding in insects. *Theoretical Population Biology* 29: 438–455.

Yamamura, N., and M. Higashi. 1992. An evolutionary theory of conflict-resolution between relatives: Altruism, manipulation, compromise. *Evolution* 46: 1236–1239.

Yamaoka, K., and T. Hirao. 1977. Stimulation of virginal oviposition by male factor and its effect on spontaneous nervous activity in *Bombyx mori*. *Journal of Insect Physiology* 23: 57–63.

Yano, K., F. Sato, and T. Takahashi. 1999. Observations of mating behavior of the manta ray, *Manta birostris*, at the Ogasawara Islands, Japan. *Ichthyological Research* 46: 289–296.

Yasui, Y. 1998. The "genetic benefits" of female multiple mating reconsidered. *Trends in Ecology & Evolution* 13: 246–250.

Yekta, N., and D. G. Blackburn. 1992. Sexual dimorphism in mass and protein content of the forelimb muscles of the northern leopard frog, *Rana pipiens*. *Canadian Journal of Zoology* 70: 670–674.

Yi, S. X., and C. Gillott. 1999. Purification and characterization of an oviposition-stimulating protein of the long hyaline tubules in the male migratory grasshopper, *Melanoplus sanguinipes*. *Journal of Insect Physiology* 45: 143–150.

Ying, Y., P. H. Chow, and W. S. O. 1998. Effects of male accessory sex glands on deoxyribonucleic acid synthesis in the first cell cycle of golden hamster embryos. *Biology of Reproduction* 58: 659–663.

Yokomine, T., A. Kuroiwa, K. Tanaka, M. Tsudzuki, Y. Matsuda, and H. Sasaki. 2001. Sequence polymorphisms, allelic expression status and chromosome locations of the chicken IGF2 and MPR1 genes. *Cytogenetics and Cell Genetics* 93: 109–113.

Young, A.D.M., and A.E.R. Downe. 1987. Male accessory gland substances and the control of sexual receptivity in female *Culex tarsalis*. *Physiological Entomology* 12: 233–239.

Yun, K. 1998. Genomic imprinting and carcinogenesis. *Histology and Histopathology* 13: 425–435.

Yusuf, R. Z., and R. Naeem. 2001. Cytogenetic studies of spontaneous miscarriages: a seven year study to compare significance of primary vs. secondary culture methods for assessment of fetal karyotype yield and maternal cell contamination. *Early Pregnancy: Biology and Medicine* 5: 121–131.

Zahavi, A. 1975. Mate selection: Selection for a handicap. *Journal of Theoretical Biology* 53: 205–214.

Zahavi, A. 1977. Cost of honesty: (Further remarks on handicap principle). *Journal of Theoretical Biology* 67: 603–605.

Zaragoza, M. V., T. Surti, R. W. Redline, E. Millie, A. Chakravarti, and T. J. Hassold. 2000. Parental origin and phenotype of triploidy in spontaneous abortions: Predominance of diandry and association with the partial hydatidiform mole. *American Journal of Human Genetics* 66: 1807–1820.

Zeh, D. W., and R. L. Smith. 1985. Paternal investment by terrestrial arthropods. *American Zoologist* 25: 785–805.

Zeh, D. W., and J. A. Zeh. 1988. Condition-dependent sex ornaments and field tests of sexual selection theory. *American Naturalist* 132: 454–459.

Zeh, J. A., and D. W. Zeh. 1996. The evolution of polyandry I: Intragenomic conflict and genetic incompatibility. *Proceedings of the Royal Society of London Series B—Biological Sciences* 263: 1711–1717.

Zeh, J. A., and D. W. Zeh. 1997. The evolution of polyandry II: Post-copulatory defences against genetic incompatibility. *Proceedings of the Royal Society of London Series B—Biological Sciences* 264: 69–75.

Zeh, J. A., and D. W. Zeh. 2001. Reproductive mode and the genetic benefits of polyandry. *Animal Behaviour* 61: 1051–1063.

Zeh, J. A., and D. W. Zeh. 2003. Toward a new sexual selection paradigm: Polyandry, conflict and incompatibility (invited article). *Ethology* 109: 929–950.

Zeiss, C., A. Martens, and J. Rolff. 1999. Male mate guarding increases females' predation risk? A case study on tandem oviposition in the damselfly *Coenagrion puella* (Insecta: Odonata). *Canadian Journal of Zoology* 77: 1013–1016.

Ziegler, T. E. 2000. Hormones associated with non-maternal infant care: A review of mammalian and avian studies. *Folia Primatologica* 71: 6–21.

Zuk, M. 2002. *Sexual Selections: What We Can and Can't Learn about Sex from Animals*. University of California Press, Berkeley.

Author Index

Abdallah, G., 59
Abe, J., 131
Abenius, J., 115, 141
Abernathy, R. L., 144
Abeydeera, L. R., 124
Abplanalp, H., 128
Abraham, S., 143
Adamo, S. A., 196, 198, 199
Adams, A. E., 120
Adams, G. M., 138
Adiyodi, K. G., 156
Adiyodi, R. G., 156
Agrawal, A. F., 3, 18, 183, 206
Aguade, M., 101, 153, 154, 205
Ahnesjö, I., 5, 38, 183
Ahola, A. L., 131
Ai, N., 147
Aigaki, T., 142
Aiken, R. B., 83
Akney, P. F., 141–142
Alaghbari, A., 98
Alatalo, R. V., 5, 21, 25, 103, 105, 165, 166, 167
Albon, S. D., 184
Albone, E. S., 93
Alcock, J., 21, 44, 46, 48, 77, 96, 107, 108t, 109t, 114, 118, 132
Alleman, M., 181
Allendorf, F. W., 131
Aloia, R. C., 87
Alonzo,S. H., 170
Alvarez, E., 131
Amundsen, T., 173–174
Andersen, N. A., 181
Andersen, N. M., 46, 68, 148
Andersson, J., 108t, 110, 115, 143
Andersson, M. B., 6, 13–23, 27, 29, 38–39, 41, 44, 46, 49, 92, 103, 183, 206, 208
Andersson, S., 49, 208
Andrade, M.C.B., 51, 52
Andrés, J. A., 30, 71, 132, 147–148, 210, 211, 212f, 220
Applebaum, S. W., 100, 143, 153
Aquadro, C. F., 125, 153, 205
Arak, A., 31, 35, 50, 72, 96, 167, 207, 208, 220

Arisawa, N., 142
Arnold, S. J., 8, 10, 15, 20, 45, 66–67, 104–105, 170, 223
Arnqvist, G., 3–5, 11, 16, 28, 29, 31–34, 36, 37, 40–43, 46, 47, 49, 52, 62–65, 68, 72, 77–79, 81– 86, 91, 96, 101, 106, 107, 108t, 109t, 113, 114, 119, 120, 122, 128, 131, 136–137, 139–141, 144, 146–148, 148–149, 153, 183, 204, 205, 206, 207–211, 212f, 217t, 218, 219, 220, 221, 224, 225
Arsenault, M., 80
Arunachalam, S., 134, 135f
Asch, R., 124
Ashton, M. L., 191
Ashwell, T. A., 34
Atkin, L., 173
Aubin, T., 208
Aulner, D. A., 135–136
Aumüller, G., 93, 94, 95
Aureli, F., 219
Austin, C. R., 94, 95, 123, 127t
Axelrod, R., 189
Azmeh, S., 205

Babcock, R. C., 125, 127t
Badyaev, A. V., 8
Baer, B., 42, 108t, 113, 114f, 135
Bahuguna, A., 94–95
Bailey, D.M.D., 131
Baker, R. H., 34, 124, 140
Baker, R. L., 71
Baker, R. R., 7, 121, 216, 218
Baker, V. F., 93–94
Bakowski, B., 189
Bali, G., 99–100
Ball, M. A., 128
Balzer, A. L., 103
Banerjee, B. N., 93–94
Bangert, P. J., 42, 62
Bangham, J., 36, 37, 40, 41, 42, 81, 217t, 225
Banta, E. M., 52
Baril, C. T., 27, 32, 74, 79
Barnes, H., 98
Barnes, M., 98
Barra, A., 94, 151
Barraclough, T. G., 205

Barrett, S.C.H., 202, 203
Barrett, T., 131
Barta, Z. N., 168, 173
Barth, F. G., 51
Bartolomei, M. S., 181
Barton, D. E., 131
Barton, N. H., 25, 81, 204, 207
Basolo, A. L., 27, 28, 71–72
Bateman, A. J., 7, 38, 157–158, 164, 193, 224
Bateson, M., 71
Bateson, P.P.G., 202, 213f
Baur, A., 187, 191
Baur, B., 156, 187, 189, 190, 193, 194, 198, 199
Beaton, G., 70f
Beck, C. W., 160
Becker, E., 56
Bedford, P. M., 124
Bedwal, R. S., 94–95
Beekman, M., 223
Beerli, P., 15–16, 45
Begun, D. J., 101, 131, 148, 153, 154, 210, 211
Belec, L., 94, 151
Bell, G., 22
Benavides, E., 67
Benetton, S., 59
Bennett, P. M., 205
Bennett, T., 201, 202
Bensch, S., 165, 166
Benton, T., 56
Berger, I., 46, 132
Berglund, A., 17, 33, 46, 48, 207
Bergsten, J., 43, 69f, 83, 84
Berlocher, S. H., 203
Bernasconi, G., 131
Berrill, M., 80
Berringan, D., 15–16, 45
Beukeboom, L. W., 183–184
Bhaskaran, G., 99
Birkhead, T. R., 3, 16, 17, 18, 52, 92, 94, 107, 109t, 114, 121, 122, 123, 124, 128, 131, 133, 140, 173
Bisazza, A. G., 17, 48, 59, 60, 81, 207
Bishop, J.D.D., 191, 202
Bissoondath, C. J., 144
Björklund, M., 8, 9f
Blackburn, D. G., 67–68
Blackwell, A., 138
Blanckenhorn, W. U., 45, 47, 129, 136–137, 148, 210, 211
Blandau, R. J., 127t

Blandy, J. P., 93, 94
Bloch-Qazi, M. C., 132
Blomquist, G. J., 147
Bluhm, C. K., 105–106, 172
Blumer, L. S., 160
Boake, C.R.B., 204
Bodnar, W. M., 143
Boitani, L., 58
Boiteau, G., 101
Bollack, C., 93, 94, 95
Bonduriansky, R., 17, 26, 39, 48, 49, 143
Boness, D. J., 58
Boomsma, J. J., 184
Borg-Karlson, A. K., 108t, 110, 115, 143
Borgia, G., 21, 22, 23, 41, 108t
Born, J., 120
Borovsky, D., 98
Borowsky, B., 65, 80
Borries, C., 53
Bottoni, L., 103, 105–106, 172
Boucher, D. H., 159
Boucher, L., 141–142
Boughman, J. W., 27, 72, 205
Bourne, G. R., 160
Bouvet, J. P., 94, 151
Bowen, W. D., 58
Bradbury, J. W., 14, 44, 46
Bradbury, R. B., 215
Braden, A.W.H., 122, 123, 127t
Brady, U. E., 147
Brain, P. F., 215
Brand-Jacobsen, K. F., 219
Breed, W. G., 127t
Breene, R. G., 52
Bride, J., 191
Briskie, J. V., 121, 124, 140
Brockes, J. P., 101
Brockman, H. J., 71, 185
Brockmann, K. K., 189
Brodie, F.E.D., 77, 104–105, 183
Brooks, M. D., 3
Brooks, R., 13, 14, 17, 25, 34, 36, 37, 41, 81, 205
Brower, A.V.Z., 55
Brown, D. V., 210, 211
Brown, J., 183
Brown, L. B., 129
Brown, M.J.F., 113
Brown, S. W., 182–183
Brown, W. D., 17, 153
Brownell, P. H., 56, 191
Bryant, E. H., 119–120

Bryga, A. H., 59
Bubucis, P. M., 66
Buchamann, S. L., 48
Bullington, S. W., 68
Bulmer, M. G., 7, 21, 218
Buntin, J. D., 171
Burd, M., 202, 203
Burford, F.R.L., 71
Burghard, W., 46, 132
Burke, W. D., 183
Burke, W. H., 105–106, 172
Burley, N. T., 8, 71, 103, 104–105, 106, 200
Burns, J. R., 74, 74f
Burt, A., 23, 25
Buschhaus, N., 47, 118
Buskirk, R. E., 51, 52
Bussing, W. A., 73
Butin, J. D., 172
Butlin, R. K., 204, 207, 224
Butt, K. R., 191
Byrd, E. W., 125
Byrne, P. G., 109t

Cabada, M. O., 125
Cabrero, J., 119
Cade, W. H., 204
Cain, M. L., 109t
Callaini, G., 127t
Camacho, J.P.M., 119
Cameron, E., 21, 36, 40, 81, 210, 211, 225
Campagna, C., 58
Cappozzo, H. L., 58
Carayon, J., 87, 89, 91
Carbone, S. S., 71
Cark, A. G., 43, 154
Carlisle, T. R., 159, 162, 163
Carlson, A., 5, 166
Carlson, D. A., 98
Carner, K. R., 125
Carrel, J. S., 98
Carrier, J. C., 65, 66
Castillo-Juarez, H., 131
Castro, J. L., 66
Catchpole, C., 166
Cerolini, S., 98
Chakravarti, A., 127t
Chan, S.T.H., 124
Chan, Y. F., 124
Chandra, H. S., 182–183, 214
Chang, M. C., 127t
Chapman, T, 152

Chapman, T., 5–6, 34, 36, 37, 40, 41, 42, 43,
    81, 86, 87, 96, 98, 109t, 128, 141–142,
    150–151, 154–155, 205, 217t, 218, 225
Charlesworth, B., 23
Charlesworth, D. D., 200, 201, 202
Charnov, E. L., 188–191, 192, 198–201
Chase, I. D., 159f
Chase, R., 194, 196, 197, 198, 199
Chastel, O., 105, 106
Chen, L. L., 56
Chen, P. S., 98, 99, 118, 142, 150
Cheng, Y., 59
Cheung, A. P., 124
Chippindale, A. K., 8, 10f, 131, 203, 206,
    214–215, 217t
Cho, F., 124
Choe, J. C., 46, 96
Choffat, Y., 100, 143, 153
Chow, P. H., 127t
Chu, J., 109t
Chung, D.J.D., 198, 199
Civetta, A., 43, 108t, 128, 151, 153, 204, 205
Clark, A. G., 101, 108t, 125, 128, 131, 148,
    151, 152, 153, 205, 210, 211
Clark, D. R., 119–120
Clark, S. J., 137
Clark, T., 74f
Clavert, A., 93, 94, 95
Clobert, J., 206
Clutton-Brock, T. H., 5, 6, 29, 31, 38, 39, 47,
    56, 68, 77, 120, 156, 158, 160, 164, 184,
    219, 220, 222, 223
Cockburn, A., 157, 184
Colegrave, N., 106
Collins, D. A., 184
Collins, F. D., 125
Conlan, K. E., 65
Constantz, G. D., 59
Contreras-Garduño, J., 36, 81
Cook, P. A., 109t, 115
Cook, S. E., 71
Cooke, H. J., 131
Coppola, A., 141–142
Cordero, A., 71
Cordero, C., 36, 37, 40, 41, 81, 137
Córdoba-Aguilar, A., 36, 81
Cotter, R., 202
Cotton, S., 23
Cottrell, T., 189
Cowie, E. G., 65
Coyne, J. A., 128, 204, 207
Cranz, C., 93, 94, 95

Crean, C. S., 81–82
Crespi, B. J., 46, 96
Crouse, H. V., 182–183
Crowley, P. H., 189
Crudgington, H. S., 45, 109t, 129, 130f, 132, 136–137, 139
Crump, M. L., 40, 81, 109t, 134
Cuervo, J. J., 205
Cumming, J. M., 76, 77, 131
Cunningham, E.J.A., 105, 106
Currie, D. C., 64
Curtsinger, J. W., 94, 123
Cusso, M., 99
Cuthill, I. C., 163

Dahm, K. H., 99
Dale, S., 124, 166, 169, 173–174
Daly, M., 46
Danielsson, I., 101, 131
Darwin, C., 2–4, 14–18, 20–21, 22, 44, 48, 60–62, 80, 92, 103, 200, 204
Darwin, E., 3
David, H. A., 127t
Davies, N. B., 108t, 156, 158, 159f, 165, 168, 170, 174–178
Davis, N. T., 88, 89–91
Dawkins, M. S., 205
Dawkins, R., 4, 31, 156, 157, 159, 162, 163, 166, 207, 209, 216, 220, 224, 225
Day, B. N., 124
Day, T., 3, 18, 21, 26, 36, 40, 42, 81, 82, 206, 210, 211, 225
De Bruyn, L., 71
De La Riva, I., 67
de Lope, F., 34, 103–104, 105, 106
De Visser, J.A.G.M., 193
de Waal, F.B.M., 219
DeGrugillier, M. E., 135
Deinert, E. I., 55
Dekort, C.A.D., 100–101
DeLasena, C. A., 128
Delisle, J., 99
Deloof, A., 143
Delph, L. F., 200, 201, 202, 203
Denlinger, D. L., 147
Dennis, D. S., 68–70
Derrickson, S. R., 57, 77
Desrochers, A., 176
Destephano, D. B., 147
Dewsbury, D. A., 137
Diaz, M., 131
Diesel, R., 98

Diez-Sanchez, C., 131
Dill, L. M., 33
Ding, J., 124
Dixson, A. F., 137
Doctor, J., 181
Dodds, P., 202
Donaldson, W. E., 120
Donat, S., 131
Double, M. C., 184
Doyle, L. R., 87
Doymaz, F., 56
Dressler, R. L., 72
Drevon, T., 166
Dubois, A., 67
Duellman, W. E., 67
Dunn, D. W., 82
Dunn, P. O., 16, 167
Duvall, D., 45, 170
Duvoisin, N., 108t, 113, 135
Dybas, H. S., 140
Dybas, L. K., 140
Dyer, K. A., 128

Eady, P. E., 139, 148, 207, 210, 211
Eberhard, W. G., 5, 16–17, 36, 37, 40, 41, 59, 62, 64–65, 81, 92–96, 98, 100, 101, 105, 107, 108t, 109t, 110, 118, 121, 123, 129, 133, 136–137, 138, 146, 204, 205, 222
Eberly, L. E., 184
Edvardsson, M., 96, 205
Eens, M., 168
Eggert, A. K., 42, 62, 168, 169f
Eibl-Eibesfeldt, I., 157, 173
Eickbush, D. G., 183
Eickbush, T. H., 183
Eipper-Mains, M. A., 43, 109t, 151
Eisner, M., 141–142
Eisner, T., 141–142, 215
El Halawani, M., 105–106, 172
Elgar, M. A., 51, 52, 136–137, 148–149
Elinson, R. P., 127t
Ellegren, H., 8, 215
Elliott, D. J., 131
Eltz, T., 72
Elwood, R. W., 51, 52
Emlen, S. T., 170
Endler, J. A., 15, 27, 28, 59, 71–72, 205, 223
Engel, H. N., 98
Enquist, M. A., 31, 35, 72, 96, 106, 158, 167, 172–173, 207–209, 220
Enriquez, J. A., 131
Epel, D., 125

Epplen, C., 53
Epplen, J. T., 53
Epstein, M. S., 67
Ericsson, R. J., 93–94
Eriksson, D., 166
Eroglu, A., 124
Estoup, A., 113
Etges, W. J., 205
Evans, J. P., 59, 125
Ezcurra, I., 202

Fahey, B. F., 51f
Fairbairn, D. J., 46, 64, 77, 83, 132
Fairfull, R., 215
Falls, J. B., 49
Faltin, S., 142, 143
Fan, Y. L., 100, 143, 153
Farley, R. D., 52
Farr, J. A., 59
Fatinikun, T., 143
Fechheimer, N. S., 128
Feltmann, A. J., 131
Fenton, M. B., 55
Fernandez, N., 93, 94, 123, 151
Fernandez-Cano, L., 127t
Ferraretti, A. P., 124
Festa-Bianchet, M., 184
Fiebig, A., 202
Field, S. A., 137
Fielding, K., 66
Fincke, O. M., 70–71
Finnerty, C. M., 43, 109t, 151
Fiorention, A., 124
Fischer, E. A., 188–189
Fischer, R. L., 182
Fisher, R. A., 18–21, 22–23, 26, 27, 28, 36, 184, 207, 208, 221
Fleischmann, P., 142, 143
Fletcher, F., 128
Flores, G., 67
Floyd, P. D., 191
Folsom, M. W., 201, 202
Forbes, M.R.L., 71
Forman, D. W., 215
Formas, J. R., 67
Forsberg, J., 99, 141
Forster, L. M., 52
Forsyth, A., 109t
Fowler, K., 5–6, 23, 34, 150
Fox, C. W., 101, 139
Fox, J. H., 28
Foxcroft, G. R., 124

Frank, S. A., 109t, 122, 125, 126, 131, 201
Franke, E. S., 125, 127t
Franklin, F.C.H., 202
Franklin-Tong, V. E., 202
Franz, D., 160, 162
Fretwell, S. D., 157f
Frey, D., 47
Friberg, U., 3, 11, 28, 29, 31–34, 36, 37, 40, 42, 72, 96, 106, 153, 205–209, 217t, 220, 224
Fricke, C., 148, 210, 211
Friedlander, M., 115, 126
Friedman, M., 172
Fritsche, E., 131
Frohlich, C., 51, 52
Froman, D. P., 98, 131
Frongillo, E. A., 150
Fugo, H., 142
Fujihara, N., 97–98
Fuller, R., 46
Funahashi, H., 124
Funk, D. H., 39–40, 76

Gack, C., 42, 62
Gaffar, N., 51
Gage, M.J.G., 121, 140, 144, 205
Galan, P., 71
Galanti, V., 103, 105–106, 172
Galimberti, F., 58
Galindo, B. E., 125
Gallagher, P. D., 141–142
Galtung, J., 219
Galvani, A., 137
Ganeshaiah, K. N., 200
Ganjian, I., 147
Garcia, T., 189
García-González, F., 34, 153
Garden, R. W., 191
Garner, T.W.J., 43, 120, 131, 148, 210, 211
Gavrilets, S., 3, 11, 28, 29, 31–34, 36, 37, 40, 72, 122, 125, 126, 206, 207, 208, 209, 217t, 220, 224
Gebhard, J., 55
Gemmel, N. J., 131
Getty, T., 26, 221
Ghirlanda, S., 31, 35, 220
Gianaroli, L., 124
Gibson, J. R., 8, 10f, 49, 206, 214–215
Giebultowicz, J. M., 99, 144
Gil, D., 105, 106
Gilbert, F., 205
Gilbert, L. E., 55

Gilbert, P., 15–16, 45
Gilburn, A. S., 81–82
Gilchrist, A. S., 131
Gilchrist, G. W., 141, 154
Gileadi, C., 143, 153
Giller, P. S., 132
Gillott, C., 93, 98, 118, 143
Ginzburg, A. S., 122, 124, 126, 127
Giusti, F., 198
Gleason, L. N., 87
Gleeson, J. M., 191
Gliozzi, T., 98
Godfray, H.C.J., 40, 218, 219
Godin, J.G.J., 59
Goldberg, R. B., 182
Goldsmith, A. R., 105–106, 172
Goldstein, D. B., 180
Gomendio, M., 121, 123, 131, 137, 215
Gomez, M. I., 125
Gomot, L., 191
Gomulkiewicz, R., 209
Gordon, I., 65, 66
Gould, M., 122, 123, 124, 127, 128
Gould, S. J., 52
Gowaty, P. A., 5, 47, 118, 121, 131, 225
Grafen, A., 26, 185, 225
Graves, J., 105, 106
Gray, D. A., 204
Greeff, J. M., 122, 187, 189, 190, 192, 193, 194, 198, 199
Green, A., 5–6
Green, D. M., 194, 198
Gregory, P. G., 109t
Grether, G. F., 27, 32, 74, 79
Greve, T., 124, 127t
Griffiths, R., 215
Gross, M. R., 21, 26, 160
Gubernick, D. J., 156
Guilford, T., 71
Guinness, F. E., 184
Gupta, G. N., 105–106, 172
Gustafsson, L., 167
Gwynne, D. T., 17, 29, 39, 46, 102, 108t, 114–115, 118, 119, 141, 143, 144

Habib, F. K., 93
Hackett, J. B., 8
Haig, D., 180, 181, 182, 201
Hale, J. A., 56
Hall, K., 180
Halliday, T. R., 5, 8, 10, 17, 79
Hamaoka, D., 47

Hamilton, W. D., 23, 154, 189, 225
Hammerstein, P., 30, 34, 220, 224
Hammock, B. D., 56
Han, Y. M., 124
Hancock, J. L., 127t
Hancock, R. G., 98
Haniffa, M. A., 134, 135f
Harada, J. J., 182
Harbison, D. S., 124
Harcourt, A. H., 121, 137
Härding, R., 31, 34, 41, 42, 86, 201, 218, 219, 220, 222
Hare, H., 185
Hargreave, T. B., 131
Harmsen, R., 119–120
Harper, A. B., 191
Harrison, R. F., 131
Harshman, L. G., 101, 109t, 128, 154
Hartley, I. R., 167, 173, 175, 176, 183
Hartmann, R., 108t, 143
Harvey, P. H., 121, 205
Hassold, T., 127t
Hasson, O., 26
Hastings, A., 209
Hatch, M., 189
Hatchwell, B. J., 175, 176
Hatfield, C., 201, 202
Haupt, H., 160, 162
Hausfater, G., 53
Havens, K., 200, 201, 202, 203
Hayashi, F., 143
Hazon, N., 105
Heckman, N. E., 23, 26, 104–105, 224
Hedrick, A. V., 33
Heifetz, Y., 150, 152
Heller, K. G., 142, 143
Heming, B. S., 137
Heming-Van Battum, K. F., 137
Hendriks, H., 119–120
Henriksson, S., 52
Henry, C. S., 204
Herbeck, J. T., 132
Herberstein, M. E., 52
Herick, 183
Herndon, L. A., 152
Herzmark, P., 202
Heywood, J. S., 26
Hidreth, P. E., 127t
Higashi, M., 181, 209, 219
Hill, C. E., 15–16, 45
Hinde, R. A., 105–106
Hinds, M. J., 138

Hinnekint, B.O.N., 71
Hinojosa, I. A., 77
Hinton, H. E., 129
Hiraiwa-Hasegawa, M., 53
Hirao, T., 142
Hirschenhauser, K., 103
Ho, P. C., 124
Hoang, A., 15–16, 45
Hoback, W. W., 58
Hocini, H., 94, 151
Hockham, L. R., 40, 76
Hodl, W., 67
Hoe, H., 160
Hoekstra, H. E., 15–16, 45
Hoekstra, J. M., 15–16, 45
Höglund, J., 103, 105
Hoi, H., 162, 163, 164
Holland, B., 6, 11, 25, 29, 31, 32, 35, 40, 43,
    72, 79, 86, 96, 108t, 122, 124–125, 152–
    153, 206, 217t, 218, 221–222, 224
Holmgren, N.M.A., 209
Holsinger, K. E., 187, 204
Hongtrakul, T., 135
Honma, K., 47
Hosken, D. J., 16, 25, 43, 45, 47, 120, 121,
    129, 131, 136–137, 148, 153, 206, 207,
    210, 211, 224
Houck, L. D., 66–67
Houde, A. E., 59
Houle, D., 22–23, 25, 26, 36, 205
Houston, A. I., 158, 159f, 160, 163, 168, 173
Howard, D. J., 109t, 122, 201, 202, 203, 207
Hrdy, S. B., 5, 53–54
Huber, F., 120
Hughes, B., 131
Hughes, K. A., 27, 32, 74, 79
Hughes, R. N., 191
Hugot, J. P., 87
Huignard, J., 141–142
Hunt, D. F., 143
Hunter, R.H.F., 123, 124, 127t
Hurst, G.D.D., 183–184
Hurst, L. D., 96, 131, 181, 182, 214
Hutchinson, J.M.C., 192

Iason, G. R., 184
Ilango, K., 140
Inceoglu, B., 56
Ingram, R. S., 181
Insel, T. R., 172
Inwards, S. J., 55
Ionov, I., 143

Iscaki, S., 94, 151
Iverson, S. J., 58
Iwasa, Y., 21, 22, 26, 181, 208
Iyengar, V. K., 215

Jacobs, H. T., 131
Jacobsen, C. G., 219
Jaenike, J., 183–184
Jaffe, L. A., 122, 123, 124, 127, 128
Janetos, A. C., 25
Janson, C. H., 53
Jennions, M. D., 13, 14, 17, 21, 25, 26, 36,
    37, 38–39, 41, 81, 113, 157, 167
Jequier, A. M., 131
Jing, J., 56
Jirtle, R. L., 181
Johnson, C., 71
Johnson, J. C., 52
Johnson, M. A., 202
Johnson, S., 202
Johnstone, R. A., 21, 23, 26, 108t, 110, 119,
    120, 137
Johnstone, S. E., 202, 208, 225
Jones, C. E., 48
Jones, G. R., 42, 113
Jones, T. M., 148–149
Jormalainen, V., 31, 34, 41, 42, 47–48, 58, 59,
    65, 77, 86, 134, 135, 218, 219, 220, 222
Jungfer, K. H., 67
Jurenka, R. A., 147

Kaitala, A., 115, 141, 144
Kajiura, S. M., 65, 66, 68
Kalb, J. M., 5–6, 43, 109t, 150–151
Kane, R. E., 125
Kao, S. H., 131
Kapoor, N. N., 64
Karlsson, B., 102, 141, 144, 145, 146
Karube, F., 134
Kasuga, H., 142
Katvala, M., 139, 149
Katzourakis, A., 205
Keenleyside, M.H.A., 165f
Keller, L., 94, 108t, 109t, 110, 119, 120, 123,
    128, 131
Kelly, C. D., 59
Kempenaers, B., 168, 169
Kenagy, G. J., 121
Kennedy, J. S., 219
Ketterson, E. D., 172
Khan, A., 83
Khan, H. R., 191

Kida, R., 134
Killian, J. K., 181
Kim, H. U., 202
Kim, J. H., 124
Kingan, T. G., 99–100, 143, 144
Kingsolver, J. G., 15–16, 45
Kinoshita, T., 182
Kirchner, H., 94
Kirkpatrick, M., 5, 13, 14, 19–23, 25–28, 35–
    36, 41, 46, 71–72, 81, 104–105, 165, 204,
    207
Klepal, W., 98
Klimley, A. P., 66
Klinkhamer, P.G.L., 201–202
Klopfer, P. H., 156
Klowden, M. J., 98, 100, 118
Knisley, C. B., 66
Knowles, L. L., 148, 210, 211
Kobayashi, M., 134
Koene, J. M., 43, 140, 191, 196, 197f, 198,
    199–200
Kokita, T., 169
Kokko, H., 3, 13, 14, 17, 25, 26, 36, 37, 38–
    39, 41, 81, 104–105, 205, 225
Komatsu, S., 147
Komdeur, J., 168, 223
Kondoh, M., 181, 209
Kondrashov, A. S., 22–23, 25, 205
Koopmanschap, A. B., 100–101
Koprowski, J. L., 108t
Korpimäki, E., 166
Kostjuk, I., 143
Kotiaho, J. S., 131
Kotrschal, K., 103
Kraus, B., 109t
Krause, W., 93, 95
Krebs, J. R., 4, 31, 205, 207, 209, 216, 225
Krebs, R. A., 141–142
Kreiter, N. A., 52
Kritsky, G., 66
Krupa, J. J., 4, 5, 11, 34, 40, 42, 46, 47, 48,
    49, 78, 79, 82, 84, 108t, 120, 137, 183
Kubli, E., 100, 143, 150, 153
Kubo, I., 147
Kuhl, A., 189
Kulathinal, R. J., 125
Kulikauskas, R., 202
Kulkarni, C. V., 60, 61f
Kuroiwa, A., 181
Kurpisz, M., 93, 94, 123, 151
Kutz, E., 143
Kvarnemo, C., 38, 183

Lack, D. L., 3, 157, 158
Lambert, C. C., 125
Lampe, H. M., 169
Lamprecht, J., 169, 170
Lams, H., 127t
Lande, R., 3, 8, 19, 20, 21, 26, 204, 206, 223
Landel, H., 205
Landolfa, M. A., 194, 196, 198, 199
Lane, R. P., 140
Lange, A. B., 143–144, 147
Langkilde, T., 44, 49, 50f, 72
Langley, C. H., 101, 153, 154
Langlois, S., 127t
Langmore, N. E., 175, 176
Lango, J., 56
Lanier, D. L., 101
Lapena, A. C., 131
Larson, B.M.H., 202
Larson, S. G., 121
Lauer, M. J., 77
Launhardt, K., 53
Lavigne, R. J., 68–70
Lawes, J. S., 208
Lawes, M. J., 49
Lawrence, S. E., 52
Lazarus, J., 160, 163, 164
Le Boeuf, B. J., 47, 57, 58
Le Feuvre, E. M., 65
LeBas, N. R., 76
Lederhouse, R. C., 109t
Lee, K. E., 52
Lee, K. H., 128
Lee, P.L.M., 215
Lee, Y. H., 125
Legge, S., 184
Lehrman, D. S., 172
Leigh, C. M., 127t
Leimar, O., 141, 144, 146
Leisler, B., 166
Lentz, A. J., 98, 101, 118
Leonard, J. L., 188, 189, 193, 198
Leong, K.L.H., 47
Leonhard-Marek, S., 94
Leopold, R. A., 135
Leopold, W. R., 48, 108t, 120
Lepri, A., 198
Lessels, C. M., 7, 119, 156, 184, 216, 217t,
    218
Leugger, R. R., 129
Lewis, C. T., 129
Lewis, S. M., 132, 141–142
Leycam, J. F., 98, 101, 118

Leykin, L., 124
Li, L. J., 191
Liappala, P., 131
Liddle, L. F., 5–6, 43, 150–151
Liersch, S., 113
Lifjeld, J. T., 34, 165–166, 173–174
Ligon, J. D., 164, 165–166, 167
Liker, A., 169
Lilja, H., 94
Lind, H., 197
Linden, M., 103
Lindenfors, P., 8
Lindfors, V., 115, 141, 145f
Linley, J. R., 138
Liu, H. F., 150
Lloyd, J. E., 135, 136–137
Locher, R., 193
Loehrl, H., 163
Loher, W., 108t, 143, 147
Longino, J. T., 55
Lonsenmair, K. E., 72
Lopez-Leon, M. S., 119
Lopez-Perez, M. J., 131
Lorch, P. D., 3, 18, 26, 206
Lorenzi, M. C., 189
Lubin, Y., 53–54
Lucchesi, J. C., 127t
Luce, R. D., 220
Lukowiak, K., 189
Lundberg, A., 5, 103, 105, 166
Lung, O., 43, 109t, 150, 151
Lungberg, 167
Lutwak-Mann, C., 93–94, 95
Lyell, Charles, 2
Lyman, R. F., 8
Lytton, B., 93, 94

Macdonald, D. W., 53, 157
Mackay, T.F.C., 8
Mackereth, R. W., 165f
Macpherson, A., 143
Maddison, W. P., 205
Maddy, S. Q., 93
Magli, M. C., 124
Magnhagen, C., 46
Magrath, M., 168
Magurran, A. E., 59, 203, 210
Maheshwari, P., 202
Maier, G., 46, 132
Maile, P., 113
Maile, R., 42
Mair, J., 138

Majerus, M.E.N., 183–184
Maldjian, A., 98
Mallet, J., 209
Manfredi, S., 60, 81
Mann, T., 93–94, 95
Mannazhath, N., 143
Manriquez, P. H., 191
Marcanato, A., 34, 109t, 134
Marchington, D. R., 131
Markow, T. A., 55, 140, 141–142, 148, 210, 211
Marler, C., 105–106, 172
Marsh, N. B., 101
Marshall, D. L., 201, 202
Martens, A., 108t
Martin, O. Y., 25, 43, 45, 47, 120, 129, 136–137, 153, 206, 207, 224
Martin, T. E., 206
Martin-Alganza, A., 119
Martin-Vivaldi, M., 103
Martinez, J. G., 103
Maruska, K. P., 65
Marzetti, I., 58
Mason, R. T., 44, 49, 50f, 72
Massa, R., 103, 105–106, 172
Masta, S. E., 205
Mathis, A., 58
Matsubayashi, K., 53
Matsuda, Y., 181
Matsui, M., 67
Matthysen, E., 71
Maxwell, M. R., 52
May, R. M., 203
Mayfield, J. A., 202
Maynard-Smith, J., 17, 21, 22, 31, 79, 156, 158, 160, 225
Mayr, E., 3, 204
Mazuc, J., 105, 106
McCarrey, J. R., 214
McCormick, S., 202
McFadden, D. E., 127t
McGregor, P. K., 71
McKinney, F. S., 57, 77
Mcknab, R., 59
McLain, D. K., 101, 206
McNamara, J. M., 160, 163, 168, 173
McNay, D., 131
McVean, G. T., 181, 182
Mead, A. R., 191
Mehmet, D., 131
Meinwald, J., 141–142
Melser, C., 201–202

Melville, J. M., 56
Menezes, N. A., 72
Meredith, J. A., 144
Merilä, J., 8
Merilaita, S., 38, 42, 47–48, 59, 65
Merritt, D. J., 129, 135
Mesnick, S. L., 47, 57, 118
Mestel, R., 50
Metz, E. C., 125
Michiels, N. K., 87, 119, 146, 185f, 186, 187, 188f, 189–194, 196, 198, 199
Millam, J. R., 105–106, 172
Miller, D., 99
Miller, G. T., 153
Miller, J. R., 98, 101, 118
Miller, K. B., 84
Miller, L., 137
Miller, M. N., 70–71
Millie, E., 127t
Mineau, P., 57, 77
Misof, B., 205
Mitra, S., 205
Miyashita, N., 153
Miyatake, T., 96, 98
Mladenov, G., 134
Mochizuki, A., 181
Mock, D. W., 40, 174, 183, 218
Møller, A. P., 18, 21, 23, 25, 26, 34, 92, 94, 103–104, 105, 106, 107, 109t, 114, 121, 122, 123, 133, 173, 205, 206
Montgomerie, R., 121, 124, 140, 206
Montgomery, W. I., 51, 52
Montoro, L., 131
Montoya, J., 131
Mook, M. S., 138
Moore, A. J., 49, 77, 104–105, 225
Moore, P. J., 49, 225
Moore, S. L., 206
Moore, T., 180, 181
Moreno, J. M., 103
Moretti, R. L., 87
Morgan, E. D., 42, 113
Morgan, M. T., 192–193, 194–195, 200, 203
Morgan, T. H., 19
Moriguch, M., 127t, 134
Morley, J., 13, 14, 17, 25, 36, 37, 41, 81
Moro, Y., 134
Moroz, T. P., 191
Morris, A. W., 56
Morris, M. R., 33
Morrow, E. H., 30, 91, 119, 120, 121, 122, 128, 132, 140, 205, 206

Moshitzky, P., 100, 143, 153
Mostl, E., 103
Moulton, M. P., 206
Muhlhauser, C., 47
Mukai, S. T., 191
Mumme, R. L., 173
Munday, B. L., 181
Murphy, C., 124
Murray, A. M., 132
Murtaugh, M. P., 147

Nacheva, I., 134
Nachman, R. J., 144
Naeem, R., 127t
Nagai, T., 124
Nagoka, S., 142
Nakazono, A., 169
Nasrallah, J. B., 202
Nassal, B., 46, 132
Naylor, R., 91
Nebel, D., 176
Nee, S., 21, 205
Neiland, R., 202
Nelson, K., 73, 74
Nestor, K. E., 128
Neubaum, D. M., 150
Newberry, K., 88–89
Newman, J. A., 52
Newman, L. J., 185, 196
Nielsen, R., 125
Nijhout, H. F., 98, 99–100, 99f, 101
Nilsson, A. N., 43, 69f, 83, 84
Nilsson, T., 34, 46, 96, 101, 107, 108t, 113, 114, 139, 140–141, 146, 148, 149, 201, 205, 210, 211
Noble, R., 98, 143
Nolan, C. M., 181
Nolan, V., 172
Norris, K. J., 103
Noureddine, M. A., 56
Nuismer, S. L., 209
Nur, N., 26
Nur, U., 127t, 183
Nuutinen, V., 191
Nylin, S., 205

Oberhauser, K. S., 141, 144
Ochoa, O., 67
O'Connor, D., 49
O'Donald, P., 19, 20
Odor, D. L., 127t
Ohlsson, R., 180, 182

Öhrström, P., 160, 162
Okada, K., 202
Okuda, T., 127t
Oliver, J. H., 98
Oliveria, R. F., 71
Olsson, M., 66
Omu, A. E., 143
O'Neill, M. J., 181
Ord, T., 124
Ord, V. A., 124
Orians, G. H., 164, 165, 170
Oring, L. W., 170
Ornborg, J., 49, 208
Orsetti, D. M., 101
Ortigosa, A., 77, 82, 137
Ortiz, J. C., 67
Osanai, M., 99, 142
Osorno, J. L., 173
Ota, T., 125
Otto, S. P., 180
Overstrom, N. A., 66
Owens, I.P.F., 39, 167, 205

Packer, C., 184
Paemen, L., 143
Page, D. C., 214
Pagel, M., 28
Palomino, J. J., 103
Palumbi, S. R., 109t, 124–125
Panhuis, T. M., 204, 207
Panzella, S., 124
Park, G. A., 122
Park, Y. I., 98, 99, 100
Parker, G. A., 4–8, 11, 14, 18, 26, 29–31, 34,
    38, 40, 41, 46, 47, 50, 51, 52, 56, 58, 68,
    72, 77, 86, 92, 96, 102, 106–107, 108t, 110,
    113, 116, 118–122, 128, 132, 133, 137,
    141, 148, 158, 173, 174, 183, 187, 192,
    195, 198, 199, 203, 205, 207, 210, 211,
    216–224
Parmigiani, S., 53
Partridge, L., 5–6, 7, 8, 41, 43, 86, 96, 98,
    109t, 122, 128, 131, 141–142, 148, 150–
    151, 152, 154–155, 203, 207, 210, 211,
    217t, 218, 224
Parzefall, J., 59
Pasa-Tolic, L., 191
Pasyukova, E. G., 8
Patel, C. S., 172
Payne, R.J.H., 28
Pearse, L., 25, 206, 224
Pedersen, H. C., 171

Peloquin, J. J., 147
Pen, I., 190
Persson, O., 160, 162
Pessah, I. N., 56
Peter, M. G., 99
Peters, S. E., 135–136
Petersen, A. L., 124
Petersen, C. W., 34, 109t, 134
Petersen, H. H., 127t
Peterson, E., 76
Petitte, J. N., 181
Petkov, N., 134
Petrie, M., 21, 26, 103, 105, 113, 167, 168
Pfennig, D. W., 215
Phillips, R. E., 105–106, 172
Phillips, S.D.F., 204
Phillips, W. R., 55
Pieneman, A. W., 187
Piko, L., 127t
Pilastro, A., 17, 48, 59, 60, 81, 207
Pillot, J., 94, 151
Pinxten, S. K., 168
Pitcher, T. E., 16, 167, 205, 206
Pitnick, S., 17, 34, 119, 120, 122, 128, 140,
    141–142, 153
Pizzari, T., 41, 81, 131, 217t, 221
Plath, M., 59
Polhemus, D. A., 148
Polhemus, J. T., 148
Polis, G. A., 52, 56
Pollock, J. N., 129
Pomiankowski, A., 14, 17, 21, 22–23, 23, 26,
    34, 47, 79, 198–199, 208
Pongratz, N., 193, 194
Poston, J., 17, 48, 81
Poulton, J., 131
Pouvreau, A., 113
Prather, R. S., 124
Pratt, H. L., 65, 66, 68
Premoli, M. C., 189
Prenter, J., 51, 52
Presgraves, D. C., 124, 140
Preston-Mafham, K. G., 77, 78f
Preudhomme, J. L., 94, 151
Preuss, D., 202
Pribil, S., 165, 167
Price, C.S.C., 128
Price, D. K., 8
Price, T. D., 23, 26, 104–105, 148, 204, 207,
    208, 224
Proctor, H. C., 28, 74
Promislow, D., 25, 106, 206, 224

Proost, P., 143
Proulx, S. R., 3, 18, 26, 206, 225
Prout, T., 101, 109t, 128, 131, 148, 152, 154, 210, 211
Pruett-Jones, S., 205
Pryke, S. R., 49, 208
Punitham, M. T., 134, 135f
Purvis, A., 205

Queller, D. C., 29, 38–39, 200, 202, 203, 221
Questiau, S., 204

Raabe, M., 98, 118
Radwan, J., 131
Rafaeli, A., 100, 143, 153
Raiffa, H., 220
Raina, A. K., 99–100, 143, 144
Ramaiah, T. R., 142
Ramaswamy, S. B., 98, 99, 100
Ramnarine, I. W., 59
Rand, A. S., 28
Randerson, J. P., 214
Ratnieks, F.L.W., 223
Ratti, O., 165, 166, 167
Raven-Yoo-Heufes, A., 189
Ravigné, V., 204, 207
Raz, T., 124, 125, 126
Read, A. F., 51
Reagan, J., 153
Redfearn, T. P., 206
Redline, R. W., 127t
Reece, M., 109t
Reed, K. J., 215
Reeve, H. K., 40, 94, 123, 131, 215
Reguera, P., 198–199
Rehfeldt, G., 108t
Reik, W., 180
Reim, C., 45, 47, 129, 136–137
Reinhardt, K., 91
Reinhold, K., 214, 215
Reise, H., 192
Rembold, H., 98
Reynolds, J. D., 21, 26
Reznick, D., 59
Ribeiro, J.M.C., 88–89
Rice, W. R., 6, 8, 10f, 11, 25, 29, 31, 32, 35, 40, 41, 43, 72, 79, 86, 96, 108t, 122, 124–125, 131, 152–153, 154, 170, 194–195, 203, 204, 206, 207, 211, 213, 214–215, 217t, 218, 221–222, 224
Richards, A., 34
Richards, O. W., 61–62

Richardson, J.M.L., 71
Richter, H. J., 73
Riechert, S. E., 56
Rihimaki, J., 47–48
Ringo, J. M., 204
Rink, L., 94
Rintamäki, P. T., 103, 105
Riparbelli, M. G., 127t
Ritchie, M. G., 40, 76, 204
Ritzén, M., 180
Riva, A., 94, 95
Rivera, A. C., 71, 220
Roach, B., 141–142
Roberts, J. D., 109t
Roberts, R. J., 20
Roberts, T. R., 73
Robertson, G. W., 148
Robertson, H. M., 70–71
Robertson, R. J., 103, 164–165
Robinson, M. H., 51
Rodd, F. H., 27, 32, 59, 74, 79
Rodriguez-Molina, M. C., 144
Roff, D. A., 96, 156
Rogers, D. W., 197, 198, 199
Roldan, E.R.S., 121, 123, 131, 137, 215
Rolff, J., 45–46, 108t
Roller, H., 99
Ronchi, E., 105–106, 172
Rooney, J., 141–142
Rosenblatt, J. S., 156, 171
Rosenthal, G. G., 32–33
Ross, K. G., 51, 52
Rotheray, G., 205
Roubik, D. W., 72
Rovio, A. T., 131
Rowe, L., 3, 4, 5, 11, 18, 21, 23, 26, 33, 36, 37, 40–43, 46, 47, 49, 64, 68, 77, 78, 79, 81–86, 108t, 109t, 120, 132, 134, 137, 139, 140, 143, 144, 183, 206, 208, 210, 211, 217t, 218–221, 225
Royale, N. J., 173, 183
Rozenboim, I., 105–106, 172
Ruiz-Pesisni, E., 131
Russel, A. F., 105, 106
Rutowski, R. L., 101, 141
Ryan, M. J., 13, 14, 19, 26, 27, 28, 33, 35–36, 41, 46, 59, 71–72, 207
Rybak, F., 208
Ryckman, R. E., 88

Sadowski, J. A., 77
Sahara, K., 115

Saifi, G. M., 214
Sakaluk, S. K., 42, 62, 141, 168
Saleuddin, A.S.M., 191
Sanchez-Guillen, R. A., 71, 220
Sandell, M. I., 166, 169
Sanderson, J. G., 206
Sankai, T., 124
Sargent, R. C., 104–105, 106, 156, 160, 170, 189
Sasaki, H., 181
Sato, F., 65
Sauer, K. P., 42, 62, 109t, 135–136
Savalli, U. M., 101, 139
Schatten, G., 124
Schemski, D. W., 200, 202
Schleicher, B., 160, 162, 163, 164
Schlupp, I., 59
Schluter, D., 23, 26, 104–105, 204, 208, 224
Schmid-Hempel, P., 42, 108t, 113, 114f, 135
Schmitt, A., 51
Schneider, J. M., 52, 53–54, 136–137
Schoech, S. J., 173
Scholl, A., 113
Schoofs, L., 143
Schooneveld, H., 100–101
Schrader, F., 127t
Schradin, C., 169, 170
Schroder, I., 108t
Schuh, R. T., 87, 129
Schulenburg, H., 43, 140, 196, 197f, 199–200
Schuppe, H. C., 131
Searcy, W. A., 165, 166, 167
Sebastian, A. P., 65, 68
Seehausen, O., 204
Seger, 183
Seghers, B. H., 59
Seldon, B. C., 8
Sella, G., 189
Senar, J. C., 8, 9f
Senda, M., 202
Servedio, M. R., 32–33
Seth, P. C., 124
Shaanker, R. U., 200
Shabanowitz, J., 143
Shalgi, R., 124, 125, 126
Shapiro, D. Y., 34, 109t, 134
Sharp, W., 5–6
Shaw, K., 28
Sheldon, B. C., 103, 105, 106, 128
Sherman, P. W., 28, 40
Sherratt, T. N., 70–71
Shields, K. M., 105–106, 172

Shimizu, K. K., 202
Shinbo, H., 127t
Shine, R., 44, 49, 50f, 67, 72, 160
Shirk, P. D., 99
Shirley, S. G., 93
Short, R. V., 121
Shu, S. Q., 98, 99
Shuker, D. M., 42, 82
Shuster, S. M., 31, 38, 158, 217t, 218, 219, 221–223
Shykoff, J. A., 113
Sih, A., 4, 5, 11, 34, 40, 42, 46–49, 78, 79, 82, 84, 108t, 120, 137, 183
Sillén-Tullberg, B., 133
Siller, S., 3, 18, 206
Silsby, J., 105–106, 172
Simerly, C., 124
Simmons, L. W., 38, 92, 96, 102, 107, 108t, 114–115, 119, 121, 131–133, 141
Simon, S., 66
Sinclair, A. H., 215
Singer, F., 56
Singh, R. S., 125, 204, 205
Singh, T., 134, 153
Sirot, L. K., 71
Sissom, W. D., 56
Siva-Jothy, M. T., 45–46, 89, 91, 109t, 129, 130f, 132, 133, 136–137, 139
Skeet, J., 176
Slabbekoorn, H., 204
Slagsvold, T., 34, 165–166, 169, 173–174
Smedley, S. R., 141–142
Smid, H. M., 100–101, 142
Smith, C. C., 157f
Smith, E. A., 25, 206, 224
Smith, H. G., 31, 34, 41, 86, 166, 169, 218, 219, 220, 222
Smith, H. K., 96, 98
Smith, P. H., 129
Smith, R. D., 191
Smith, R. L., 156
Smith, T. B., 204
Smith, V.G.F., 7, 107, 121, 216, 218
Smuts, B. B., 47, 53, 68, 77
Smuts, R. W., 47, 53, 68, 77
Snook, R. R., 41, 81, 217t, 221
Snow, A. A., 201
Snowdon, C. T., 156, 171
So, W.W.K., 124
Sokolowski, M. B., 59
Soler, M., 103
Solignac, M., 113

Soltis, J., 53
Sorci, G., 105, 106, 206
Sork, V. L., 200, 202
Speake, B., 143
Speicher, B. R., 127t
Spencer, J. L., 98, 101, 118
Sperling, F.A.H., 215
Spicer, G. S., 140, 141–142
Spielman, A., 88–89
Springer, S., 65
Srinivasan, A., 99, 100
Stallmann, R. R., 47, 49, 108t, 132
Stanley-Samuelson, D., 147
Stanton, M. L., 201
Stearns, S. C., 96
Steel, E., 105–106
Stephenson, A. G., 201
Sterck, E.H.M., 53
Stewart, I.R.K., 77
Stewart, N., 181
Stitch, S. R., 93
Stjernholm, F., 102, 145
Stockel, J., 144
Stockley, P., 107, 121, 131
Stokes, B. J., 189
Stoks, R., 71
Stone, G. N., 48
Stout, C. D., 125
Streng, A., 189
Stricker, S. A., 124, 126
Strong, D. R., 58, 59
Stutt, A. D., 87, 89
Styan, C. A., 125, 127t
Stys, P., 87, 129
Sugawara, T., 147
Sundermann, G., 191
Sundströ, L., 184
Surai, P., 98, 143
Surani, A., 180
Sureau, G., 208
Surti, T., 127t
Sutherland, S., 201
Sutherland, W. J., 5, 94, 109t, 122, 123
Suzuki, H., 134
Suzuki, N., 127t
Svärd, L., 99, 140, 141, 144
Svensson, B. G., 76
Swaddle, J. P., 103–104
Swanson, L. V., 42, 62
Swanson, W. J., 6, 125, 150, 153, 202, 205
Sweedler, J. V., 191
Sweet, M. H., 52

Syred, A., 88f
Szecsi, P. B., 94
Székely, T., 160, 163, 168, 169

Takahashi, T., 65
Takeda, Y., 181
Takemura, Y., 115
Takenaka, O., 53
Tallamy, D. W., 39–40, 76
Tallarovic, S. K., 56
Tanaka, K., 181
Tanaka, Y., 134, 205
Tang, W. H., 202
Taylor, E. H., 67
Taylor, P. D., 23
Teal, P.E.A., 144
Ter Maat, A., 193
Terkanian, B., 141
Termaat, A., 187
Terranov, A. C., 135
Teuschl, Y., 45, 47, 129, 136–137
Thibault, C., 124, 127t
Thiel, M., 77
Thomas, D. W., 55
Thomas, G., 119–120
Thomas, M. L., 136–137
Thompson, D.B.A., 39
Thompson, J. N., 209
Thomsen, R., 53
Thorne, M. H., 128
Thornhill, R., 21, 40, 41, 42, 44, 46, 51, 62,
    68, 77, 80, 92, 96, 101, 105, 107, 109t, 114,
    132, 135–136, 141
Thorson, B. J., 135
Tidemann, C. R., 55
Tilghman, S. M., 181, 182
Ting, C. T., 153
Tips, A., 143
Tobe, S. S., 147
Todd, B. L., 153
Todte, I., 160, 162
Tompa, A., 196
Toner, M., 124
Torgerson, D. G., 125
Torres-Vila, L. M., 144
Tosti, E., 124
Toth, T. L., 124
Toyra, A., 43, 69f, 83, 84
Tram, U., 43, 109t, 151
Travers, S., 46, 49
Tregenza, T., 120, 131, 132, 202, 204, 207,
    224

Trevitt, S., 5–6, 109t, 141–142
Tricas, T. C., 65, 66, 68
Trillmich, F., 169, 170
Trivers, R. L., 4, 7, 14, 21, 29, 38, 96, 102, 156, 157–158, 159, 163, 164, 166, 174, 180, 183, 185, 218, 224
Trombulak, S. C., 121
Trowbridge, C. D., 87
Truckenbrodt, W., 127t
Trumbo, S. T., 168
Tsaur, S. C., 153
Tsudzuki, M., 181
Tubman, S. C., 148
Tufton, T. J., 139
Tuomi, J., 31, 34, 41, 59, 86, 218, 219, 220, 222
Turelli, M., 204, 207
Turner, G. F., 204
Turner, J.R.G., 209
Turri, F., 189

Ueshima, N., 88
Ulfstrand, S., 5, 166
Urrieis, A., 131
Usinger, R. L., 88, 89, 90f
Utzeri, C., 71
Uyenoyama, M. K., 187

Vaccari, G., 59
Vacquier, V. D., 6, 124–125, 202, 205
Vahed, K., 101, 102, 114, 140, 141–142, 145, 147
Valck, F., 71
Valera, F., 160, 162, 163, 164
Van Delden, W., 119–120
Van Den Berg, M. J., 119–120
Van Duivenboden, Y. A., 187
Van Gossum, H., 71
van Schaik, C., 53
Van Valen, L., 209
Vandamme, J., 143
Vandenbroeck, J., 143
Vangerwen, A.C.M., 129
Vanhandel, E., 98
Veelaert, D., 143
Veiga, J. P., 169
Vermette, R., 64
Verner, J., 164, 170
Vernon, J. G., 71
Vieira, C., 8
Vignieri, S. N., 15–16, 45
Vincent, A., 5

Viriyapanich, P., 124
Vitullo, A. D., 123
vom Saal, F. S., 53
Von Helversen, O., 142, 143
Vrana, P. B., 181
Vreys, C., 189
Vueille, M., 65

Wachtmeister, C. A., 31, 35, 105, 106, 158, 172–173, 220
Wade, M. J., 15, 31, 38, 158, 217t, 218, 219, 221, 222, 223
Wagner, W. E., 33
Waldrip-Dail, H. M., 125, 153, 205
Walker, E. D., 98, 101, 118
Walker, W. F., 134
Waller, D. M., 187
Wallin, W. G., 225
Walpole, D. E., 88
Walsh, N. E., 201
Walter, B., 169, 170
Wang, P. J., 214
Wang, W. H., 124
Wang, Z., 172
Ward, P. I., 43, 45, 47, 58, 120, 129, 131, 136–137
Ward, S., 98
Warner, R. R., 34, 109t, 134, 170
Watanabe, M., 141
Watson, P. J., 47, 49, 108t, 109t, 117–118, 132
Watts, D. P., 53
Watts, T. D., 141–142
Waxman, D., 122, 207
Weatherhead, P. J., 20, 103, 164–165
Webb, J. N., 160, 163
Wedell, N., 102, 104, 109t, 115, 120, 131, 132, 137, 140, 141, 145, 202, 224
Wei, Y. H., 131
Weigensberg, I., 77, 83
Weir, T. A., 148
Weissing, F. J., 190
Weitzman, S. H., 72, 74, 74f
Wells, A., 105
Wells, M.L.M., 204
Wells, M. M., 204
Werren, J. H., 183–184
West-Eberhard, M. J., 27, 35, 71–72, 81, 204, 205, 207
Westcott, D. A., 26
Westlake, K. P., 64
Westneat, D. R., 77, 104–105, 106, 156, 170
Westoby, M., 180, 182, 201

Wheeler, J., 49, 143
Wheeler, M. J., 202
White, J. M., 189
Whitfield, P., 166
Whitley, P., 153
Whitlock, M. C., 206
Whitten, W. M., 72
Whittingham, L. A., 16, 167
Wichmann, L., 131
Wickler, W., 4, 73, 74, 102, 141, 146
Wiens, B. L., 191
Wiersma, P., 168
Wiklund, C., 99, 108t, 110, 115, 140, 141, 143, 144, 146, 205
Wilcock, C., 202
Wilcox, R. S., 118
Wilczyski, W., 28
Wiley, R. H., 17, 48, 81
Wilkins, J. F., 180, 181
Wilkinson, G. S., 124, 140
Willard, H. F., 215
Williams, A., 103, 105
Williams, G. C., 21, 22, 23
Williams, T. D., 103, 160
Willson, M. F., 164, 170, 200, 201
Wilson, K., 206
Wilson, N., 139, 148
Wing, S. R., 132, 135
Wingfield, J. C., 105–106, 172, 173
Winkler, H., 166
Winkler, P., 53
Wise, D. H., 52
Wisenden, B. D., 157
Wishart, G., 143
Wolf, J. B., 104–105
Wolfenbarger, L. L., 28
Wolff, J. O., 53
Wolfner, M. F., 5–6, 43, 109t, 125, 150–151, 152, 153, 154, 205
Wu, C. I., 153, 207

Xu, H. F., 56
Xu, X., 124

Yadegari, R., 182
Yamaja-Setty, B.N., 142
Yamamoto, J. T., 105–106, 172
Yamamura, N., 133, 219
Yamaoka, K., 142
Yanagimachi, H., 125
Yang, F., 214
Yang, Q. F., 125
Yano, K., 65
Yasui, Y., 113
Yekta, N., 67–68
Yeung, W.S.B., 124
Yi, S. X., 143
Ying, Y., 127t
Yokomine, T., 181
Yolov, A., 134
Yonezawa, Y., 142
Yoshikawa, Y., 124
Young, D. K., 141–142
Young, P., 52
Yun, K., 182
Yusuf, R. Z., 127t
Yuval, B., 137

Zahavi, A., 3, 21, 22, 23, 225
Zahiri, N., 64
Zaragoza, M. V., 127t
Zeh, D. W., 6, 41, 132, 156, 202
Zeh, J. A., 6, 41, 132, 202
Zeiss, C., 108t
Zeltner, N., 129
Zeng, F. R., 98, 99
Zeng, Z. B., 8
Zhang, Z. H., 153
Ziegler, T. E., 172
Zonneveld, C., 193
Zuk, M., 5, 23, 204, 207, 225

Subject Index

abiotic cues, stimulating hormone production, 172

accessory reproductive gland substances, 150–152; deposition of, 129; effects of on female behavior and physiology, 151t; sequential transfer of, 134–135; transferal of in hermaphroditic snails, 198–199

accessory reproductive glands: function of, 95; secretions of, 92–94, 100

Acp genes, divergent evolution of, 153–155

Acps, in fruit fly seminal fluid, 150–155

adaptations: accumulated antagonistic, 4; antagonistic, costs of, 34; "hidden," 41; for persistence and resistance, 55–78. *See also* female adaptations

adaptive evolution: constraints on, 206; intralocus conflict limiting, 7–8

aggression, egg burial and, 163–164

aggressive ejaculates, 121; to avoid polyspermy, 122–128; costs and benefits of, 109t

allatotropins, 98

alloparental care, 157

allopatric male signal, reproductive response to, 213f

allopatric populations: crosses in, 210–212; sexual conflict in, 209–210

allozyme markers, 198

alternating gamete trading, 189–190

amino acids, in ejaculate, 143

amphibians: grasping structures in, 67; humeral spines in, 67

amplexus: grasping structures during, 67; humeral spines used in, 67–68

andromorphs, damselfly, 71

angiosperms: hermaphroditic, 203; imprinted genes in, 181–182

anisogamy, 7, 14; evolution of, 218

antagonistic sexual coevolution. *See* sexually antagonistic coevolution

antiaphrodisiac, 115; costs and benefits of, 108t

antigrasping traits, 68–71; evolution of in diving beetles, 83–84; in water striders, 84–86

antisperm antibody responses, 122–123

apophallation, 192

apyrene sperm, 115

arms race hypothesis, 84–87, 216, 220–222; avoidance-of-polyspermy hypothesis and, 124–125; coevolutionary, 209; frequency-dependent, 154–155. *See also* female resistance; harassment; male persistence; sexual conflict; sexually antagonistic coevolution

artificial insemination, increased polyspermy risk with, 124

attractiveness: male display traits and, 103–106; paternal care and, 104–105

avoidance-of-polyspermy hypothesis, 122–128

banana slugs, reciprocal copulation and sperm exchange in, 191–192

Batesian mimicry, 209

bats, male copulation with torpid females in, 55

battle of sexes, 4, 216, 220–222

bedbugs: paragenital systems in, 90f; sexual antagonistic coevolution in, 87–91; sperm insemination in, 88f

bees: fragrances in mating of, 72; male harassment rates of, 48

bill morphology, selection of in male versus female finches, 9f

biparental species: in dunnocks, 175; partial mate desertion and sexual conflict in, 164–170; sexual conflict over care effort in, 170–174; signals regulating parental care in, 172–174

birds: female aggressive behavior during pair bonding in, 168–169; remating among, 96; seminal fluids in, 97–98; socially monogamous, 172–173

biting: copulatory and noncopulatory, 66; as grasping strategy, 65–67; in male mating behavior, 65–66

black widow spiders, multiple mating of males in, 52

blowflies, secondary ejaculate substances in, 129

body coloration, in male mimicry, 71

Bombyx mori: female reproductive fitness in, 135f; mating duration in, 134–135; spermatophore in, 142

brood defence behaviors, hormones in, 171–172

bumblebee colonies, with polyandrous versus monandrous queens, 113–114

bursa copulatrix, 196–197

butterflies: male sperm competition adaptations in, 114–115; spermatophore of, 144–145

Callosobruchus maculatus: female kicking in, 138–139; sclerotized spines on male genitalia of, 130f

cannibalism: female, 50–52; of male spiders in mating, 1

carotenoids, in seminal fluid, 143

Centronella spinosa, prepollical spines on, 67

choice, costs of, 47

chromosomes, parent-of-origin-specific behavior of, 182–183

cladogenesis, 203–204

clutch size, evolution of, 96

coevolution. *See* sexually antagonistic coevolution

Colorado potato beetle, seminal substances in, 100–101

condition, in sexually selected traits, 24f

conflict of interest, female reproductive effort and, 96–106

control, sexually antagonistic coevolution and, 222–223

convenience polyandry, 77–78

copulatory biting, 65–66

copulatory setae, 190–191

corpus allatum: gonadotropin production in, 100, 142–143; seminal substances stimulating, 99–100

Corynopoma, sexual dimorphism and mating in, 72–73

Corynopomini, phylogeny of, 75f

counteradaptations: "hidden," 41; to male postmating adaptations, 137–139; rapid evolution of, 147–148; risk of infanticide and, 53

courtship, 44; evolution of signals in, 173; in socially monogamous birds, 172–173

crickets, gin trap structures in, 62–63

crustaceans: courtship in, 120; male clasping devices in, 65

cryptic female choice, 16; conflicts over, 129–132; definition of, 92

Ctenophorus maculosus, copulatory biting in, 66

damselflies, sex-limited female di- or polymorphism in, 71

dance fly (*Rhamphomyia longicauda*): exploitive mating behavior of, 74–77; female flattened leg scales in, 76f

Darwinian evolution, sexual selection in, 14–18

death-feigning behavior, 70

deceptive alarm calls, costs and benefits of, 109t

defensive adaptations, male, 107–120

The Descent of Man and Selection in Relation to Sex (Darwin), 2, 15

Desmognath wrighti, copulatory biting in, 66–67

differential allocation hypothesis, 103, 104–105

dimorphic dentition, 66

direct benefits models, 26–27

direct selection, 26–27, 36; in Fisher process model, 21

dishonest signaling, 225

display traits, evolution of in males, 103–106

divergence: coevolutionary, 209; conflict and, 13. *See also* evolutionary divergence

divergent populations: magnitude of, 211; mismatches in, 210–211

diversity: natural selection and, 15; sexual conflict in, 226–227

diving beetles: antigrasping traits in, 68; polymorphic females of, 69f; rough and smooth morphs in, 69f; sexual antagonistic coevolution in, 83–84

Drosophila (fruit fly): pupal mating in, 55; seminal substances in, 98; sex peptide in, 100; sexually antagonistic coevolution in, 149–155; sperm displacement in, 128

Drosophila melanogaster, intralocus conflict in, 8–10

dual venom system, 56

dunnocks, mating system of, 174–178

earthworm, copulatory setae in, 190–191

economic studies, 42

ectospermalege, evolution of, 89–91

Efferia, female resistance traits in, 68–70

egg burial behavior, 163–164

egg laying: aggression during, 163; mating stimulating, 98; in penduline tit, 1

egg-laying hormones, 191
egg maturation rate: seminal substances stimulating, 100; stimulation of, 101
egg production rate: evolution of, 96; in hermaphrodites, 194; lifetime peak of, 134; regulation of, 99–100; seminal substances increasing, 150–152; substances increasing, 142–143
eggs: ejaculate proteins in, 141–142; gamete recognition system in fertilizing, 124–125; mediation of, 126; size of, 101; trading of, 188–189
ejaculate: elaborated, 140–146; large and proteinaceous, 114–115; protein-rich, 146; proteins in, 143; rapid processing of, 145f
ejaculatory substances: rapid evolution of female response to, 147–148; rapid processing of, 145f; secondary, 129
elasmobranches: biting and holding in mating behavior of, 65; sexual dimorphism in, 66; sexually dimorphic skin thickness in, 68
elephant seals: group raids among, 58; male harassment of resistant females in, 57–58
Empis opaca, using fake nuptial gifts, 77, 78f
endosperm, early growth of, 182
energy expenditure: in female resistance, 47–48; in mating, 46
equilibrium: evolutionary, 30–31; line of, 20, 21; low rate of convergence to, 209; of male traits and female preference, 37; of sexual and natural selection, 26
ESS models, 30–31
Euplectes ardens, male ornamentation in, 48–49
eupyrene sperm, 115
evolutionary divergence: rapid and gradual, 213f; sexual conflict as engine of, 207–210
evolutionary power, in conflict resolution, 222–223
experimental evolution techniques, 43
extinction, sexual selection and, 205–206

fecundity, 16f; males contributing to, 26–27; sexual cannibalism and, 52
female adaptations, evolution of, 147
female aggressive behavior: in polygynandrous dunnocks, 176; against secondary females, 168–169
female choice, 17; Darwin's description of, 18; origin and exaggeration of, 18–19; rapid and gradual divergence in, 213f;

sexual conflict and, 225. *See also* cryptic female choice; female preference
female counteradaptations, rapid evolution of, 147–148
female deception hypothesis, 166–167
female foraging: male exploitation of in courting, 72–74; reduced rates of, 82; remating rates and, 120; sexual cannibalism in, 52
female gonadotropin, 100
female infertility, male offensive adaptations and, 122–128
female longevity, mating costs to, 45–46
female lures, evolution of in fish species, 72–74
female mating behavior, sperm competition and, 106–132
female perception bias, for novelty, 207–208
female preference: coevolution of male traits with, 22–25; definition of, 17; direct benefits of, 26–27; direct selection working against, 26–27; driving male traits, 25–26; evolution of, 19–21, 207; evolutionary modification of, 36; for exaggerated male traits, 22; factors in evolution of, 36–37; in Fisher process, 18–21; hidden, 207–209; indirect and direct selection forces in, 35–36; male trait exaggeration and, 35; origin of, 27–29; preexisting biases in evolution of, 27–29; selective forces shaping, 19; sensitivity of, 211
female productivity, copulation duration maximizing, 134
female remating: seminal substances reducing, 118; sexual conflict over, 111f, 113–115; sperm competition adaptations and, 112f
female reproduction: mating duration and, 135f; postmating conflict over, 92; regulatory pathway of in insects, 99f
female reproductive effort: conflicting interests of sexes and, 96–106; optimal, 102; sexual conflict over, 97f
female reproductive rate: conflict over, 96–97; male attractiveness and, 103–104, 105–106; nuptial feeding and, 102; seminal substances regulating, 98; sexual conflict in, 10–11
female reproductive receptors, evolution of, 146–147
female reproductive tract: functional morphology of, 137; hostile to seminal proteins, 94;

female reproductive tract *(cont.)*
penetration of walls of during copulation, 129; sperm viability in, 93
female resistance, 30–32; adaptations in, 12f; alleles for, 221–222; antigrasping traits in, 68–71; in bedbugs, 89–91; costs of, 37, 46, 47–48, 57, 78–79; in diving beetles, 83–84; driving exaggerated male traits, 32–33; evolution of, 207; in evolution of male grasping devices, 61–68; gamete recognition systems and, 125; generalized, 81–82; grasping devices in male and, 61–62; male response to, 35; male trait exaggeration and, 35; manipulation of, 77–78; as mate screening, 81–82; mating duration and, 138–139; phenotypic studies of, 42; potential results of, 79; repeatability of, 82; in selection of male persistence, 82–93; sensitivity of, 211; widespread occurrence of, 58
female sensory system, direct selection in, 27
female sexual cannibalism, evolution of, 50–52
female sexual traits: deceptive secondary, 76–77; manipulative, 39–40
female viability: male traits reducing, 116; sperm competition adaptations and, 112f
fertilization: doubling of, 200–201; sexual conflict and efficiency of, 10–11
fertilization success, 18; in hermaphrodites, 194
fertilization system: internal versus external, 159–160; mate desertion and, 159–160
fish: male paddles as female lures in, 75f; paddle extensions as "female lures" in, 72–74
Fisher process model, 18–21, 25–26, 36
Fisher school. *See* Fisher process model
flowering plants: genomic imprinting in, 181–182; reproductive parts of, 201
foraging efficiency, reduced, 77–78
funnel-web spider, male mating strategy of, 56

game-theory model, 30; demonstrating parental care systems, 162–163; of male trait evolution, 119; winners and losers in, 221–222
gamete recognition system, 124–125
gametes, selected for rapid fusion, 125
garter snakes: courtship in, 72; male competition among, 50f
genetic diseases, imprinting in, 182
genetic diversity: in plants, 202; processes of, 203–205

genetic experiments, sexual conflict, 43
genetic monogamy, 3, 96
genetic sexual conflict models, 31–34
genetic variance: additive, 25; maintaining high level of, 23
The Genetical Theory of Natural Selection (Fisher), 19
genital barbs, 129; on male genitalia, 129, 130f
genital spines: costs and benefits of, 109t; mating duration and, 136–137
genitalia: evolution of, 205; spines flanking, 68, 136–137
genitonogamy, 203
genomic imprinting, evolution of, 179–183
gin trap, 62–63
glandular secretions, in mating, 66–67
gonadotropic substances, in seminal fluid, 97–102
good father hypothesis, 104–105
good genes hypothesis, 19, 21, 22–25, 26, 30, 36–38, 106; male attractiveness and, 105; screening of, 81–83; sexual selection in, 81
grasping devices, 56, 59–68; coevolution of resistance adaptation with, 220; costs and benefits of, 109t; in diving beetles, 83–84; evolution of in water striders, 84–86; female traits to resist, 68–71; for male-male competition, 80; mating duration and, 135–137; selective value of, 80–81; traits reducing efficiency of, 68
gray seals, lactating, male harassment of, 58
growth promoters (GPs), embryonic/fetal, 180–181
growth promoters/suppressors, mismatch of, 209
growth suppression, embryonic/fetal, 180–181
guppies, harassment in, 59–60

handicap model. *See* good genes hypothesis
harassment, sexual, 47–48, 56–59; avoidance of, 68; convenience polyandry as response to, 77–78; female responses to, 68–70; female viability and ability to resist, 119–120; resistance and, 57–60; in water striders, 84–86
harems: female aggression in, 169–170; peripheral males surrounding, 57; reduced levels of harassment in, 57–58
hatching asynchrony, 174
hatching failure rate, polyspermy rate and, 126–127

hermaphrodites: conflict in, 13; postmating conflict in, 190–192; premating conflict in, 187–190; relative interests of, 188f; sexual conflict in, 186–200; sexual selection and antagonistic coevolution in, 192–196
heterogametic sex, 214–215
heterospecific ejaculates, toxic effect of, 88–89
hormones: regulating parental care behavior, 171–172; stimuli triggering production of, 172
horns, intrasexual selection of, 17
humeral spines, 67–68
hybridization: avoidance of, 3; fitness effects of, 210
hypodermic insemination, 186, 189–190, 196

immunosuppressants, in female reproductive tract, 94–95
incestuous mating, 224
indicator mechanisms, 22–25. See also good genes hypothesis
indicator models: versus Fisher models, 22–23; male trait exaggeration in, 25–26
indirect selection, 20–21, 26–27, 35–38; versus direct benefits models, 27
infanticide, by males, evolution of, 53–54
infertility: in plants, 201–202; with sperm shortage, 107–110
insects: copulatory biting in, 66–67; female responses to male harassment in, 68–70; genomic imprinting in, 182–183; male clasping structures in, 62; male mimicry in, 71; pupal mating in, 55; regulatory pathway of female reproduction in, 99f; remating among, 96; sexual conflict among, 5–6
insulin-like-growth-factor/insulin pathway, 181
interlocus conflict, 10–11, 221–222
interploidy studies, 181–182
intralocus sexual conflict, 7–10, 203, 206
invasive male genitalia, harmful, 139
isogamous species, reproductive conflict in, 6–7

juvenile hormone (JH), 98, 99–100

lactogenic hormones, 171–172
Led-MAGP, 100–101
lineages, splitting of, 203–205
litter size, evolution of, 96
"love dart," 196–200; evolution of, 199–200

Malabar ricefish: barbed spermatophores of, 61f; male harassment in, 60; mating behavior of, 1
male accessory reproductive glands. See accessory reproductive glands
male-attracting pheromones, 117–118; blocking of, 143
male attractiveness, indirect genetic benefits of, 105
male combat, grasping structures used in, 67
male courtship behavior, 44
male deceptiveness, 166
male defensive adaptations: direct costs of to females, 118–120; indirect costs of to females, 116–118
male display traits, 103–106
male-female coevolution, 30–31; alternative forms of, 226–227; multiple traits in, 208–209; postmating conflicts and, 139–140; pre-existing biases in, 27–29
male-female conflict of interests, postmating, 92–155
male fertility, zinc in, 94–95
male fragrance signaling, 72
male infanticide, evolution of, 53–54
male-male competition, 3, 17; costs of to females, 48–50; in evolution of grasping/clasping devices, 80–81; use of spines in, 67
male-male conflict: sperm competition and, 110f; ubiquity of, 216–218
male manipulative adaptations: in hermaphrodites, 190–191; novel, 207
male manipulative substances, 142–143
male mimicry, 71
male offensive adaptations, 121–132
male persistence, 30–31, 41f; adaptations for, 195; alleles for, 221–222; female resistance and, 47; female resistance to, 4; overcoming female resistance, 32; sexual selection for, 81–83
male postmating adaptations, 146–149
male size: mating success and, 81–82; takeover success and, 80–81
male traits: coevolution of female preferences with, 22–25; evolution and covariance of, 25–26; exaggerated, 19–20, 22, 23, 24f, 25, 28, 37–38, 42; female fitness spread and, 35; female responsiveness to, 32–33; genetic covariance with, 20–21; as honest indicators of nonmating fitness, 23; manipulative, 38–39; multiple, 208–209; preexisting

male traits *(cont.)*
  bias for, 28; in response to female resistance, 35; secondary, 72–74
mammary glands, 16
mandibles, sexually dimorphic, 66
marine flatworms: hypodermic insemination in, 186; penis fencing duels between, 185–187; sperm trading in, 189–190; unidirectional insemination in, 185–186
mate: low quality, 46–47; recognition of, 204; screening of, 80–83
mate acquisition, in hermaphrodites, 193–194
mate attracting strategies, 167–169
mate choice: evolution of, 44–45; sexual conflict over, 224–225. *See also* female choice; resistance adaptations
mate desertion: partial, in biparental species, 164–170; in penduline tits, 160–164; in uniparental species, 158–160
mate guarding: costs and benefits of, 108t; grasping devices in, 80; indirect costs to females of, 118; postcopulatory, 133; precopulatory, 134; reducing paternal care time, 172
maternal investment, optimal, 179–180
mating: conflicts over duration of, 132–139; costs and benefits of, 224; costs of resisting, 47–48; deleterious, 128–129; economy of, 45; female thresholds for, 32–33; genetic interests in, 29–30; injurious interactions in, 185–187; multiple, in hermaphroditic snails, 198–199; patterns of, 14; sexual conflict after, 92–155; sexual conflict before, 44–91
mating biases, 79; preexisting biases in evolution of, 72–77
mating costs: direct, 45–46; ecological, 137; female resistance and, 82–93; for females, 5–6, 87–88; in hermaphrodites, 196; of low quality mate, 46–47; reduced foraging efficiency in, 77–78; time-dependent, 132–133
mating duration: female counteradaptations and, 137–139; male adaptations for, 135–137; optimal, 134
mating frequency, 45; nuptial gifts and, 140–141; selection of, 7
mating plugs, 108t
mating rate: reduction of, 79; sexual conflict in, 10–11
mating scars, in bedbugs, 90f, 91
mating strategies, coercive, 196
mating success, 18; variance in, 15–16

mating systems: reproductive output and success with, 177t; sexual conflict over and partial mate desertion, 164–170; sexual selection of, 205; traditional views of evolution of, 171f; variability of in dunnock, 174–178
MEDEA gene, imprinted, 182
Medea gifts, 146
mesospermalege, artificial inseminations into, 89–91
mitochondria, in sperm, 131
monogamous species: genetic contributions in, 179; mutual sensory exploitation in, 172; sexual conflict over care effort in, 170–174
monogamy, 3; biparental care and, 164–166; in dunnocks, 177t; female fitness and, 152–153; in hermaphrodites, 190; imposed on males, 167–168, 169–170; true genetic, 96
Müllerian mimicry, 209
mutual gamete exchange, 188–189
mutualistic mimicry, 209

Nabidae bugs, cryptic intrusive copulation in, 129
natural selection, 14–15; checking exaggerated male traits, 19–20; evolutionary divergence and, 204–205; of female resistance, 79; fitness and, 16f; interaction of sexes and, 3; pleiotropic effect of, 79; of primary sexual traits, 16; sexual selection and, 3–4, 15–16, 26; sexual selection conflicting with, 20–21, 24f; sexual selection promoting, 18
Nephila plumipes, female sexual cannibalism in, 51f
neuropeptides: affecting female reproduction, 144; in male ejaculate, 142–143; oviposition-stimulating, 143–144; regulating female reproductive rate, 98; stimulating egg production in hermaphrodites, 191
nonequilibrium models, 35
Notophthalmus viridescens, landular secretions in mating of, 67
nuptial feeding, 102; sex-role-reversed, 148–149
nuptial gifts, 140; benefits of, 114–115; in dance fly mating, 74–76; deceptive non-nutritional, 77, 78f; evolutionary origin and maintenance of, 141–142; postmating sexual conflict and, 140–146
nuptial pads, 67

offensive adaptations, male, 121–132
offensive sperm competition adaptations, 128–129
offspring: age and size of in parental care, 173–174; attractive male mates in survival of, 34; costs of male harassment to, 58; equal genetic contribution of parents in, 179; male attractiveness and survival of, 105; sex ratio of, 183–185; sexual conflict over provisioning of, 180–183; unequal genetic expression in, 180–183
offspring fitness: parental care and, 156, 176; total parental expenditure and, 157f
offspring quality: male attractiveness and, 103–104; parental care and, 102
opponent-independent cost game, 31
Orconectes rusticus, evolution of grasping devices in, 80–81
Origin of Species by Means of Natural Selection (Darwin), 14–15
orthopterans: prostaglandin synthease in male ejaculate in, 147; spermatophore composition in, 143–144
oviposition, 98; seminal substances stimulating, 100; stimulation of, 99–100
ovulation, seminal substances stimulating, 98–102

pair bonding: female aggressive behavior during, 168–169; strengthening of, 172–173
parent relatedness asymmetries, 184–185; in plants, 200–201
parental care: age and size distribution of offspring and, 173–174; amphi- or ambisexual, 160; antagonistic negotiations over, 173; biparental, 164–170; conflict over, 92; costs and benefits of, 159–160, 164–165; in dunnocks, 176; hormones regulating, 171–172; male contributions in, 27; offspring fitness and, 156; offspring quality and, 102; sexual conflict over, 156–178; total expenditure for and offspring fitness, 157f; uniparental, 158–164
parental investment: asymmetric, 74–77; asymmetrical, 14, 38–39, 96; in biparental monogamous species, 170–174; female versus male, 159f; gender differences in, 29; sexual conflict in, 10–11. *See also* maternal investment; paternal investment
parental risk taking, prolactin-mediated, 171–172

parental roles: distribution of among taxa, 158–159; fixed, 160
paternal care: costs of, 104–105; decreased, 34
paternal investment: nuptial feeding and, 102, 141–142; sexual cannibalism as form of, 51–52
penduline tits: breeding system of, 160; egg-laying behavior of, 1; mate desertion in, 160–164; roofed and pendulous nest of, 161f
penis fencing duels, 185–187
persistence adaptations, 11; adaptations for, 55–78; costs and efficacy of, 222–223; phenotypic manipulations in, 42; selection for, 12f. *See also* harassment; male persistence
phenotype-dependent costs, 34
phenotype-independent costs, 34
phenotypic manipulations, 42
phenotypic selection gradient approach, 223–224
pheromone blockers, in male ejaculate, 143
pheromone-emitting webs, 117–118
pheromones: in burying beetles, 168, 169f; female-attracting, 168; male-attracting, 117–118
physiological resistance adaptations, 89
pinnipeds, mating systems of, 57
plants: double fertilization in, 200–201; genomic imprinting in, 181–182; infertility rates in, 201–202; sexual conflict in, 200–203
plasticity, adaptive phenotypic, 42
pleiotropic effect, 27–28
Poeciliid fish, male harassment in, 59
pollen, 201, 202
polyamines, in seminal fluid, 95
polyandry, 38; bee colonies of, 113–114; convenience, 77–78; in dunnocks, 175, 177t; indirect benefits of, 167; sexual conflict and, 164
polygamous mating systems, postmating sexual conflict in, 152–153
polygamy: in dunnocks, 175; prevalence of, 2–3
polygynandry, in dunnocks, 176, 177t
polygynous signaling, 168
polygyny: costs of, 166–167; in dunnocks, 176, 177t; parental care and, 164; in secondary females, 164–168
polygyny threshold model, 164–166
polymorphism, Acp genes, 154

polyspermy, 122–128; barrier to in plants, 202; blocking of, 126, 127–128; cost of, 126–127; occurrence of under natural conditions, 127t

population crosses, 210–212, 212f

postmating events, 16, 18; in humans, 92–93. *See also* cryptic female choice; sperm competition

postmating fertilization success, 198–199

postmating sexual conflict, 139–140; in polygamous mating systems, 152–153

precopulatory biting, 65–66

precopulatory guarding, 58, 80

predation, 58, 132; mating and susceptibility to, 46; mating duration and, 133f

preference: collective selection of, 20–21; costs of, 21, 23; definition of, 17; in male-female coevolution, 31–35. *See also* female preference

preferred traits, handicapping function of. *See* good genes hypothesis

prehension organs, diversity of, 60–62

premating asymmetries, 38–39

prion proteins, 100–101

Prochyliza xanthostoma, selection conflict in, 49

prolactin: abiotic cues stimulating production of, 172; in parental risk taking, 171–172

promenade à deux, 56

prostaglandin biosynthesis, 147

protease inhibitors, in seminal fluids, 151–152

proteolytic enzymes: in female reproductive tract, 151; seminal substances interfering with, 95

Pterobrycon, sexual dimorphism and mating in, 73–74

pupal mating, evolution of, 55

Rana blanfordii, spines of, 67

Rana pipiens, humeral spines in, 67

reciprocal copulation: in banana slugs, 191–192; in hermaphroditic snails, 196–197

reciprocal population crosses, male-female interactions with, 212f

relatedness asymmetries, 184–185

remating, 96; arms race in, 144–145; delayed, cost of to females, 111–118; indirect benefits of, 113–114; sexual conflict in, 10–11

remating inhibitors: costs of to females, 111; female susceptibility to, 110; in male ejaculate, 143

reproduction: as cooperative endeavor, 4; costs of, 96; economics of, 6; evolving views of, 2–6; female versus male rate of, 38–39; variation of in dunnocks, 176, 177t

reproductive behavior, synchronizing, 172–173

reproductive conflict, primordial, 6–7

reproductive investment, evolution of exploitative traits and, 74–77

reproductive outcome, conflict over control of, 222–223

reproductive output, variation of in dunnocks, 176, 177t

reproductive responses: to allopatric male signal, 213f; weak, 212f

reproductive roles, conflict of in hermaphrodites, 187–189

resistance adaptations, 11, 55–78; alternative explanations of, 80–83; costs and efficacy of, 222–223; costs of, 46, 47–48; female, 12f; to grasping devices, 220; harassment and, 57–60; natural selection in evolution of, 78–79; phenotypic manipulations in, 42

resistance/persistence polymorphism, 220

Rhamphomyia longicauda (dance fly), sex roles of, 39–40

Rheumatobates, male grasping devices in, 64–65

robber flies: female resistance in, 70f; mating season activity of, 1

round stingray, mating behavior of, 66

salamanders, copulatory biting in, 66–67

satyr effect, 88–89

scorpion, dual venom of, 56

scorpionflies: clasping structures in, 62; dorsal clamp in males of, 136f

sea slugs, sperm trading in, 189

seals, copulatory biting in, 66

seaweed flies, generalized female resistance in, 81–82

selection: direct, 26–27, 36; indirect, 26–27, 35–36, 37–38. *See also* natural selection; sexual selection

selective resistance hypothesis, 138–139

self-fertilization, harmful effects of, 203

self-incompatibility, evolution of, 202

self-sacrifice, potential benefits of, 51–52

seminal fluid: adaptations of, 118; complexity of, 97–98; composition of, 143; defenses against hostile environment in, 94–95; evolution of, 205; female adaptations to, 152;

female antibodies to proteins of, 94; function of substances in, 142–143; with gonadotropic effects, 97–102; in male fruit flies, 150; polyamines in, 95; promoting sperm displacement, 128; proteins of penetrating vaginal wall, 93–94; remate-reducing substances in, 118; substances enhancing pregnancy rates in, 93–94; substances of, 93; traditional view of, 95; zinc concentration in, 94–95

seminal receptivity inhibitors, 108t

seminal toxins, 109t

sensory biases: in evolution of female traits, 79; exploitation of, 27–28, 38–39, 71–77, 79, 173; in female preference evolution, 27–29; preexisting, 72

sensory drive, 205

sensory exploitation, 36–37

serranid sea bass, egg trading in, 188–189

sex allocation: in hermaphrodites, 193–196; sexual conflict and, 183–186

sex-biased gene transmission, 213–214

sex chromosomes: autosomes and, 183–184; sexual conflict and, 212–215

sex-determining genes, 213–214; architecture of, 214–215

sex functions, in hermaphrodites, 193–196

sex-limited gene expression, evolution of, 7–8

sex peptide, 100

sex ratios, sexual conflict and, 183–186

sex roles, 14; reversal of, 17; in sexual conflict, 38–40

sexual cannibalism: before copulation, 52; evolution of in female, 50–52

sexual conflict, 14, 29–30; after mating, 92–155; definitions of, 217t; "economy" of, 41; empirical approaches to, 40–43; in evolution, 4–6; evolution of sexual cannibalism by female in, 50–52; in evolutionary divergence, 207–210; in framework of sexual selection, 35–38; generating diversity of adaptations, 226–227; genetic models of, 31–34; genomic imprinting and, 179–183; in hermaphrodites, 186–200; hidden, 218; increasing interest in, 5f; intensity of, 223–224; interlocus, 10–11, 12f; intralocus, 7–10, 203, 206; levels of, 216–219; lineage divergence and, 13; male defensive adaptations and, 107–120; male infanticide and, 53–54; male offensive adaptations and, 121–132; in nature, 1–11; nonequilibrium models of, 35; over control of interactions, 222–223;

over duration of mating, 132–139; over mate choice, 224–225; over parental care, 156–178; over parental effort in biparental monogamous species, 170–174; Parker's initial models of, 30–31; phenotype-dependent and phenotype-independent costs of, 34; in plants, 200–203; premating, 44–91; resolution of, 219–220; resolution of, evolutionary power in, 222–223; sex chromosomes and, 212–215; and sex ratios and allocation, 183–186; sex roles in, 38–40; in sexual selection, 78–79; sexually antagonistic selection and, 5–11; in speciation and extinction, 203–212; winners and losers of, 220–222. See also male-male conflict; sexually antagonistic coevolution

sexual dimorphism, 14; in crustaceans, 65; Darwin's explanation of, 14–15; in elasmobranches, 66; evolution of, 206; female choice in, 20; Fisher process explaining, 20–21; male/female heterogameity and, 214–215; not related to sexual selection, 16–17; in prepollical spines, 67; sexual selection of, 205; in skin thickness, 68

sexual harassment. See harassment

sexual selection: adaptive, 18; adaptive versus maladaptive, 18; balanced by natural selection, 26; categorization of, 19; classes of, 17; conventional theory of, 225; currency of, 18; Darwinian views on, 14–18, 92; definition of, 15; fitness and, 16f; frequency-dependent, 214–215; good genes process in, 106; in hermaphrodites, 192–196; indirect, 22; modes of, 210; natural selection and, 3–4; operational sex ratio and, 183–184; for persistence traits in males, 81–82; in plants, 200; promoting extinction, 205–206; sexual conflict and, 35–38, 78–79; in speciation, 204–205; through female choice, 19–21

sexual signal/response systems, evolutionary divergence of, 204–205

sexual traits: diversity of, 216–218; primary, 16–17; secondary, 16–17, 39–40

sexually antagonistic coevolution: case studies in, 83–91; conflict over parental care in, 157–178; control as dynamic result of, 222–223; costs of, 34; direct selection forces in, 36–38; divergence and, 210–212; empirical approaches to, 40–43; in fruit flies, 149–155; in hermaphrodites, 192–196; postmating conflicts and, 139–140;

sexually antagonistic coevolution *(cont.)*
research in, 226–227; selection in, 5–11; sexual conflict in, 216–219; theoretical framework for, 12–13; theory of, 14; winners and losers in, 221–222

sexy son benefit, 20, 30–32, 36–38, 81, 103, 132, 195

Sierra dome spiders, destructive male strategies of, 117–118

snails, love dart in, 196–200

speciation, 203–204

sperm: defense of, 110f; displacement of, 18, 128; donation of, costs and benefits of, 187–188; dumping, prevention of, 133; exchange of in banana slugs, 191–192; female barriers against, 123–124; fertile eupyrene and nonfertile apyrene, 115; infertile, costs and benefits of, 109t; mitochondria in, 131; shortage of, 107–110; storage of in hermaphroditic snails, 199, 200; stored during hibernation, 55; trading of, 189–190; viability of, seminal substances enhancing, 93

sperm allocation: costs and benefits of, 109t; mating duration and, 137

sperm competition, 16; aggressive ejaculates and, 121; definition of, 92; female mating behavior and, 106–132; heritable genetic variation for success in, 131; male adaptations in, 107, 108–109t; mating duration and success of, 133; selecting for aggressive ejaculate, 122, 123f

sperm competition adaptations: costs of to females, 112f; in hermaphroditic snails, 198

sperm digestion: in hermaphrodites, 192; in hermaphroditic snails, 199; in snails, 197

sperm-egg interactions, mediation of, 125–126

sperm offense, 110f

sperm transfer: facilitation of, 59–60; sequential, 134–135

sperm transport, alternative mechanics of, 87–91, 129

spermalege, 89–91

spermaticide, in hermaphrodites, 192

spermatophores, 142; barbed, 61f; composition of, 143–144; in egg production rates, 99; exchange of, 196–197; production of, 60; release of, 1

spermatozoa, female antibodies against, 94

spiders: courtship among, 1; female sexual cannibalism in, 50–52; male infanticide in, 53–55

spines: elongated and upturned connexival, 68; humeral, 67–68; on male genitalia, 129, 130f; prepollical, 67

starvation: female remating rates and, 119–120; mating duration and, 137

Stegodyphus lineatus, male infanticide in, 53–55

stingrays: precopulatory biting of, 65; sexually dimorphic dentition in, 66

successively reciprocal mating, 188–189

supernumerary sperm nucleus, 127–128

takeover attempts, 49–50; prevention of, 80

Telmatobius, humeral spines in, 67

territorial defence, reducing paternal care time, 172

thanatosis, in female resistance, 70

theoretical framework, 12–13

traumatic insemination, 87–91; homosexual, 91

uniparental species, mate desertion in, 158–164

vaginal wall, permeability of, 93–94

vitellogenesis, 98, 102, 142; accumulation of, 99f; increased rate of, 142

water striders: abdominal clasping processes on males among, 63–64; antigrasping traits in, 68; convenience polyandry in, 77–78; degree of antagonistic armaments in, 85f; duration of mating in, 132; female resistance among, 47; male grasping structures in, 63f; male versus female armament levels in, 86f; mating costs to, 46; selection conflict in, 49; sexual antagonistic coevolution in, 84–87; sexual selection of male persistence in, 82–83

web destruction behavior, 117–118

Zeus bugs: dorsal glands in females of, 148f; mating system of, 148–149

Zoonomia (Darwin), 3